From Observations to Optimal Phylogenetic Trees

Species and Systematics

The *Species and Systematics* series will investigate the theory and practice of systematics, phylogenetics, and taxonomy and explore their importance to biology in a series of comprehensive volumes aimed at students and researchers in biology and in the history and philosophy of biology. The book series will examine the role of biological diversity studies at all levels of organization and focus on the philosophical and theoretical underpinnings of research in biodiversity dynamics. The philosophical consequences of classification, integrative taxonomy, and future implications of rapidly expanding data and technologies will be among the themes explored by this series. Approaches to topics in *Species and Systematics* may include detailed studies of systematic methods, empirical studies of exemplar taxonomic groups, and historical treatises on central concepts in systematics.

For more information visit:
www.crcpress.com/Species-and-Systematics/book-series/CRCSPEANDSYS

Species and Systematics

Phylogenetic Systematics: Haeckel To Hennig, *by Olivier Rieppel*

Evolution By Natural Selection: Confidence, Evidence and the Gap, *by Michaelis Michael*

The Evolution of Phylogenetic Systematics, *edited by Andrew Hamilton*

Molecular Panbiogeography on the Tropics, *by Michael Heads*

Beyond Cladistics, *edited by David M. Williams and Sandra Knapp*

Comparative Biogeography: Discovering and Classifying Bio-Geographical Patterns of a Dynamic Earth, *by Lynee R. Parenti and Malte C. Ebach*

Species: A History of the Idea, *by John S. Wilkins*

What Species Mean: Understanding the Units of Biodiversity, *by Julia Sigwart*

What, if anything, are species? *by Brent D. Mishler*

Biological Systematics, *by Igor Ya. Pavlinov*

Species Problems and Beyond: Contemporary Issues in Philosophy and Practice, *by John S. Wilkins, Frank E. Zachos, Igor Ya. Pavlinov*

From Observations to Optimal Phylogenetic Trees: Phylogenetic Analysis of Morphological Data: Volume 1, *by Pablo A. Goloboff*

For more information about this series, please visit:
www.crcpress.com/Species-and-Systematics/book-series/CRCSPEANDSYS

From Observations to Optimal Phylogenetic Trees

Phylogenetic Analysis of Morphological Data, Volume 1

Pablo A. Goloboff

CRC Press is an imprint of the
Taylor & Francis Group, an **informa** business

First edition published 2022
by CRC Press
6000 Broken Sound Parkway NW, Suite 300, Boca Raton, FL 33487–2742

and by CRC Press
2 Park Square, Milton Park, Abingdon, Oxon, OX14 4RN

CRC Press is an imprint of Taylor & Francis Group, LLC

Library of Congress Cataloging-in-Publication Data
Names: Goloboff, Pablo A., author.
Title: From observations to optimal phylogenetic trees : phylogenetic analysis of morphological data / Pablo Goloboff.
Description: First edition. | Boco Ratan : CRC Press, 2022. | Series: Species and systematics | Includes bibliographical references and index. |
Summary: "This book outlines the steps in a phylogenetic analysis that follow the generation of most parsimonious trees. In addition, character reliability approaches and methods of analysis for morphometric characters are summarized. The algorithm used throughout the book is TNT, a freely available software package able to summarize and compare multiple trees produced by ambiguous datasets, or analyses of different datasets. Unstable taxa (wildcards or rogues), which may obscure the relationships of the other taxa, are discussed extensively, as well as their identification and handling with several options implemented in TNT"– Provided by publisher.
Identifiers: LCCN 2021047215 | ISBN 9781032114859 (volume 1 ; hardback) | ISBN 9781032114873 (volume 1 ; paperback) | ISBN 9781003220084 (volume 1 ; ebook)
Subjects: LCSH: Phylogeny–Data processing. | Cladistic analysis.
Classification: LCC QH367.5 .G65 2022 | DDC 576.8/8–dc23/eng/20211025
LC record available at https://lccn.loc.gov/2021047215

ISBN: 9781032114859 (hbk)
ISBN: 9781032114873 (pbk)
ISBN: 9781003220084 (ebk)

DOI: 10.1201/9781003220084

Typeset in Times
by Apex CoVantage, LLC

Contents

Preface

Why This Book?

The field of phylogenetic reconstruction has seen tremendous advances in the last few decades. The sources for those advances have come from different subdisciplines of biology (evolution, genetics, or molecular biology) and even outside biology (computer science, mathematics, philosophy). But a number of the contributions to phylogenetic methodology have come from workers interested mostly in systematics and taxonomy. Systematists were instrumental in establishing phylogenetics as an important field in the late 60s and early 70s. The general outlook of researchers in systematics and taxonomy, and the kinds of problems they are interested in, tend to differ from those of people approaching phylogenetics from evolutionary biology or computer science. Systematists focus more on the relationships between species, are more interested in diversity than in model organisms, and tend to rely much more heavily on morphological characters—which are used as part of standard taxonomy. More evolutionarily oriented phylogeneticists, instead, tend to focus on the use of trees for testing specific hypotheses. Both are different kinds of problems and legitimate fields of inquiry; personal preferences lead to focusing on one or the other type of problem. As someone with an upbringing rooted in systematics and taxonomy, I felt that no recent book summarized the main advances in this field from this perspective. This is the first reason to write this book.

Molecular phylogenetics exploded in recent decades and became the dominant approach in the field. There is only one "systematics," and the goals and principles are ultimately the same for all of it; the principles for "morphological" and "molecular" systematics are exactly the same. However, it also true that different types of data pose different challenges and attract different groups of people. The enormous interest in molecular data is probably the combined result of molecular evolution being much more amenable than morphology to modeling and mathematical treatment, the fascination of many scientists with new technological innovations, and the intrinsic importance and interest of molecular biology and evolution. But despite that predominance, phylogenetic reconstruction using morphological data continued being developed in recent decades and still presents many interesting challenges. Phylogenetic analysis of morphological data continues to be very important for researchers working in the systematics of groups where sequencing of most species is unfeasible (not just fossils, but also groups that are hard to collect or rarely represented in collections by fresh material). New molecular results continue to be tested against morphology—the main taxonomic groups, taken as reference in any analysis, have been established, and continue being recognized, on the basis of morphology. And morphology (in the wide sense of phenotypic characteristics, including behavior and their associated adaptations) will forever continue being the main reason behind the interest and fascination that animals and plants exert on us. It is thus logical to continue focusing on how to use morphological data for making inferences about phylogeny, for understanding morphology in a phylogenetic context, and for complementing the results of molecular data.

Despite the fact that morphology continues to be an important source of data for phylogenetic analysis, however, no recent book on phylogenetics devotes much space to the peculiarities and problems intrinsic to the analysis of morphological datasets. No other book summarizes the analysis of morphological datasets, certainly not including a balance between concepts and more technical developments of the last 20 or 30 years. This is the second reason to write this book. Problems specific to molecular data are discussed in this book only insofar as they serve to illustrate differences with, or improve understanding of, the problems of morphological data.

In addition to covering methods for morphological phylogenetics in more depth than usual, I also aspire to provide practical guidance for taxonomists in a way most books do not. This is the third reason to write this book. Some books help translate theoretical concepts into practice by discussing real implementations in detail, but all of these deal almost exclusively with molecular sequences. No book dealing with morphological phylogenetics and systematics includes an in-depth, authoritative treatment of computer implementations. Discussion of abstract concepts is important, but I also aim at facilitating application of those concepts to real-world problems. In recent decades, computational applications for phylogenetics have evolved toward a few major programs and a host of small, highly specific programs, useful for different parts of the analysis. While it is, of course, impossible for a single program to keep up with the ever-increasing repertoire of methods in phylogenetics, implementing most of the tasks required for a phylogenetic analysis in a single, general-purpose program has some definite advantages: the internal coherence of assumptions, not being limited by the possibilities of interconnection between programs, etc. Given my belief that the detailed assumptions required for model-based (or "statistical") phylogenetics are not warranted in the case of morphology, I consider parsimony as the best justified method of analysis for morphological datasets. The practical parts of this book thus use TNT—which focuses on parsimony. It is not the only program for phylogenetic analysis, but (having developed it over the course of more than 20 years and having taught its use in numerous courses and seminars) it is one I know very well.

A few words on what this book is *not*. It is not a neutral summary of the literature on phylogenetics; other types of books, and primary literature, must be consulted for that—I take a position on many issues. The fact that this book takes positions is (hopefully) one of the aspects that make it interesting. As a consequence, this is not by any means a complete, comprehensive treatment of phylogenetic analysis; even within its focus, many books, papers, or approaches go unmentioned. It is not intended, either, to provide a full historical account on the origin of all the different ideas presented. To some extent, many omissions of alternative methods and proposals are purposeful and (barring involuntary oversights) implicitly provide an overview of what I consider the higher and lesser contributions to the field. As in any communication process, what is not there *is* also part of the message. So, I take responsibility for omissions.

The book includes two volumes, and the general development of themes follows from my experience teaching courses in cladistics for almost three decades. Teaching taught me (among many other things) how people who are starting to work in systematics and phylogenetics view the problems, as well as common misunderstandings and prejudice. Much of the flow in the different chapters of this book is the result of my experience teaching and explaining these problems to people who hear them for the first time. The sequence of topics loosely follows that of an analysis, from gathering of data to deciding an optimality criterion and finding appropriate trees (Chapters 1–5, in the first volume) to summarizing results and more carefully considering the evidence (weighting) and non-standard types of data, such as continuous and morphometric characters (Chapters 6–10, in the second volume). Hopefully I have managed to achieve at least a fraction of the clarity I aimed for and this book helps upcoming systematists in their first phylogenetic steps. All the educational scripts accompanying this book can be downloaded from http://www.lillo.org.ar/phylogeny/eduscripts/.

Finally, debts of gratitude. My approach to systematics has been shaped in interactions with many colleagues; I thank all of them both for their agreements and disagreements, but mostly for what I have learned from them: Lone Aagesen, Victor Albert, Dalton Amorim, Salvador Arias, Ronald Brady (†), Andrew Brower, James Carpenter, Santiago Catalano, María M. Cigliano, Joel Cracraft, Jonathan Coddington, Jan De Laet, Julian Faivovich, Joseph Felsenstein, Gonzalo Giribet, Norberto Giannini, Charles Griswold, Peter Hovenkamp (deceased), Jaakko Hyvönen, Mari Källersjö, James Liebherr, Melissa Luckow, Camilo Mattoni, Rudolf Meier, Marcos Mirande, Martín Morales, Jyrki Muona, Gareth Nelson, Kevin Nixon, Norman Platnick (†), Diego Pol, Martín Ramírez, Robert Raven, Randall Schuh, Mark Siddall, Mark Simmons, Michael Steel, David Swofford, Claudia Szumik, John Wenzel, Ward Wheeler, Kipling Will, and Mark Wilkinson. I only imposed on a few the task of reading the manuscript; I both apologize for the burden and thank them for their useful comments: Andrew Brower, James Carpenter, Camilo Mattoni, Diego Pol, Christiane Weirauch, and Ward Wheeler. I also thank Chuck Crumly for help and advice during the editorial process. Last but most important, I especially acknowledge María E. Galiano (†), world-renowned salticidologist who (in the late 70s) introduced me to taxonomy, arachnology, and critical thinking, and James "Steve" Farris, who led phylogenetic systematics into the modern era and (in the late 80s) helped me jump onto that train when it was already moving fast. Steve also originated the TNT project in the late 90s (approaching Kevin Nixon and me with the proposal to join forces and write a comprehensive phylogeny program). Steve never wrote a book (he was always too busy developing new methods and ideas), but his influence all along this book is in fact so strong that—if my vanity can be excused—I'd like to think this book is not too different from the one he might have written.

Author Biography

Born in Buenos Aires, Pablo A. Goloboff became interested in spider biology and systematics in the late 70s in the Museo Argentino de Ciencias Naturales. His first papers (published during the 80s) were on spider systematics, but he soon became more interested in systematic theory and phylogenetic methods. He graduated with a Licenciatura in Biology in 1989 from Universidad de Buenos Aires and then pursued doctoral studies in Cornell University and the American Museum of Natural History, in New York, between 1989 and 1994. He published his first methodological papers in the early 90s, gradually switching his research from spider systematics to systematic theory. During his stay at Cornell University, he became more involved with quantitative methods for parsimony analysis and wrote his first computer programs. He moved to Tucumán in 1994 to work for the CONICET and continued working on theory and methods for systematics and historical biogeography. He has published over a hundred scientific papers and about a dozen computer programs, the best known of which are Nona, Piwe, TNT (for phylogenetics), and VNDM (for biogeography). He is a Fellow Honoris Causa of the Willi Hennig Society and served as president of the society from 2004 to 2006. Since 1995, he has been regularly teaching courses on phylogenetics in Argentina and about a dozen countries.

Author Biography

Abbreviations

Acctran	A character state reconstruction that, in the face of ambiguity, implies that transformation occurs as close to the root as possible
AIC	Akaike information criterion, a method to choose among statistical models based on minimizing loss of information
AFD	Absolute fit difference (used in the standard Bremer support)
CPP	Clade posterior probability
CSS	Constrained sectorial searches
Deltran	A character state reconstruction that, in the face of ambiguity, implies that transformation occurs as far from the root as possible
DC	Distortion coefficient, a measure of tree (dis)similarity
DMC	Discrete morphological characters
DSS	User-defined sector
FWR	Frequency within replicates, a method to summarize results from resampling
GM	Geometric morphometrics
GUI	Graphical user interface
HTU	Hypothetical taxonomic unit, reconstructed internal node of a tree
ILS	Incomplete lineage sorting
LBA	Long branch attraction
MAP	Maximum a posteriori tree (resulting from an *MCMC*)
MCMC	Monte Carlo Markov chain (Bayesian analysis)
$\mathbf{M}_{DG}$	In model-based analyses, model under which the data truly evolves (known with certainty only when performing a simulation or formal mathematical analysis)
ML	Maximum likelihood
MLT	Most likely tree
MP	Maximum parsimony
$\mathbf{M}_{PI}$	In model-based analyses, model assumed to evaluate trees for phylogenetic inference (as opposed to M_{DG}, model for generating data)
MPR	Most parsimonious reconstruction; a set of parsimony-optimal ancestral state assignments
MP-set	Set of states that occur in an *MPR*, at a given node
MPT	Most parsimonious tree
NCM	No Common Mechanism
NFL	No Free Lunch theorem, stating that no optimization method can perform best across all possible datasets
NNI	Nearest neighbor interchange
NP	Number of parameters
$\mathbf{NPT}_{(n)}$	Number of possible trees for *n* taxa
$\mathbf{NPTr}_{(n)}$	Number of possible rooted trees for *n* taxa, equal to $NPT_{(n+1)}$
O(s)	Number of operations needed to calculate results as function of the size s of the problem; computational complexity

OTU	Operational taxonomic unit
PP	Posterior probability (in Bayesian statistics)
RAS	Random addition sequence
RF	Robinson-Foulds distance
RFD	Relative fit difference
RSS	Random sectorial search
SLM	In model-based analyses, a standard likelihood model (i.e. a Poisson model where branch lengths differ across tree but are similar for all characters)
SPR	Subtree-pruning regrafting
SS	Sectorial searches
SVG	Scalable vector graphics
TBR	Tree-bisection reconnection
TF	Tree fusing
TSA	Transformation series analysis
XSS	Exclusive sectorial search

1 Introduction to Phylogenetics

This is a book about systematics—that is, establishing relationships among taxa—with emphasis on morphological data. With the acceptance (starting in the 70s) of the intimate connection between classification and phylogeny, it became evident that systematics is central to all of biology. Classifications are thus not only a way to name things (scientific names being just a convention), but in their function of hypotheses of relationships and statements about the existence of natural groups, they are also devices that facilitate making comparisons, transmitting information about organisms, and making predictions about as yet unobserved characteristics. This remains as true today as it was in the days of Aristotle, when he produced the first general classifications, absurd and arbitrary if judged from the optic of today's knowledge but which at the time surely enhanced his own way of understanding relationships and making better sense of the biological world. Classifications provide a means for abbreviated transmission of information through the association of characteristics with groups. In biology, the goal of explaining is often contrasted with that of merely describing, but the distinction between these two is blurred. The distinction is blurred not only in biology; the equations describing the speed and position of a falling body are also predictive—of where the body will be at any given time. Are those equations descriptive or predictive? In generalizing the description beyond a single observed instance (and instant), they can be seen as making predictions and, in a sense, providing explanation. The same is true of biological classification.

The features of organisms which classification helps summarize and understand are all kinds of intrinsic attributes, or *characters*: morphological in the broad sense (i.e. physiological, behavioral, and anatomical), as well as molecular. Some authors (e.g. Nelson and Platnick 1981) have argued, correctly in my opinion, that extrinsic attributes should not be considered in deciding classifications. Extrinsic attributes are, for example, those related to time (what is the geological horizon at which we find this species?) and space (what is the geographical distribution of this taxon?). Such attributes no doubt can provide information about phylogeny, and thus, in the final decision of which scheme of phylogenetic relationships is best, there is no reason to prescribe ignoring them (as correctly pointed out by proponents of so-called "stratocladistics"; Fox et al. 1999; Fisher 2008). Systematics, however, has traditionally been concerned mostly with finding evidence from intrinsic attributes only; determining the patterns of distribution of intrinsic attributes is a legitimate field of inquiry and is based on considerations and mechanics rather different from those applicable to extrinsic attributes. The discipline concerned with the study of intrinsic characters deserves a name, and the name best associated with such endeavor is *systematics*. Systematics is thus not concerned with phylogeny per se; rather, systematics

DOI: 10.1201/9781003220084-1

is concerned with a phylogenetic understanding of intrinsic characters. A classification is, then, strictly speaking, a hypothesis based solely on features observable on the organisms themselves; the explanation for many of those features being associated with groups is the standard evolutionary or phylogenetic account: the features characterizing groups have originated in the most recent common ancestor of the group and were inherited by the descendants. A classification is then interpretable as a phylogeny, and among different classifications—phylogenies—we should select those that best explain observed (and observable) features.

1.1 LOGICAL AND CONCEPTUAL ASPECTS OF PHYLOGENETIC ANALYSIS

The main goal of systematists is in producing classifications which—in their interpretation as phylogenies—help understand character distributions. Given a series of observations, the problem is then in how to best summarize that information. More properly: given that the summary is (at least in the case of morphology) always made by means of trees and resorting to genealogical explanation of observed similarities, the problem is how to measure which tree(s), among all possible ones, best allow summarizing and understanding observations about characters in genealogical terms.

Those goals of taxonomy and systematics naturally lead to preferring the trees that minimize the requirements of *homoplasy*—homoplasies are similar but independently evolved similarities. The criterion of comparing the requirements of homoplasy is widely used for measuring the degree to which different trees fit the data and is also known as the *parsimony* criterion (often abbreviated as *MP*, for *maximum parsimony*). Systematists have often used the term *cladogram* as a near synonym of a phylogenetic tree and *cladistic analysis* as a synonym of phylogenetics; although this usage is not universal, the term *cladist* in practice generally denotes a researcher who infers phylogeny using parsimony, and this is how the term is viewed in this book. Evolutionary biologists who use methods other than parsimony (often, molecular or evolutionary biologists who tend not to be taxonomists) use the term *phylogenetic* almost exclusively. Cladistic analysis is thus a subset of phylogenetic analysis, but since this book focuses on the use of parsimony, the two terms will be used more or less interchangeably here. Readers should keep in mind the distinction, however.

The term *parsimony* can be used with different meanings, and this has caused some confusion regarding the reasons to use and justify that criterion in phylogenetics. In philosophy, parsimony is commonly taken to measure the simplicity of hypotheses. A hypothesis that Pluto moves in a circular orbit is simpler than the hypothesis that it moves in an elliptical orbit: it can be specified with fewer parameters. Note that the hypothesis that Pluto moves in a circular orbit is simpler, but observations do not fit that hypothesis; the more complex ellipse does fit the observations and is thus preferable despite being more complex. Given several alternative hypotheses, other things being equal (i.e. if all hypotheses fit the data equally well), hypotheses that are in themselves simpler are generally to be preferred, and there is no big controversy over this principle. It is the same idea implemented in model testing, under statistical approaches to phylogenetic reconstruction (see Chapter 4). In a sense, any *theory* produced by a phylogenetic analysis is equally "complex" (a binary tree); all phylogenetic hypotheses are under this view equally complex, and thus this sense of the word *parsimony* is not too helpful at the present point.

In phylogenetics, however, parsimony is taken to have a specific meaning: the *minimization of homoplasy*. It was stated prior that the goals of taxonomy and systematics naturally lead to the criterion of parsimony, which is to be interpreted in the sense that those goals naturally lead to the idea of minimizing homoplasy, not to the idea of philosophical parsimony. It is never known whether the similarity in a given feature for two different species is truly a homology, but it can be interpreted as such—when that similarity is traceable to a common ancestor or node of the tree (more on this in Chapters 2 and 3). Observations are fixed when effecting a phylogenetic analysis, but different trees display different groups of species, thus differing in the degree to which similarities can be considered as homologies or homoplasies. That is, some trees require larger numbers of independent originations of similar features, and those are less useful in understanding character distributions in genealogical terms. Note that homology and homoplasy are complementary concepts; thus, by definition, minimizing homoplasy amounts to maximizing homology.

What needs to be justified then, in the case of phylogenetic analysis, is not the general philosophical principle but parsimony *qua* minimization of the number of independent originations of similar features required by the tree—that is, homoplasy. The most important discussions of the logical basis of parsimony in phylogenetics have been provided by Farris (1979b, 1982b, 1983; see also Farris 2008), and the treatment that follows is based completely on his. He has defended parsimony a.k.a. minimization of homoplasy on a number of interrelated properties: maximization of explanatory power and conformity to evidence, minimization of ad hoc hypotheses (i.e. those required to defend a theory against falsifying evidence), and maximization of descriptive power. Although the treatment that follows does not add much to that from Farris (1983, 2008), many workers (from de Queiroz and Poe 2001 and Rieppel 2003 to more recent ones such as Wright 2019) seem to have been oblivious of most aspects of this justification for parsimony, so the points bear repetition.

1.1.1 Explanatory Power

The sense in which trees can provide explanation is limited to a single aspect: genealogical explanation. A tree by itself, in fact, does not explain anything; it does so only by conjoining the scheme of descent embodied in the tree with a specific mechanism for transmission of features. That mechanism of transmission is descent with modification: descendants inherit their features either unmodified from their ancestors or in a modified form. An example is the tree in Figure 1.1a: the species A–F are observed to have a given feature (e.g. *black*) and G–L to have the alternative (e.g. *gray*), but the ancestors of the tree (internal nodes) have not been assigned any states; thus, the tree of Figure 1.1a by itself (i.e. without a coloring of the ancestors) makes no concrete statements on the reasons for the similarities. With these distinctions, the prior meaning can be clarified: the tree is a *vehicle* to provide genealogical

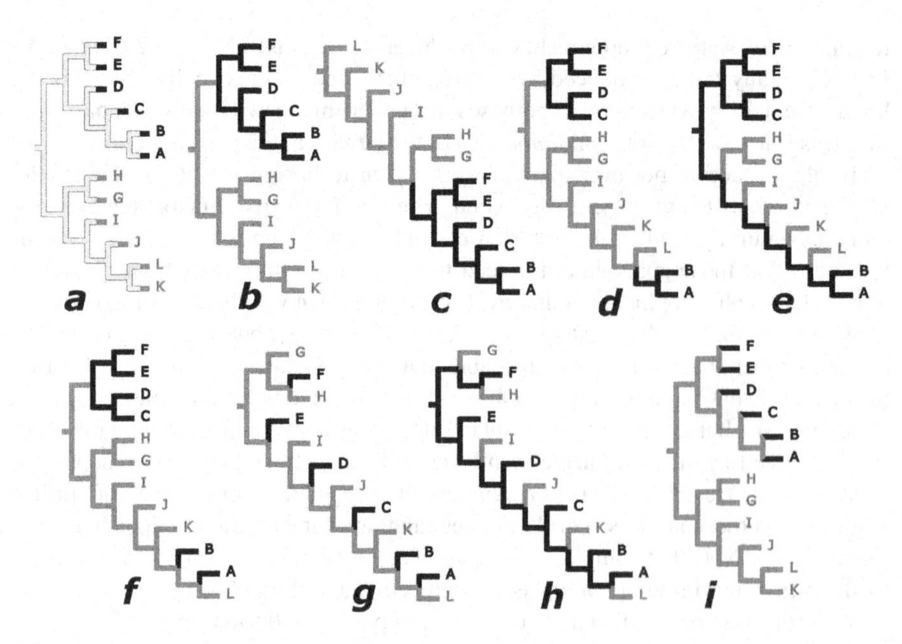

FIGURE 1.1 Examples of possible interpretations of a given character state distribution on alternative trees. Case **(a)** makes no interpretation; only the terminal character states are given. **(b)** Explanation of the observed similarities in terms of common ancestry is provided only when ancestors are assigned a color. **(c)** Similarities in both the *black* and *gray* state can be completely explained by common ancestry even if taxa with one of the colors do not form a monophyletic group. **(d)** Ability to explain similarities by common ancestry decreases as the taxa with the *black* states are separated in two groups. **(e)** Given the same tree as in (d), the similarity in *black* state can be explained by common ancestry, but this requires that similarities in the *gray* state are not due to common ancestry. **(f, g)** The ability to explain observed similarities by common ancestry continues decreasing as taxa with the same color are placed in more separate groups of the tree. **(h)** On the same tree as (g), an alternative equally parsimonious explanation attributes to common ancestry all similarities in *black* color (instead of *gray*). **(i)** Even when all the taxa with *black* color form a group in the tree, it is possible to assign ancestral colors such that no similarity is due to common ancestry; this explains nothing but is a logical possibility.

explanation, via common descent, of similarities observed among different species. Such an explanation is only made when we assign states to the ancestors (*gray* or *black*) in the tree of Figure 1.1a, thus producing tree 1.1b; the tree itself is the same, but now there is a particular account of how the character evolved.

The question to be asked (the "why" part) is then, "why are these two species, A and B, alike in being *black*?" The explanation (the "because" part) consists of responding, "because they inherited that color from a common ancestor". That is the *genealogical* explanation; other explanations may also be possible and—depending on the point of view of the investigator—more relevant. For example, answering that A and B are alike because of stabilizing natural selection may also be correct (for the evolutionary biologist), as well as explanations invoking genetic (for the geneticist) or developmental causes (for the embryologist). The systematist is interested mostly in the genealogical explanation, and this is what trees help provide.

An important point is that trees contribute to understanding character distributions by explaining similarities, *not* differences. On the question of why taxa A–F differ from taxa G–K in being *black* instead of *gray*, even a tree placing A–F and G–K in two separate groups (such as Figure 1.1b) contributes nothing. Again, different levels of explanation may be possible and desirable; an ecological or evolutionary theory may shed light on the causes for the difference. The tree described (i.e. Figure 1.1b) can only help explain why A–F are similar and why G–K are similar.

A tree that places taxa A–B separate from C–F (for example, nested within the group with *gray* color, as in Figure 1.1d) can still explain *some* of the color similarities between taxa but fewer than the tree with *black* and *gray* in two separate groups. For example, on tree 1.1d we can explain the color similarity between A and B, and that between C and D, by reference to common inheritance, i.e. homology. However, the color similarity between A and D, or that between B and E, cannot be explained by reference to common ancestry. Note, however, that those two comparisons are not independent. As discussed in more detail by Farris (1983: 19), the count of cases of unexplained similarity must consider only independent pairwise comparisons. Comparing A with D seems independent from comparing B with E, but for any assignment of ancestral states where A shares its color with B as homology, and D its color with E as homology, then it follows that the color in A being non-homologous with that in D implies by logical necessity that the color in B is not homologous with that in E. The count of independent instances that are not attributable to common ancestry then changes directly with the number of independent originations of similar features implied by a set of ancestral state assignments.

In the example of Figure 1.1d, the colors are assigned to the ancestors so that the largest possible number of similarities are considered as due to common inheritance. Were one convinced, by divine revelation for example, that the color in A and D is indeed shared by common ancestry, it is still possible, given that tree, to force such interpretation; and more generally, any set of taxa with a similar condition can be considered to have the shared condition by common ancestry. This can be done simply by hypothesizing that all the ancestors connecting the taxa with the shared common state have that same state, as in Figure 1.1e—the same tree as in Figure 1.1d, but with a different set of ancestral state assignments. The problem with such a combination of ancestral states should be obvious: while it attributes all similarities in *black* color to common ancestry, it attributes almost no similarity in *gray* color to common ancestry (i.e. only between G and H). Therefore, given the tree topology in Figs. 1.1d and 1.1e, a tradeoff is necessary, and attributing to common ancestry more similarities in one state requires unavoidably considering more similarities in the other state as unexplained. The largest number of similarities explained, given the tree topology, is when we hypothesize the character to have evolved as in Figure 1.1d; the explanation of the observations provided by Figure 1.1e is clearly inferior in that regard.

It should be clear, at this point, that not all trees allow attributing to a common origin the same numbers of features. For the tree in Figure 1.1a, for example, we can explain every single instance of similarity as due to common ancestry by hypothesizing ancestors to have had the colors indicated in Figure 1.1b. But for the tree topology of Figure 1.1d, no matter how hard we try, explaining every similarity by common ancestry is not possible. And other trees can leave unexplained even greater numbers of similarities, as in Figure 1.1f, where we no longer can explain the color shared by A and B as inherited from a common ancestor. The lower limit is given by tree in Figure 1.1g, where no similarity in *black* can be attributed to common ancestry and six independent derivations of *black* occur; no tree can do worse than that. The ancestral colors on that tree could have been reconstructed as all *black* (Figure 1.1h), in which case six independent derivations of *gray* would have occurred, but (assuming *black→gray* transitions as equivalent to *gray→black*) this is exactly the same number of unexplained similarities. This could change if there are grounds to prefer explaining the similarity in some of the features over those of others; in that case, the counts should take into account the relative importance of each of the alternative conditions. A typical example is in the presence or absence of insect thoracic wings: it is often considered that, given alternative possibilities, explaining by common ancestry the shared presence of wings is preferable to explaining the absence of wings. The logic of the counts is exactly the same; in the example, equivalence between all conditions of a character is assumed for simplicity, but the criterion of parsimony requires no such equivalence.

The preceding discussion shows, then, that the degree to which observed similarities can be explained by reference to common ancestry on a given tree depends directly on the minimum number of independent derivations of similar features required by the tree. Every independent derivation, or *step*, amounts to a hypothesized *transformation* from one condition to the other.

1.1.2 Ad Hoc Hypotheses

Scientific hypotheses or theories make predictions (about what observations should be made), and this is what makes them both testable (amenable to empirical rejection) and useful (accepting a hypothesis changes what we expect to observe in the real world). But it is clear that empirical rejection can never be absolute. When a theory appears falsified by non-conforming evidence, there are always two possible courses of action: retain the theory and dismiss the evidence or reject the theory and accept the evidence. As theories—particularly in biology—are rarely of a level of generality and precision such that all the evidence seems in conformity with the theory, the choice is typically between theories that are falsified more or less—any theory will be unable to explain some of the evidence, and falsification is never absolute (cf. Lakatos's 1980 "sophisticated falsificationism"). The rationale for dismissing apparently falsifying evidence can take different forms (e.g. "incorrect observation", "defective microscope"), but the important aspect is that such dismissal is being made solely for protecting the theory against rejection, hence the name: the dismissal is made only on the ad hoc grounds that the hypothesis must be true and thus the observation must be incorrect.[1] Of course, evidence previously thought to reject a hypothesis and discarded only for the sake of protecting it against rejection may later be found to be dismissible on separate grounds (e.g. a subsequent inspection showing that the microscope is indeed defective and produces incorrect observations), but in that case there is no longer a falsifying observation. The term *ad hoc hypothesis* is reserved for those cases where the observations must be hypothesized to be incorrect only for preserving the theory, *without any external grounds to do so.* It is clear that to the extent to which ad hoc hypotheses are allowed freely, the connection between theories and observations is destroyed. Thus, ad hoc hypotheses must be minimized, and this is one of the senses in which the term *parsimony* is often used in philosophy. Note that minimization of ad hoc hypotheses is not the same thing as logical simplicity or the posing of few parameters, already discussed. This is a source of confusion in some claims (e.g. Steel and Penny 2000; Steel 2005; Huelsenbeck et al. 2008) that parsimony (as that criterion is used in phylogenetics) resorts to an "unparsimonious" model of evolution; the criticism stems from mixing alternative senses of the term *parsimony*.

Every instance of homoplasy can be seen as a hypothesis ad hoc. When two taxa in separate parts of the tree have exactly the same feature, we must assume (if we are to keep the tree) that the similarity, despite all appearances, is in fact due to homoplasy. For example, by logical necessity, in Figure 1.1d the similarity between taxa A and C is a parallelism (or the similarities between G–L are, as in Figure 1.1e, but that interpretation requires an even larger number of ad hoc hypotheses, so it is not considered further). If the similarity between A and C is truly due to inheritance (and so is the similarity between G–L), then it follows that the tree in Figure 1.1d is an incorrect depiction of relationships. Of the trees shown in Figure 1.1, only trees 1.1b or 1.1c could be correct (along with many other possible trees on which the color can evolve with a single step). The idea that the similarity between A and C is due to a parallelism, then, if we are to accept the tree in Figure 1.1c, stems exclusively from our desire to retain the tree; no external evidence supports that hypothesis of a parallelism—it is required by the tree itself, and the hypothesis is thus ad hoc.

1.1.3 Logical Asymmetry

Farris (1983) pointed out an important logical asymmetry in how trees behave relative to homoplasy. One might be tempted to think that, just like concluding homoplasy is required to keep the tree 1.1d, concluding homology is required to keep the tree in 1.1b—that is, keeping the tree in 1.1.b requires that we consider *black* as truly shared by inheritance. This is definitely not so. Given the tree topology in Figs. 1.1a–b, it is logically possible for the character to have evolved as in Figure 1.1i, where every single species with *black* acquired its color independently of the others (such scenario may not be likely, or credible, but it is *logically* possible). Put differently: if *black* was indeed acquired independently by every species with that color, this does not make the tree in Figure 1.1b false; it instead makes color irrelevant for choosing among trees. But if *black* was indeed acquired through common ancestry (as well as *gray*), then this makes trees compatible with that possibility (as the one in Figure 1.1.b) the only viable alternatives, making a tree like the one in Figure 1.1g logically impossible: on tree 1.1g there is no way to have the *black* color shared by common ancestry (at least, not unless we postulate that the *gray* color originated in every species independently, as in Figure 1.1h). Incorrectly identified homologies do not make the trees they support automatically false, but properly identified homologous similarities support some trees by truly making the alternatives impossible. Thus, parsimonious trees are compatible with the idea of low or high amounts of homoplasy; unparsimonious trees, instead, are compatible only with high amounts of homoplasy. While parsimony is often criticized on the grounds that it assumes evolution proceeds parsimoniously (e.g. Camin and Sokal 1965; Felsenstein 1979, 1981), the truth is that the situation is the opposite: unparsimonious trees force us to believe evolution proceeds unparsimoniously, but parsimonious trees require nothing in that regard.

1.2 TREES AND MONOPHYLY

A tree accepted as the current hypothesis of relationships defines a series of *monophyletic* groups—that is, groups which correspond to a branch of the tree. A tree could equally well be represented as a series of internested subsets—every branching of the tree defining two (or more) subsets of the more inclusive group. All groups, or *taxa* (sing. *taxon*), named in a classification need to correspond to a group in the tree, but not all groups in the tree need to be named—this depends on whether establishing a name for the group is considered useful or practical. The term *monophyletic*, in actual phylogenetic practice, is relative to the problem at hand: of all the species being considered in the dataset, those that are more closely related among themselves than with any other species not included in the group. That is, the group which includes all the descendants of the most recent common ancestor of the group. Groups defined artificially on the basis of independently derived characters are termed *polyphyletic*, while groups containing some but not all descendants of the most recent common ancestor of the group are termed *paraphyletic* (typically defined on the basis of shared primitive features, see the following). The group (or species) most closely related to another group X is said to be the *sister group* of X (note that the sister group may be a single species). Hennig (1966) was one of the first authors to insist that a classification should contain only monophyletic groups and that sister groups should have equal rank in the classification; while this view was controversial in the early days of phylogenetics, it is accepted today without discussion. The main differences in alternative approaches to phylogenetic reconstruction lie on the methods used to hypothesize the genealogical groups themselves.

Defining groups in the classification only from monophyletic groups has several important consequences in terms of what "evolution" is considered to be and the kind of statements that can be logically made regarding ancestry. A requirement of a phylogenetic system is that sister groups must have equal rank; as evolution proceeds, subdivisions of a group must always be subordinate to the original group. Prior to Hennig (1966), statements about some taxon being ancestral to another were common; for example, "Reptiles are ancestral to Birds". At the level of higher taxa, when group membership is defined on the basis of monophyly, such a statement is illogical and self-contradictory: if some "Reptiles" are more closely related to Birds than to other "Reptiles", then "Reptiles" is not a group. There is divergence among authors when it comes to whether species-level taxa (Wilkins 2009 provided a historical review of the "species problem") can be ancestral. Some authors (notably Lovtrup

1987, with this view apparently also embraced by Nelson 1989b and some points in common to that of de Queiroz and Donoghue 1988) maintained that species differ from other taxa in the hierarchy only by virtue of not being (or not having been found to be, at least yet) further subdivisible and that taxa in general (species included) are not evolutionary units. That is, taxa simply subdivide, and there is no origination of new taxa in the phylogenetic process (only new sub-taxa). On this view, which is perhaps the most logically consistent one, a species may become divided in two *sub*species, never in two species; alternatively, the original "species" would now be called a "genus" and the two new forms "species". This also means that the members of the original undivided form, if they could be found and identified, would now be classifiable only to the level of genus, belonging to no nameable species. From this perspective, evolution is seen only as the origination and modification of characters (or character states) in the actual process of descent among individuals, not as the origination of taxa. The phylogenetic tree (that is, the classification) provides the proper scaffold to understand such modifications along time. I assume that the reader has some familiarity with these ideas and the existing alternative views on this topic; otherwise, the reader is referred to books that deal in more depth with such problems (Eldredge and Cracraft 1980; Wägele 2005; Wiley and Lieberman 2011; Baum and Smith 2013; Hamilton 2014).

1.2.1 Tree Terminology

Figure 1.2 illustrates the basic components of trees. The trees connect (via *branches*, sometimes called *edges*) the observed taxa to *hypothetical ancestors*, postulated to have been intermediate between the observed taxa (and whose morphology is reconstructed as part of the phylogenetic analysis, at least in some of the methods for inferring phylogenetic relationships). Reconstructed ancestors (i.e. the character states that can be parsimoniously assigned to ancestral nodes) are often called hypothetical taxonomic units, or *HTU*s, by analogy to the "operational" taxonomic units of the pheneticists, *OTUs*, or observed taxa. The entities connected by branches are known as *nodes*; observed taxa are often referred to as *terminal nodes* (sometimes simply as *terminals* or *leaves*) and hypothetical ancestors as *internal nodes*. Tree branches themselves are sometimes classified as *internal* (connecting two internal nodes) or *terminal branches* (leading to a terminal taxon). The node at the base of the tree is known as *root* and has no ancestor specified; all other nodes of the tree have a single ancestor (a tree where some node has more than two "ancestors" amounts to a tree with loops, or network, which can be used to represent events of hybridization or horizontal gene transmission, beyond the scope of this book). Internal nodes lead to two or more *descendants*; observed taxa have no descendants. When no internal node leads to more than two descendants, the tree is said to be *binary* (or *bifurcating*, or *fully resolved*); otherwise, the tree is *polytomous*. An individual node that leads to two descendants is said to be *dichotomous*; numbers of descendants of an internal node (often called the *degree* of the node) can be used to classify the nodes in *trichotomous* or *tetrachotomous* (higher degrees of furcation rarely receive a name). Trees can be represented as sets of relationships or groups; a common way to represent

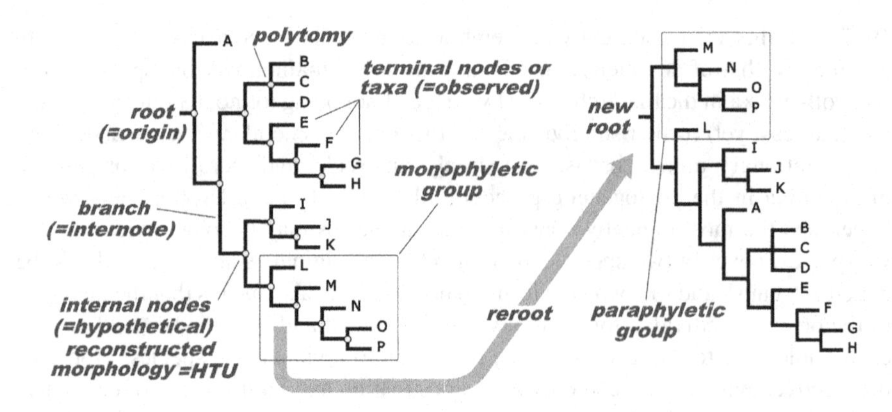

FIGURE 1.2 A tree for taxa A–P, indicating the terminology. The tree in the left can be rerooted by "pulling down" the tree from the branch indicated, producing the tree in the right. Some groups that are monophyletic in the first rooting (e.g. L–P) become paraphyletic with the new rooting.

them is by parenthetical notation, where each of the groups of the tree is represented by a set of nested parentheses. Thus, the notation (A(B(CD))) represents a tree where C and D share a most recent common ancestor not shared in turn with A or B, B is the sister group of C + D, and A is the sister group of the three other taxa. To better understand parenthetical notation, you can study the example TNT script ***parent-rees.pic***, which reads and saves trees in parenthetical notation, bypassing TNT's internal functions. Internally, TNT uses lists of ancestors and descendants to handle trees; these can be interconverted into and from parenthetical trees or converted into tree diagrams (e.g. as in the TNT script ***treeplotting.pic***).

A tree can be *unrooted* when it has no direction, simply establishing interconnections between observed and internal nodes but without specifying which of the two connected nodes is the ancestor and which the descendant. A tree which is unrooted establishes no groups; only designating a node as root ("pulling" the tree down from that node so that now the root is "below" all the other nodes in the tree) imposes a direction on the tree and allows for some groups to be monophyletic. For some combinations of terminals, no possible rooting of the tree will make them monophyletic [for example, in ((AB)(CD)), there is no way to root the tree such that B+D form a monophyletic group]. Thus, an analysis producing an unrooted tree constitutes a partial test of the monophyly of a group. Each one of the branches of an unrooted tree establishes a *partition* of the taxa (i.e. the taxa in one and the other side of the branch in question); the partition established by a branch connecting an internal and a terminal node is trivial (terminal nodes are established before the analysis). Of the two groups in a partition determined by each internal branch, only one will be monophyletic in most rootings, with the other one paraphyletic (except when the root is chosen to be in the same branch connecting the two partitions, in which case both partitions are monophyletic). Partitions in an unrooted tree are often denoted with a vertical bar, indicating the groups of taxa in each partition at both sides of the line.

Thus, the partition AB|CD can be rooted so as to produce the tree (A(B(CD))), but the partition AC|BD cannot.

The numbers of branches, nodes, and possible distinct trees are determined, as logical possibilities, by the number of leaves in the tree. In the case of unrooted trees for three terminals, there is a single tree, with three branches interconnecting the observed taxa through a single internal node. As a fourth taxon is added to this unrooted tree, it could be connected to each of the three existing branches (each connection producing a distinct unrooted tree), so that for four taxa three different unrooted trees are possible (AB|CD, AC|BD, and AD|BC). As each taxon is added, a new node must be created, dividing in two one of the existing branches, and a branch must be added to connect the taxon and the new node. Thus, for each terminal taxon added to the tree, an additional node is created (starting with one node for three taxa, so that the number of nodes in binary trees for T taxa always equals $T - 2$), and two branches (starting with three branches for three taxa, so that the number of branches in binary trees of T taxa always equals $2T - 3$, determining the same number of partitions, of which $T - 3$ are non-trivial). As another taxon is added to an unrooted tree with N taxa, it can be placed in any of the $2N - 3$ branches, so that the number of unrooted trees for T taxa, $NPT_{(T)}$, is the double factorial, $(2T - 5)!!$.

The progression of the number of binary rooted trees, $NPTr_{(T)}$, is similar to that for unrooted trees, except that rooting the unrooted tree in each of its branches produces a distinct rooted tree. Thus, the number of binary rooted trees for T taxa equals the number of unrooted trees for $T + 1$ taxa, or $NPTr_{(T)} = (2T - 3)!!$. The earliest derivations of numbers of trees were done by Schröder (1870); Felsenstein (1978a) provided more general equations for numbers of labeled trees (including formulae for polytomous trees as well). As the number of taxa grows, the number of possible trees grows superexponentially. For example, for even a modest $T = 15$, the number of possible binary trees is $3 \times 5 \times 7 \times 9 \times \ldots \times 27 = 2.135 \times 10^{14}$; for $T = 100$, the number of possible binary trees is 3.350×10^{184}. The number of possible trees to consider is overwhelming, and constitutes the major challenge of any phylogenetic analysis (see Chapter 5).

If non-binary trees are counted as well, the number is even larger. Computer programs for phylogenetic analysis carry out most operations on binary trees, eliminating the branches that lack character support to produce polytomies only in a subsequent step (see Chapters 5–6). In phylogenetic analysis, polytomies are generally understood as an expression of ignorance (i.e. we do not know how to resolve the relationships of the groups connected to the polytomous node) rather than as a positive statement that multiple speciation occurred. The impossibility of finding support leading to prefer a polytomy over its alternative resolutions is further discussed in Chapter 3 (see Section 3.10.1).

1.3 PARSIMONY, SYNAPOMORPHIES, AND ROOTING

The criterion of parsimony is sometimes characterized (e.g. Felsenstein's 1982 version of "Hennig's dilemma"; see discussion in Farris and Kluge 1997) as needed only when characters are in conflict—as if a set of perfectly congruent characters univocally determined a tree. Even for such a set of congruent characters, it will be possible to postulate trees which imply some homoplasy. The reason why the tree implying a single step for each of the characters is preferred is precisely because that tree is most parsimonious, not because of any external criterion independent of explanatory power. When Hennig (1966) introduced his methods of phylogenetic analysis, he proposed that groups are recognized by the shared possession of derived features. He called these *synapomorphies* (apomorphy shared by several taxa, from the greek *syn*, "together", and *apo*, "from" or "off"); synapomorphies are often contextually distinguished from *autapomorphies* (apomorphies found in a single species, hence defining no groups; this depends on the context of the analysis, so that adding more species to the data may well change some *aut*-apomorphies into *syn*-apomorphies). Hennig (1966) famously argued that similarity in unmodified characters, or plesiomorphies, is no evidence of monophyly. It had long been recognized that homoplasies (often distinguished in *convergence* and *parallelism*, a distinction that is not relevant in the context of phylogenetic analysis) are no evidence of monophyly, but the idea that plesiomorphies need to be distinguished from apomorphies was controversial. One of Hennig's (1966) main contributions was to convince the taxonomic community that plesiomorphies (even if originating only once) cannot serve to identify monophyletic groups (he was the most effective but not first to argue that point; see Moody's 1985 discussion of Camp 1923; Szumik's 1996 discussion of Davis 1938, 1940).

In that view of Hennig's (1966), for a case like the one shown in Figure 1.1b, only *one* of the colors could legitimately be used to recognize a group. Suppose that the tree is rerooted so that L is now at the base (and GH, I, J, K are then the successive sister groups of the *black* group, as in Figure 1.1c). In that case, the change from *gray* to *black* in the group of A–F constitutes a synapomorphy; placing some of A–F among separate *gray* species would make it impossible to explain the similarity in *black* as due to common ancestry. The notion of grouping by synapomorphy, therefore, is a consequence of the parsimony criterion and not a tenet of cladistic analysis that can be maintained separately from the goal of explaining shared similarities. Note that both trees (the ones shown in Figure 1.1b and 1.1c) can account for every instance of color similarity in terms of common ancestry, yet the tree 1.1c does not display the taxa with *gray* as a monophyletic group. On that tree, the similarity in *gray* color acts as a plesiomorphy, thus not defining a group. Note that from Hennig's (1966) point of view, plesiomorphies do not define groups; this was often erroneously interpreted to mean that plesiomorphies are considered *uninformative* in a Hennigian system. Plesiomorphies *are* informative: they inform us that the bearer of the plesiomorphic condition does *not* belong to the group defined by the alternative state.

Given that cladistic methodology was initially premised on the principle of grouping by synapomorphy (e.g. Brundin 1966; Wiley 1975; Platnick 1979; Eldredge and Cracraft 1980), early criticisms of cladistics attacked the idea that one could determine prior to an analysis which of the alternative conditions of a character is primitive and which derived (a problem that became known as character *polarity*), as well as claiming that grouping by synapomorphies meant that classifications completely ignored plesiomorphies. These are two strawmen, even from the perspective that grouping by synapomorphy (instead of parsimony) is the fundamental tenet of cladistics, as shown by some papers of the time (e.g. Wiley 1975), but after Farris (1983) the points can be more easily refuted on the perspective that the guiding principle of cladistics is parsimony and maximization of explanatory power, not grouping by synapomorphy.

Rereading preceding sections, on connecting parsimony and explanatory power, observe that the goal is explaining observed similarities between taxa: *all* similarities, regardless of whether they are similarities in primitive or derived characters. The question of whether a feature is primitive or derived never enters the picture when measuring the degree of explanatory power of a tree. A similarity that cannot be accounted for by common ancestry—i.e. any two similar taxa that are not connected via ancestors with that same state—worsens the parsimony score of the tree, regardless of whether that similarity is in a primitive or derived state. In fact, the very idea of a state being in itself "primitive" or "derived" is incorrect. Consider the case of a group with several species, at the point of origin of which a character changes from state x to y, and then (several nodes down the tree) a single species A reverts back to the condition x. Which condition is derived, x or y? In that hypothetical situation, the question is simply unanswerable without further specification: we need to specify the point of the tree to which we refer—that is, the specific hypothesized transformation: in the transformation at the point of origin of the group, x is derived relative to an ancestral y; at the transformation in the single reverting species A, y is derived relative to an ancestral x. The states themselves, x and y, cannot be characterized as primitive or derived in the abstract; they can be so classified only by reference to a specific transformation in the tree. Thus, classifying the states of a character into primitive or derived prior to the analysis is neither required nor logically possible. This is further discussed in Chapter 3, after description of specific algorithms for ancestral reconstruction which facilitate understanding additional subtleties.

The emphasis on synapomorphies as exclusive indicators of monophyly, following Hennig's (1966) formulation, is a useful approximation when datasets are relatively small and congruent and phylogenetic trees can (or need to, as in the 60s) be done by hand. If most characters are congruent, then thinking of character states as "plesiomorphic" and "apomorphic", and trying to think of groups defined by the shared possession of apomorphies, is an expedient way to approximate the problem and facilitates assembling the tree by eye. The state believed to be apomorphic then can be indicated as 1, the plesiomorphic as 0, and the 1 states can be considered as defining groups. An example is shown in Figure 1.3, where most such groups are compatible (i.e. either disjoint, or containing one another), and character 5 can be easily identified as the discordant character since it "crosses" the boundaries of several other compatible characters. This expedient approach required polarizing the characters before the analysis, which was most often done by means of the outgroup criterion (i.e. establishing the character states outside of the study group; Watrous and Wheeler 1981; Maddison et al. 1984; this is in fact just an application of parsimony, see Farris 1982a) or the ontogenetic criterion (Nelson 1978). But having a quick approximation is not the same as considering the approach to be the conceptual foundation of cladistics; the conceptual foundation is the criterion of parsimony and maximization of explanatory power, as outlined prior.

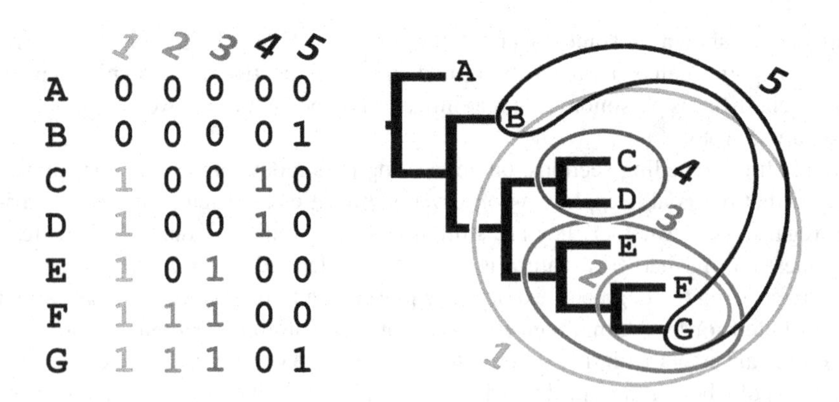

FIGURE 1.3 Although the ultimate justification for selecting trees is parsimony (i.e. preferring trees where the maximum possible similarities can be explained by common ancestry), Hennig's notion of "grouping by synapomorphy" facilitates mental analysis, thinking about groups defined by the apomorphic state.

The confusion of considering the useful quick approximation of grouping by synapomorphies as the foundational basis of cladistics has led some authors to make proposals that are hard to defend, elevating synapomorphies to an absolute evidentiary status they cannot possibly have. Two related examples are the proposals that (a) only synapomorphies represent positive evidence and need to be explained (and thus accounted for by groups), and (b) the notion that homology equates to synapomorphy so that plesiomorphies cannot be considered homologs. These two views originated in the early 80s, mostly from Nelson and Platnick (1981) and Patterson (1982b) respectively, but were maintained by several other authors as well (e.g. Ebach et al. 2013; Brower and de Pinna 2012). In this view,

> the fact that both lice and ticks lack spinnerets [as present in spiders] is no more in need of explanation than is the fact that they lack mammary glands, or the fact that they lack Cadillac engines in their stomachs. Regarding such shared 0 entries as meaningful, and in need of explanation, is simply a remnant of phenetics, and one that systematists would do well to discard.
>
> (**Platnick et al. 1996;** the Cadillac example is taken from **Nelson and Platnick 1981: 29** and repeated by **Platnick 2012).**

On that view, what the group "spiders" serves for is explaining why spiders have spinnerets, but that was never Farris's (1983) argument on how genealogical explanation is achieved. Genealogical reasoning can only explain why all spiders are *alike* in having spinnerets, not why they have them in the first place; there is a difference, and not a subtle one. What genealogy can explain is not why lice and ticks (and all the rest of life) lack engines in the abstract (as in Platnick's 2012 *reductio ad absurdum*); what it can explain is instead why they are all *similar* in lacking engines: because they all descend from ancestors that did not have such engines. As for plesiomorphies not being homologies, paraphrasing Nixon and Carpenter (2012a), all the taxa with *gray* color in Figs. 1.1c, 1.1d, and 1.1f share that condition by having inherited it from their most recent common ancestor—thus, *homology*. In other words: all synapomorphies are homologies, but not all homologies are synapomorphies. While Farris (1983) did not discuss these ideas explicitly, viewing the problem of phylogenetic inference as one of explaining observed similarities by invoking common ancestry easily leads to the proper conclusions, as clarified and expanded by Nixon and Carpenter (2012b) and Farris (2014b).

1.4 MAXIMUM LIKELIHOOD AND ASSUMPTIONS IMPLICIT IN PARSIMONY

The previous formulation of the parsimony criterion focuses on the degree to which a tree allows rationalizing observations about character distributions—that rationalization being made on the basis of explaining similarities by common ancestry. While some of the arguments concern whether the most parsimonious tree can be considered the historically correct tree, the main argument concerns a logical relationship between evidence and conclusions, which is not a probabilistic argument. This applies to some degree to individual propositions of ancestral states but much more obviously so to hypotheses of relationships between taxa. The proposal of monophyly of a group supported by multiple characters (or lines of evidence) continues providing explanation even if some (or all) of the supporting characters are in fact homoplasies. Explanation is the process by which a common cause can be attributed to a series of otherwise independent observations so that those observations can be understood in terms of a unifying theory, rather than astonishing coincidences. The quality of the explanation is given by the degree to which a larger number of observations (or aspects of the observations, more generally) can be subsumed under this common theory, not by whether the explanation itself is true or not. As so aptly put by Farris (1982b: 414), “theories are not taken as explanations because they are known to be true, but rather truth is attributed to them precisely because they provide explanations”.

But it is always possible that the explanation is wrong. That the tree which best allows explaining in terms of common ancestry the observations—necessarily identical with the most parsimonious tree—can be different from the historically accurate phylogeny may seem surprising at first. Felsenstein (1978b) was the first to propose a formal mechanism for the characters to evolve in such a way that parsimony may produce an incorrect tree. He considered a model of evolution where:

a) all characters could take only two alternative states, 0 or 1;
b) the probability of transformation for each of the characters along a given branch of the tree is exactly the same;
c) the probability of transformation for the characters changes along the different tree branches.

This type of model is called a common mechanism model (e.g. Tuffley and Steel 1997) because it assumes (at point b) that there are factors that can make the probability of transformation along a given branch of the tree higher, or lower, for all characters simultaneously. Evidently, under such model, the final states for the descendant of a branch with low probability of change will be very similar to those of its ancestor; the final states for the descendant of a branch with a very high probability of branch will instead be randomized relative to those of its ancestor—approximating a random series of 0s and 1s. Consider now the tree shown in Figure 1.4a; this tree is rooted only to facilitate description of the phenomenon (the root of the tree could be changed without affecting the behavior of the example). The lengths of the branches indicate the probability of transformation along that branch; the branches leading to B and C thus have a high probability of change and the rest of the branches a low one. It is easy to see that, if the characters evolve according to the probabilities indicated in the tree of Figure 1.4a, most of the characters will change along the branches x–B and y–C, and just by chance, many characters will end up in the same state in both B and C (as their branches become longer and longer, the probability that B and C end up in the same state for a given character tends to 0.5, as there are only two possible states). At the same time, there will be very few changes along the branches D–y, y–x, and x–A, so that A and D are likely to have the same states for most characters. Under those conditions, there will be almost no character states in which AB are similar and different from CD—that is, there will be no apparent synapomorphies between AB, even if AB form a monophyletic group on the true phylogeny. In contrast, half the characters for which BC happen to be similar (≈0.5) will in turn be different from the state found in AD—that is, apparent synapomorphies for BC in roughly $0.5 \times 0.5 = 0.25$ of the characters. As the differences in branch lengths increase and larger numbers of characters are evolved on the tree in Figure 1.4a, an analysis by means of parsimony is guaranteed to converge onto the wrong tree, shown in Figure 1.4b. This is known as *long branch attraction* (*LBA*) and is a case where parsimony can be inconsistent. The property of statistical consistency is that of converging to the correct value of the parameter being estimated—in this case, the phylogenetic tree—as the amount of data increases and is, in the abstract, a desirable property. Felsenstein (1978, 1981) suggested that analyses under maximum likelihood would achieve consistency in such case. The *likelihood* of a hypothesis is the probability that the data are observed—or evolved, in the case of phylogeny—should the hypothesis and model of evolution be true. Although Felsenstein had no formal proof of consistency, the consistency of maximum likelihood under simple models of evolution was subsequently proven (by Chang 1996; Rogers 1997). But the likelihood of a phylogenetic tree can be calculated only when a specific model of evolution is assumed (e.g. as specified by conditions a–c), and the property of consistency can be guaranteed only when the data truly evolve under the model subsequently used for evaluating the trees. So, maximum likelihood is not just consistent in the abstract but only consistent for a specific model of evolution; there are models of evolution (i.e. ways in which the data could evolve, for example, by not assuming the common model of point b in the list prior) that make phylogenetic estimation inconsistent even if the true model is assumed. In part because its sophistication appears attractive to many phylogeneticists, maximum likelihood (or derivative methods like Bayesian analysis) are today very common for the analysis of molecular sequences and are becoming more

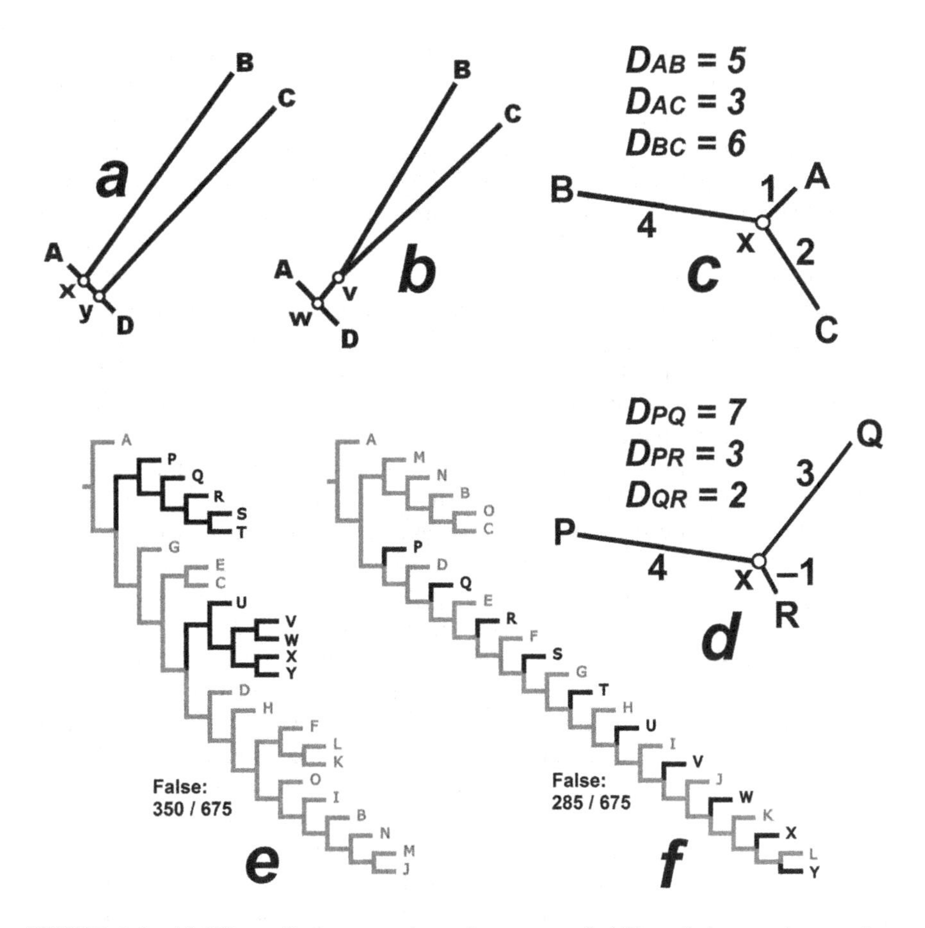

FIGURE 1.4 (**a**) When all characters have the same probability of change along a given branch, and differing among branches, the taxa at the end of long branches (B, C) are likely to share more apparent synapomorphies in common than actually related taxa (A, B). This forms the basis of Felsenstein's (1978b) long branch attraction. (**b**) For the data generated on tree (a), this tree is more parsimonious than the correct tree. The changes along segment v–w represent the parallelisms between B, C. (**c**) Distances obeying the triangle inequality can be fitted on a three-taxon tree so that original distances are perfectly retrievable, with all three branches having a positive distance. (**d**) When distances do not obey the triangle inequality, adjusting branch lengths so that they sum up to observed distances requires the use of negative distances—which are physically uninterpretable. (**e, f**) Using three-taxon statements leads to prefer the tree that places every *black* taxon separately and can explain no similarity in *black* color as due to common ancestry (tree f) over the tree that separates *black* taxa in only two groups and can explain the similarity among most *black* taxa by common ancestry (tree e).

popular for the analysis of morphological datasets; Chapter 4 discusses model-based methods in detail.

In the present context, what needs to be stressed is that the model which Felsenstein (1978b) constructed is one where a basic assumption of parsimony is violated: it is a model where the most likely explanation of observed similarities (at least, between some of the taxa in the tree) is *not* common ancestry. A method, like parsimony, which tries to attribute as many similarities as possible to common ancestry is then bound to fail. In the evolutionary model posed by Felsenstein (1978b), the extent to which common ancestry is or is not the most likely cause of similarities depends on the parameters of the model; for example, when the number of alternative character states is a larger number N (instead of only 0 and 1), then the chances that taxa B and C will independently end up in the same state are decreased (to 1/N). And observed similarities will be more probably due to common ancestry (thus leading parsimony to produce correct phylogenetic trees) when the transformations in all characters do not occur preferentially on some branches—that is, when character state transformations can be located on different tree branches with (about) the same probability.

The correct phylogenetic tree, in Figure 1.4a, would be able to attribute to common ancestry fewer similarities than the tree in Figure 1.4b. It is then necessary to be more careful in specifying the goal of a cladistic analysis. Is it that of finding the true phylogenetic tree? Or is it that of finding the tree that best allows explaining observed similarities by common ancestry? In the face of common mechanism models like the one posed by Felsenstein (1978b), these two goals are no longer identical. Which of the approaches is preferable depends, in good part, on whether there are reasons to believe that characters evolve mostly in accordance with one or the other model. From this point of view, the use of parsimony or probabilistic methods based on common models becomes an empirical question, and one that needs not be answered in the same way for different types of data. This is in contrast with the approach of some authors on both sides of the likelihood vs. parsimony controversy, some of whom think that parsimony is a method to apply universally, regardless of any empirical evidence, just because it makes few assumptions about evolution and is philosophically sound, and some of whom think that statistical methods are always superior just by virtue of making use of probabilistic reasoning. Further discussion of these problems can be found in Chapter 4, where it is argued that the common mechanism assumed by standard probabilistic models seems to deviate rather strongly from the way in which morphological characters appear to evolve, thus providing (to some extent at least) an empirical justification for the use of parsimony in the case of morphological datasets.

1.5 DISTANCES, PHENETICS, AND INFORMATION CONTENT

As Hennig was developing his methods for phylogenetic classification, an entirely different approach, called *phenetics*, or sometimes *numerical taxonomy*, started being developed in the late 50s as a response to perceived inadequacies of then contemporary systematics; the first book summarizing phenetics was by Sokal and Sneath (1963). The development of phenetics coincided with the introduction of computers into biology and taxonomy, and the hope of pheneticists was that the use of algorithmic methods would by itself bring systematics into a new era of objectivity. The main tenets of phenetic taxonomy (Sokal and Sneath 1963: 50–52, 55) were that:

a) groups should have "the greatest content of information" and be based on as many characters as possible;

b) overall similarity (or affinity) between two taxa is a function of the similarity in all the characters being considered;
c) a classification should establish groups of mutually similar taxa;
d) taxonomy can be a strictly empirical science only if affinity is estimated independently of phylogenetic considerations, evaluating it "purely on the basis of the resemblances existing *now* in the material at hand".

Although phenetic methods are no longer widely used in systematics, some of the ideas that were embodied in that approach are still encountered occasionally, and some discussion of phenetic methods is useful for better understanding the properties of phylogenetic methods in general.

1.5.1 Distances and Their Properties

Phenetic methods were based on first calculating the *similarity* (or its complement, *distance*) between taxa. A distance measures "how far away" two taxa are. While the pheneticists eschewed phylogenetic considerations, distances in themselves are not intrinsically antithetical to evolutionary considerations. As changes accumulate along a line of descent, the distance between the ancestor and its descendant increases. Thus, a distance measures (on a certain scale) the number of evolutionary changes occurring along a branch of phylogeny. In fact, one of the formulations of the parsimony criterion (i.e. that employed by Farris 1970) is based on calculating ancestral reconstructions such that the distances between each node of the tree and its ancestor sum up to a minimum. This formulation of parsimony uses the Manhattan distance, which is the lineal sum of differences along all the characters; for simplicity, subsequent examples will consider Manhattan distances. Other measures of distance could consider transformations in non-lineal scales (e.g. log, squared, or squared root), but such distances will necessarily pose logical difficulties (Farris 1967) to the interpretation of amounts of evolutionary change along paths between nodes of the tree. That is, only *lineal* distances fulfill the requirement that, in going through nodes (i.e. reconstructed ancestral states) x, y intermediate between two terminal taxa A and B, the sums of distances $D_{Ax} + D_{xy} + D_{yB}$ will equal the distances observed between the terminals D_{AB} (Farris 1967).

Pheneticists proposed (their tenet c, see list a–d in previous section) creating groups composed of taxa that were, overall, most similar to each other. The *clustering level* in phenograms indicated the degree of similarity shared by the terminal taxa included in the group. Consider the example in Figure 1.5a, where the tree on the left represents the true (unknown) phylogenetic tree, with numbers of changes (synapomorphies, with no homoplasy) indicated along each branch. The distances between all the terminal taxa—the sums of numbers of changes for the branches in the connecting path—can be represented in a matrix of distances (Figure 1.5b). A phenogram for such data, based on creating groups with most similar (least distant) taxa, is in this case a tree which perfectly represents in their clustering levels the observed distances between taxa, Figure 1.5c. The data about distances from Figure 1.5a can be represented in a phenogram with absolutely no distortion. That will be possible whenever the distances between taxa obey a very special condition:

note (comparing Figs. 1.5b and c) that the distances between any two members of every sister group (at any clustering level) are exactly identical. The observed distance from any one of P, Q, to any one of N, O, R, is exactly the same, 12. To the extent that the data can be so represented, the clustering level will allow "predicting" the observed distances with no distortion. Distance data obeying that condition are called *ultrametric*; for those distances, given any triplet of terminal taxa, i, j, k, D_{ij} is no greater than the maximum of D_{ik} and D_{jk}. This will be the case when the amount of evolution from a common ancestor to all the tips descended from it is exactly the same. This can be observed in Figure 1.5a, where seven characters change along the branches from the root of the tree to each of the terminal taxa. This will generally occur when evolution proceeds with a constant rate over time (and all the terminal taxa are contemporaneous), and then the ultrametric condition is equivalent to having an *evolutionary clock*, by which change accumulates along time at a constant rate. Of course, it would be very nice if evolution actually approximated a clock; in that case, one could interpret phenograms as phylogenetic trees, and (via a proper calibration) one would be able to assign actual dates to all the splits of the tree. A clock in molecular evolution (first proposed by Zuckerkandl and Pauling in 1965) could be expected if molecular evolution is mostly neutral (Kimura 1968); such a neutral model is not generally accepted in the case of molecules, but it is much less justified in the case of morphological evolution. Perhaps because the existence of a morphological clock would be so beautiful, making it possible to estimate so much more from morphological data, the hypothesis of a clock continues to be entertained by some authors (mostly in connection to models; see Chapter 4). But beautiful hypotheses are often slain by ugly facts (Huxley *dixit*). It is plain that the rates of accumulation of morphological change in the two phyletic lines leading to a coelacanth and a bat, since the divergence from their most recent common ancestor, have been vastly different—such a difference falsifies the existence of a morphological clock. In concordance with that, Puttick et al. (2016) and Parins-Fukuchi and Brown (2017) have shown that, predictably, the results in those papers applying a morphological clock depend more on the assumptions ("priors") than on properties intrinsic to the data. And the situation is worse when fossils are considered, because then taxa which underwent extinction at different time periods (from millions of years ago to the present) did not have the same time to evolve, so ultrametricity is not to be expected even if a clock governs the rates of evolution and time determines degree of morphological separation (as noted by Goloboff et al. 2017 in their criticism of the ultrametric simulations of Puttick et al. 2017).

When changes accumulate at different rates in different branches of the tree, a phenogram no longer allows retrieving the observed taxon-to-taxon distances without distortion. An example is in Figure 1.5d (for simplicity, continue assuming no homoplasy). The distance matrix and the resulting phenogram (with its corresponding clustering levels) are shown in Figure 1.5e. The clustering levels in that phenogram are the best one can do to store in the clustering levels the observed data on distances, but they do not allow retrieving those distances without distortion. For example, the two pairs of most similar taxa in the data are BC and DE (with $D_{BC} = D_{DE} = 1$), so that they form two groups. The phenogram predicts that the distance between B and D should be exactly the same as that between C and D (both B–D and C–D are separated at a level of seven), yet C is in fact a little closer to D (D_{CD} = 6) than B is (D_{BD} = 7). It is thus impossible to simultaneously reflect all the distances exactly by indicating the distances in the clustering levels. Pheneticists were of course aware of this problem, and they used the cophenetic correlation coefficient (Sokal and Rohlf 1962; Sokal and Sneath 1963: 141, 296, similar to Pearson's correlation coefficient) to measure the ability of the resulting phenogram to "predict" observed distances. One is left wondering, however, why the insistence on trying to store information of the degrees of similarity in the clustering levels, when it is plain that differences in the rates of evolution along different branches are bound to make that approach ineffective. It must be emphasized that the present criticism of the phenogram in Figure 1.5e is *not* that it displays a group BC when in fact C is genealogically closer to DE than it is to B. The criticism is instead that the distances—which according to pheneticists are what we should be striving to reflect

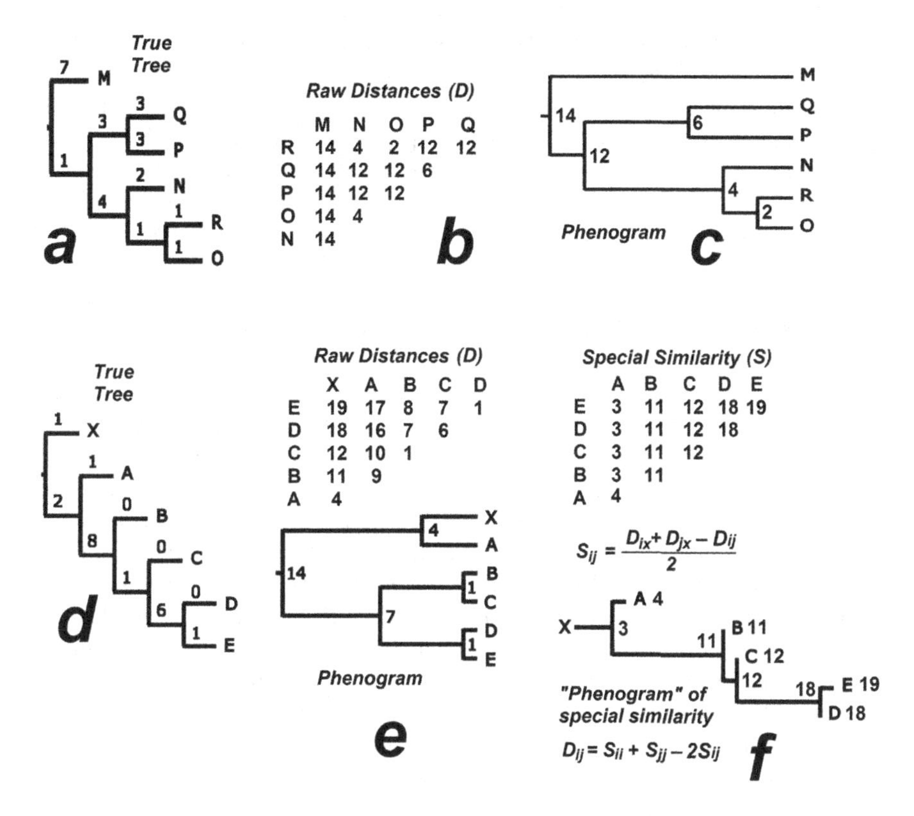

FIGURE 1.5 **(a)** When the distances between taxa are ultrametric (i.e. clocklike), the associated matrix of distances **(b)** can be represented exactly by the linkage levels of a tree **(c)**. When rates of evolution vary and the data are not ultrametric **(d)**, clustering on the original distances necessarily distorts some of the observed distances **(e)**. In that case, clustering by special similarity (which transforms the original distances into a count of "shared apomorphies"; Farris 1977) provides a perfect fit to the data **(f)**, allowing the distances to be retrieved from the tree with no distortion.

in our classifications—are retrievable only with significant distortion when the data evolve without obeying a clock, as in Figure 1.5d.

To restate the obvious: similarity in many characters is no evidence that the taxa in question should form a genealogical group. The tree in Figure 1.5d shows a case where the most similar taxa (B and C) are, however, not most closely related; C is more closely related to DE than it is to B, even if its minimal distance is indeed to taxon B. The ultrametric condition is violated in this case; the rates of change along the different branches of the tree are different. The case in Figure 1.5d obeys a condition less restrictive than the ultrametric, called a *metric*. All ultrametrics are metric, but not all metrics are ultrametric. A metric fulfills (in addition to other more obvious requirements, such as $D_{ii} = 0$ for all i) the *triangle inequality*, which states that for any three taxa i, j, k, the distance between a pair of taxa cannot be greater than the sum of the distances between the other two possible pairs: $D_{ij} \leq D_{jk} + D_{ik}$. Or, put differently, the path from one taxon to another can never be shorter if done through a third taxon. Metricity is the minimum requirement for a distance to be interpretable phylogenetically; distances not satisfying this condition create problems of interpretation. A distance fulfilling metricity implies that, if A, B, C are three taxa connected by an intermediate ancestor x, then the distances can represent (i.e. be proportional to) amounts of evolutionary change, and x can be placed at distances from A, B, C such that the same occurs. Consider as example (Figure 1.4c) the case of distances obeying the triangle inequality, $D_{AB} = 5$, $D_{AC} = 3$, and $D_{BC} = 6$; it is then possible to draw a tree, with ancestor x intermediate, connecting these three taxa at distances $D_{Ax} = 1$, $D_{Bx} = 4$, and $D_{Cx} = 2$. Observe that summing the lengths of the paths between any two terminals (passing through x) produces the correct observed distances in all cases (e.g. $D_{AB} = 5$ and $D_{Ax} + D_{xB} = 5$). The differences expressed as distances, therefore, can be interpreted as amounts of evolutionary change. Contrast this with another example, where the triangle inequality is violated (Figure 1.4d), with distances between taxa P, Q, R equal to $D_{PQ} = 7$, $D_{PR} = 3$, $D_{QR} = 2$ (i.e. the path from P to Q is shorter if done through R instead of directly). In this second case, it is not possible to draw a tree and have the ancestor x connecting these three taxa so that the distances between the terminals P, Q, and R are recoverable from the tree, *unless some of those distances are negative*. The distances can be recovered exactly if x is placed at a distance of 4 from P, 3 from Q, and −1 from R. But what does it mean to have had an amount of evolution proportional to −1 in the branch connecting x and R? No change is zero distance, some change is a positive distance; change cannot be negative. Distances violating the triangle inequality thus cannot be interpreted as proportional to amounts of evolutionary change. Some methods for reconstructing phylogenies (e.g. DNA hybridization, immunological distances) often produced distances violating the triangle inequality.

1.5.2 Phenetics vs. Cladistics

The defining aspect of phenetic methods is the notion that taxonomic groups should always be composed of the taxa most similar at the corresponding level. The most important comparisons between cladistic and phenetic classification were carried out by Farris (1977, 1979a, 1979b, 1980a, 1980b, 1982b). Although phenetic and cladistic methods were sometimes compared on the basis of whether they attempted to incorporate genealogical information, the differences between phenetics and cladistics can be resolved much more profitably by considering both methods in terms of common goals. In particular, pheneticists did not claim that their methods would be better for establishing phylogenetic relationships; they defended their methods on the premise that they were better as systems for retrieving information about observed similarities. On the one hand, as the cladistic approach was characterized prior, it also has a focus on similarities (i.e. on the degree to which similarities can be attributed to ancestry by different trees), but similarities between several taxa can often be explained by common ancestry even if the taxa in question do not form a group in the tree—that is, when the similarities correspond to plesiomorphies (as the *gray* color in Figure 1.1c). On the other hand, cladistic classification is also based on the characters that taxa have now, thus being no less empirical than phenetic groupings, and then we can also view a cladistic classification as a system for storing and retrieving information.

While the pheneticists admitted (e.g. Sokal and Sneath 1963: 216) that the best explanation of observed degrees of similarity between taxonomic groups of different levels was that of evolution and descent with modification, they contended that phylogeny was known with certainty in only a few cases. Farris (1982b: 416) noted that such an idea "amounts to the claim that description and explanation are contradictory goals. That is a most curious notion indeed, inasmuch as no theory can very well be said to explain what it cannot describe". Both cladistic and phenetic classification are based on observable characters; what differs in both methods is how observed characters are considered to bear on the conclusions. Farris (1977, 1979a, 1979b, 1980a, 1980b, 1982b) then compared both phenetic and cladistic classifications in terms of their ability to store retrievable information about similarities. His findings—that cladistic methods better store information about similarities and differences—were in a sense surprising, given that they ran against what at the time was considered common wisdom, but fully expected once it is realized that efficient description and explanation should come together.

1.5.3 Retrieving Distances

One of the aspects of information retrieval in which phenetic and cladistic classifications can be compared is that of storing information about distances. While phenetics is sometimes equated with the use of distances (instead of character-based methods), distances can themselves be interpreted phylogenetically (e.g. as in the example of Figure 1.4c) and be used without necessarily creating groups with most similar taxa.

The term *phenetics* is used here in its strict connotation, that of grouping taxa on the basis of overall similarity, which in turn is inextricably bound to ultrametricity. The distances between taxa could be treated in such a way that the ultrametric condition is not forced onto the results, and such treatment would not be "phenetic". One of the earliest examples is Farris's (1972) method for creating Wagner trees from distance matrices, which assumed no rate constancy at any point during the creation of the tree. Farris (1972) noted that operating with distances between taxa, an estimator (which he subsequently termed *special similarity S*; Farris 1977) of the number of apomorphies shared by two taxa i, j relative to a "reference" taxon r could be easily constructed, as $S_{ij,r} = (D_{ir} + D_{jr} - D_{ij})/2$. If taxon r is thought as ancestral of i, j, then $S_{ij,r}$ varies with the number of characters that are shared by i and j and also different from r (i.e. apomorphies, if r is considered, purely formally, as ancestral for all its character states); if we think of x as an ancestor intermediate between i, j, and r, $S_{ij,r}$ gives the distance between r and x which produces—at least for simple binary characters—an exact fit to the observed distances along the paths of the mini-tree. There is thus a definite relationship between distances and shared apomorphies, at least for binary characters. While the phenetic method was based on grouping by overall similarity (i.e. placing together the taxa least distant according to the original distance measure), Farris (1977) proposed creating "phenograms" grouping by special similarity; as special similarity measures resemblance in shared "apomorphies", this is equivalent to the cladistic method. Farris (1977) found that the cophenetic correlation coefficient calculated for classifications produced by grouping with special similarity invariably led to better correlation coefficients than classifications produced by grouping with overall similarity (i.e. raw distances). This is unsurprising when one realizes that the transformation of distances into special similarity turns distances that are only *metric* into a measure of similarity which is also *ultrametric* (Bandelt 1990), and phenograms are designed to recover ultrametric patterns. To understand how a metric is transformed into an ultrametric, consider that the distances from root (reference taxon) to the tips (terminal taxa) are additive: the distances from root to tips of the tree are the summation of the distances between the intermediate nodes (=ancestors). Additional clarifications on the use of special similarity in this way, and discussion of possible objections, were provided by Farris (1979a, 1979b, 1980a). An example where the Farris transform takes distance data that could not fit well into an ultrametric pattern, such as the data corresponding to the tree in Figure 1.5d, and converts it into an ultrametric matrix of special similarity, thus producing a tree with perfect fit to the matrix of special similarities, is shown in Figure 1.5f. Note that the terminals have their own values of special similarity, defined as $S_{ii,r} = D_{ir}$ (the distance to the reference taxon). Algebraic operations on the (perfectly predicted) special similarities thus allow retrieving the original distances with perfect fidelity by inverting the formula for special similarity (as shown in the bottom of Figure 1.5f). Just as in the case of the phenogram of Figure 1.5e the criticism was the distortion to the retrieved distances (not the fact that the tree is phylogenetically incorrect), the advantage of the tree in Figure 1.5f is not that it is phylogenetically correct (which doesn't hurt) but instead that it allows storing and retrieving distances with no distortion. Farris (1979b) noted that using the special similarity in this way amounts to considering that reversals provide no grounds for grouping (1979b: 494) and that assigning branch lengths to the tree in such a way that differences between observed and predicted distances are minimized (e.g. as done by Fitch and Margoliash 1967; also in neighbor-joining, Saitou and Nei 1987) produces results that are closer to those of parsimony, both in the genealogical interpretation and fit to the original distance observations.

That distances can be analyzed in a way which approximates the results of parsimony does not mean that distances can always be used phylogenetically, interchangeably with character data; Chapter 3 (Section 3.12) discusses some problems of interpretation that may arise in the case of distances, particularly when the data depart from simple, homoplasy-free binary data. The most common justification for eschewing distances is that, unavoidably, they imply loss of information—obvious when one considers that it is possible to go from character data to distances but impossible to return from distances back to character data. It is that loss of information that can be responsible for paradoxical situations such as those discussed in Chapter 3. The import of the present discussion is in showing that there is nothing intrinsically phenetic about distances and that distances can be treated in a way analogous to phylogenetic methods—that is, without producing classifications where ultrametricity is forced onto the results. Of course, if ultrametricity is (for whatever valid and relevant reason) to be generally expected for the data at hand, then using ultrametric methods may be justified. Phenetic methods are still in use in fields other than systematics (e.g. in psychology, ecology, etc.), and perhaps their use is justified in those contexts. If there are no grounds to expect—at least in principle—that the data generally obey ultrametricity, then the use of ultrametric methods (i.e. grouping most similar things!) is totally unwarranted. It amounts to trying to fit a square into a triangle.

1.5.4 Transmitting Character Information

In subsuming all character differences under a single value, distances are prone to problems. How do classifications produced by cladistic analysis thus differ from those produced by other means in terms of their ability to transmit character information—that is, information about specific features? It was Farris again (in 1979b) who showed that the goal of maximizing the ability of classifications to allow transmitting information about features leads directly to parsimony. Note that this is not to say that, among alternative methods for producing trees from character data, parsimony is the best in terms of transmitting information about character data; rather, Farris (1979b) provided a demonstration that maximizing information directly leads to the parsimony criterion so that no other criterion can possibly outperform parsimony in this regard.

Classifications and trees, regardless of the method or criterion by which they have been produced, contain no character information in themselves. The only information contained in the classification itself is that some groups are included within others. A classification can help convey information about characters only when the groups of a classification are associated with diagnoses. In this way, groups of a classification can be associated with those characteristics that distinguish them, and some classifications will be better at allowing such associations. Consider, as example, the groups Arthropoda, Insecta (*sensu strictissimo*), Diptera, and *Drosophila*. Arthropoda is a natural group, and the group serves to transmit information about an exoskeleton. The association between the exoskeleton and the group is perfect: no arthropod lacks one, and no species outside of arthropods has one. Every group in the hierarchy is best associated with some features: body divided in three segments, wings, and a peculiar set of buccal pieces in Insecta; posterior wings in the form of halteres in Diptera. The groups in the hierarchy serve to abbreviate description: in the diagnosis of *Drosophila*, there is no need to specify that it has halteres, mandibles, wings, or exoskeleton; knowing that *Drosophila* is a Diptera, an Insect, and an Arthropod suffices for us to know that. The diagnosis of *Drosophila* is best done by enumerating only the characters that distinguish it from other Diptera. A classification is then most useful in aiding diagnoses when it allows diagnoses to be kept as succinct as possible.

To be useful in transmitting character information, groups need to be homogeneous with regards to the feature in question, but it is also necessary that some group in the hierarchy can be distinguished by the feature—that is, a group which is heterogeneous with regards to their counterpart. Otherwise, the group cannot be associated with any diagnosis—cannot be associated with any distinguishing characters. The criterion preferred by the pheneticists—groups defined by high similarity—focused exclusively on the intragroup homogeneity, but a classification becomes more useful when its constituent classes are also distinguished by unique features. That the within-group homogeneity is desirable is obvious, but this needs to be balanced with regard to the between-group heterogeneity. Evidently, in a hierarchical classification, any group that is subdivided and associated with features unique to each subgroup will be heterogeneous in those features distinguishing the subgroups. Farris (1979b) noted that in such a case, it is necessary to resort to complement groups—presence of mammary glands is best described by the taxon Mammalia, absence of mammary glands is best described by referring to all of life *except* Mammalia (i.e. the complement of Mammalia). A feature associated with the definition of a given group is what—in the phylogenetic setting—is normally termed a synapomorphy, and the alternative feature described by the complement group is what is normally termed a plesiomorphy.

Thus, for a given grouping, every instance where some unrelated species also presents a feature used to define a taxon, and every instance where a group includes some species lacking the feature used to distinguish it, decrease our ability to use the classification to transmit information about the feature. Evidently, not all classifications are equally useful in aiding abbreviated description. The descriptive power of classifications is thus maximized when the sum of cases where the defining feature is found in unrelated taxa, and absent from members of the group, is minimized—that is, parsimony. Every "step" required by the classification amounts either to an instance where the feature defining some group is found outside the group as well or an instance where the feature defining a group is absent from a member included in the group. When classifications are most parsimonious, the original data can be reconstructed (via the association of features with groups) with maximum efficiency. Thus, for each character, the plesiomorphic state at the base of the entire tree needs to be specified, then indicating every group (=branch of the tree) where there is a change to a different state. The fewer the symbols (group names) that have to be used to reconstruct the original observations, the higher the information content of the symbols, and it is clear that the minimum number of indications will always correspond to most parsimonious trees—the two criteria, descriptive efficiency and explanatory power on the basis of genealogy, are exactly the same thing.

Farris (1979b) also discussed the idea of predictivity of classifications—that is, the ability to infer as yet unobserved features from jointly considering the classification and some observations of the distribution of the feature in question.[2] In this sense, the classification that provides the best description of existing observations can be expected to provide the best prediction of *future* observations. In practice, group membership is not known but only concluded from the observation of other group-defining features. We do not observe an individual to be a bird, we only observe that it has wings and feathers. It is from this observation that we can infer that the winged and feathered individual in question probably has a four-chambered heart, pneumatic bones, and reproduction with eggs. A classification, in other words, functions well as a predictive scheme to the extent that different suites of characters exhibit a good correlation—so that all, or most of them, can be summarized simultaneously by the same classification.

1.6 PATTERN CLADISTICS AND THREE-TAXON STATEMENTS

In the late 70s and early 80s, with the realization that phylogenetic classifications could be seen as statements about patterns of character distributions as much as they could be seen as representing the branching in phylogenetic trees, a movement called *pattern cladistics* (or sometimes *transformed cladistics*) began to emerge. These authors (Platnick 1979, 1982; Nelson and Platnick 1981; Patterson 1982a, 1982b; Brady 1982) emphasized the idea that cladistics is fundamentally about identifying patterns and that the identification of these patterns should not be tainted with prior notions of how evolution proceeds. Thus, a cladogram could be thought of as merely summarizing patterns in character distributions. Those patterns would, of course, call for a causal explanation, and there is no doubt that that causal explanation is (at least for the majority of patterns) an evolutionary one: descent with modification. Brady (1985) discussed how pre-Darwinian, pre-evolutionary biology had indeed perceived strong hierarchical patterns in nature and how this observation of a natural hierarchy remained unintelligible until Darwin (1859) provided an explanation in terms of common descent. Brady (1983, 1985) argued that, to the extent that the theory of common descent is well established, it is so because of the very observations that support it, and then the fact that distributions of features among different organisms follow certain patterns (the *explanandum*, or thing to be explained, in Brady's 1985 account) has as much "reality" as the explanatory notion that biological diversity results from a process of evolution (the *explanans*, in Brady's 1985 terminology). Good summaries of the ideas of pattern cladistics (and replies to common misunderstandings) were given by Brower (2000, 2019).

In part, pattern cladistics can be seen as a reaction against some excesses of the early days of cladistics, such as using preconceptions of how evolution proceeds to guide the choice of phylogenies; Platnick (1979) argued that, to test mechanistic theories of evolution, phylogenies should play a central role, and then those phylogenies—if they are to bear on the decision to accept or reject a theory of mechanism—should ideally be independent of the theory they are supposed to test. While understandable in principle under ideal circumstances, that notion of whole independence is, however, not shared by all authors; for example, from the likelihoodist perspective, a phylogenetic analysis (via model testing; see Chapter 4) may well determine the best evolutionary model and parameter values.

The notion of transformed cladistics caused a very strong reaction on the part of some authors (e.g. Beatty 1982; Kluge 1997, 1998, 1999), including two books devoted solely to criticizing pattern cladistics (Ridley 1986; Scott-Ram 1990). Outrage at the idea that taxonomy could be based on considerations not directly involving evolutionary reasoning continues to this day, as recently exemplified by Quinn (2017).[3] However, the philosophical decision of seeing a scientific theory as a sort of description certainly extends well beyond taxonomy—all of the empirical sciences could be conceived as identifying patterns. Empirical science, in a sense, can only deal with the how, not with the why—"cause" being deferred eventually to subsequent "hows", to theories that explain (remember: describe) phenomena at a higher level. While the law of universal gravitation "explains" *why* a body falls in a certain way, that law itself is merely a theory describing *how* bodies interact with each other. And "explanations" can have strong empirical support only to the extent that the patterns which they explain are indeed real phenomena. Consider the following quotation, from an author who could not possibly be suspected of any allegiance to transformed cladistics:

> The incentive for contemplating a scientific hypothesis is that through it we may achieve an economy of thought in the description of events, enabling us to enunciate laws and relations of more than immediate validity and relevance. . . . The greater the regularity or pattern in a sequence of events, the more we feel compelled to seek an "explanation" in terms of a law. . . . It is our task to detect regularity in the presence of confusion, order in the presence of chaos.
>
> **(Edwards 1992: 1)**

The position of pattern cladists is then much less controversial than the literature attacking their position would suggest and without many deep consequences for the daily practice of the taxonomist. It is a way of seeing; as put by Rieppel (2013), it concerns the metaphysical interpretation of what taxa really *are* more than anything else. Or, it can be viewed (with Brady 1985) as a call for keeping evidence and conclusions logically separate in our minds. From the pattern cladistics point of view, groups can be associated with character states; similarity in a given feature can then be attributed, via group membership, to a common cause (whatever the invoked cause is, not necessarily a genealogical one). To the extent that different (suites of) characters can be described or explained by the same (or compatible) sets of groups (=trees), the character pattern that the groups discern seems indeed to be real. Although I know of no actual attempts to do so, it seems quite likely that even model-based methods could be examined in terms of their ability to predict and describe character distributions, rather than in terms of the probability of finding the true tree. The extent to which parsimony or model-based methods are preferable in terms of making character distributions intelligible will then depend in good part on the nature of the data—which of the two types of summary, one based on parsimony or one based on common mechanisms, works better in practice, will depend mostly on the intrinsic properties of the data, which is anyway how the controversy between different approaches usually is (and should be) presented.

1.6.1 Three-Taxon Statements

What is problematic about invoking an epistemological separation between classification and evolutionary reasoning is when that divorce is (ab)used in attempts to justify otherwise unjustifiable methods. A case in point is the method known as *three-taxon statements* (also called *three-item analysis*). Formally proposed by Nelson and Platnick (1991), it is more or less akin to ideas on *component analysis* earlier presented in Nelson and Platnick (1981). Component analysis consisted of first identifying the monophyletic group(s) separately identifiable by the "synapomorphic" state in each of the characters in the matrix and then combining them into a single tree. Nelson and Platnick (1991) thought that the atomic, irreducible component of relationship is the proposal that two taxa are more closely related to each other than they are to a third. They viewed statements on homology as a proposition about relationships between *taxa* (instead of being a proposition about sameness, or comparability, in *characters*). As a consequence, they proposed to represent all the possible combinations of homology statements in the original matrix with "characters", having only two 1s (for two of the taxa in the group) and two 0s (for the third taxon left outside of the group and in an all-zero root taxon). A given statement thus represented can then be true (no extra steps for the corresponding "character") on a given candidate tree, or it may be false (one extra step). Note that no more than a single step can ever be required, so a statement can only be completely true or completely false on a given tree (no intermediate options are possible). The recoded matrix can be analyzed with any standard parsimony program, with the result then being neither the tree which minimizes the number of independent originations of similar features, nor the tree which allows explaining in terms of common ancestry the maximum possible number of similarities, but instead the tree in which the maximum number of atomic statements about relative kinship that could be derived from each of the characters in the original dataset are recovered correctly (i.e. the tree where the maximum number of three-taxon statements have a single step). There are many problems with such an approach, in addition to not maximizing the degree to which observed similarities can be accounted for by common ancestry (or common cause). The individual statements are not independent: e.g. if we know that both A(DE) and C(AD) are true, then so must be—by logical necessity—the statement C(AE). The approach requires an absolute concept of apomorphy, which, as stated earlier (Section 1.3), cannot logically exist. It also loses information; the original matrix can be transformed into a series of three-taxon statements, but it is not possible to go back from the recoded matrix to the original matrix, at least not unless a record of which statement came from which character is kept. Since such records, even if kept, have no influence on the results, the information is effectively lost when selecting trees.

Cogent criticisms of the approach can be found in Kluge (1994), Farris et al. (1995), De Laet and Smets (1998), and Farris (2012, 2014a, 2014b, 2014c). The defense on the part of supporters of three-taxon analysis has centered on the following: the conflation of particular algorithms for evaluating minimum requirements of homoplasy (which they call "Farris optimization"; see Platnick et al. 1996; Williams and Ebach 2008, 2012b, 2017) with the logic behind explaining observed similarities in terms of common cause by placing taxa in the same group and attributing the shared feature to the common node; confusing the goal of explaining observed similarities with "phenetics"; and posing that similarity in "primitive" conditions need not be explained.

It is important to note that the reasons to reject three-taxon statements have in fact nothing to do with whether or not one thinks that cladistics is about phylogeny or purely about patterns. Brower (2000, 2018, 2019), for example, had a strong position in favor of pattern cladistics, yet he considered standard parsimony—not three-taxon analysis—as the best method for establishing classifications (Brower 2015). Even if (with Williams and Ebach 2018, defenders of three-taxon statements) one does not adhere to the "commonly preconceived notion, namely that systematics requires evolution as a prior condition", the classifications resulting from three-taxon analysis provide neither the best rationalization of the data in terms of common causes (i.e. the best association of specific characters states with group membership) nor the best means for describing the original observations (which, as shown by Farris 1979a, 1979b, 1982b, requires a specific association between groups and states, i.e. parsimonious optimization).

Of interest for the present section is the dismissal of any criticism of three-taxon statements based on states assigned to common ancestors, on the grounds that cladistic analysis is not about evolution or ancestors:

> A cladogram differs from other types of phylogenetic trees in placing all organisms, both fossil and present, in terminal positions, implying that ancestral taxa are artifacts. Cladistics may possibly be improved if parts of organisms were treated in the same fashion in character (state) trees, with the implication that ancestral characters, too, are artifacts.
>
> **Nelson (1994: 137)**

According to this view, optimization (i.e. the process by which optimal ancestral assignments are found; see Chapter 3) is only

> a numerical recipe whereby the characters of organisms and taxa are fitted to a tree at various nodes, which are themselves reified as ancestors, in part or in whole, some older and some younger, so that, at least in one's imagination, some evolve into others
>
> **(Nelson 2004: 134)**

Nelson (2004: 134) thought that the problem is that

> optimization does formally derive characters from one another, in an idealistic morphology of all formal transformations imaginable, and their reversals, in the hope that some might intelligibly reflect, and thereby approximate, if not reveal, the real historical process of evolution.

Williams and Siebert (2000: 205) have also stated that parsimony analysis is problematic because it implies "a particular model of character evolution (a linear transformation of one state into another)". Perhaps these supporters of three-taxon analysis would want methods compatible with the idea that evolution consists of something other than the transformation of characters along descent or where ancestors had no characters at all?

Williams (2004: 213) was of the opinion that three-taxon analysis "has been mistakenly considered to be simply a method of analysis, and criticism has so far dwelt only on technical matters". However, there is nothing particularly "technical" about the most damning examples to three-taxon statements. Consider the character distribution of *gray* and *black*, among taxa A–Y, shown in the two trees[4] of Figs. 1.3e–f. Assume that one wants only to compare those two trees, to determine which of the two better fits the data. For the tree in Figure 1.4e, every triplet combining one of P–T with any one of U–Y (25 combinations), with any one of the *gray* taxa other than A (B–O, or 14 pairings for each combination of two *black* taxa placed apart), is false on the tree. For the tree in Figure 1.4e, therefore, a total of 25 × 14 = 350 statements in the original data are not recovered. For the tree in Figure 1.4f, the count is more involved; note that the only false statement for taxa X–Y (*black*) and a third (*gray*) taxon is with taxon L; 2 combinations (any of XY with W) are false against 2 *gray* taxa, K and L; 3 combinations (any of WXY with V) are false against 3 *gray* taxa (J, K, L); the progression is then such that the number of false statements equals $\sum_{i=1}^{i=9} i^2 = 285$. The tree that does not place any two *black* taxa together (Figure 1.4f) is considered by three-taxon statements to be much better than the tree which places the *black* taxa in two groups (Figure 1.4e). Among statements that are "true" on the tree of Figure 1.4f is, for example, the statement that P and Y are more closely related to each other than to taxon M. The approach of counting statements, that is, considers that "some" of the original homology hypothesized to be shared among P and Y is recovered by the tree of Figure 1.4f, even if P and Y are in distant positions of that tree and separated by taxa with the alternative state so that there is no way they could present their shared state by common cause (not unless all the *gray* taxa are similar in color because of independent causes). There is no way all those taxa could share their common state (as counted with three-taxon statements) by common ancestry simultaneously. In other words, if evolution has indeed occurred by transformation of characters in some descendants, the counts the three-taxon method effects to evaluate trees simply require ancestors with impossible combinations of states—the state the ancestors should have had to produce the counts used in three-taxon statement are mutually incompatible. There is nothing other than the number of false statements one could use to defend the tree in Figure 1.4f over the tree in Figure 1.4e; advocates of three-taxon analysis have certainly provided no such reason. Understanding the original data in terms of common causes, or by attributing observed similarities to common ancestry, is much better accomplished by means of the tree in Figure 1.4e, and—contra Williams (2004)—there is nothing "technical" in this example. No specific algorithm for optimization was used, just the obvious pairings with groups and colors. As defenders of three-taxon statements acknowledge themselves (Nelson 2004; Williams and Siebert 2000, in the quotations prior), their method is not just agnostic regarding evolution—it is incompatible with evolution.

1.7 PHYLOGENY AS AN ASSUMPTION; LIMITS OF PHYLOGENY

Previous sections build from the fact that the true phylogeny is unknowable and that to reconstruct it, detective-like methods are needed. That the true phylogeny is unknowable applies to almost every empirical case, with exceptions only in extraordinary circumstances, such as cultured strains of viruses or bacteria (e.g. Cunningham et al. 1998; Bull et al. 1997; Hillis et al. 1992; Boden et al. 2012), or perhaps cultivars (e.g. wine varieties). It is universally true for taxonomists working on wildlife, extant or extinct. That the true phylogeny is unknowable requires consideration of basic principles of how we know what we know and consideration of the problem of whether the ultimate goal of a phylogenetic study is finding the true phylogeny or making sense of the data in the best possible way. These two goals are interrelated, but they are not exactly the same, and they pervade many discussions (past and present) of the relative merits of parsimony vs. model-based methods (and without these difficult and subtle distinctions always made explicit, so that researchers on the two sides often just talk past each other).

Of the assumptions made in a phylogenetic analysis, the very notion that there is something out there that can be strictly called a "phylogeny", at least one that is tree-like and strictly hierarchical, is the assumption with more consequences for all the subsequent steps of an analysis. It is here where one can see some differences between the goal of finding the true phylogeny and that of rationalizing data: one can hope the properties of the data themselves will provide clues as to whether the problem of relationships in the group under question is best analyzed under the assumption of a unique, well-defined phylogeny. Obviously, genealogical relations between closely related, sexually reproducing (or conjugating) organisms are not hierarchical—an individual does not have a single originator. Hennig's famous figure (1966, figure 4) explaining *tokogeny* (the mesh of kinship within interbreeding populations, as opposed to phylogeny) illustrates the level at which relationships are clearly not hierarchical. This is the realm of *phylogeographic* relations, the interface between populations and phylogenetically distinct clades. As populations split and polymorphism is maintained within each population, an individual in one of the two populations may have a copy of a gene that is more closely related (i.e. a direct copy of the gene from a single individual in the original pre-splitting population) to the copy in an individual of the *other* population than to the copy in an individual of the *same* population. In diploid organisms, this could even happen not just within a population but also between gene copies of the same individual. If time passes without further splits in the populations, the continued interbreeding within each population will eventually fix one variant in each population separately from the other, and this situation where some individual genes do not follow the same separation as the populations will occur for fewer and fewer genes. This discordance can potentially persist after subsequent events separating populations (especially if those events are separated only by brief periods of time or if population sizes are very large and there is no selection for one or another gene variant). When this happens, the pattern of descent of individual copies of some gene(s) may differ from the pattern of descent in the populations: gene trees and species trees may be different. This is known as *incomplete lineage sorting*, *ILS*, and can be expected to occur for successive cladogenetic events happening while differentiation between populations still remains incomplete. There is a large body of theory on phylogenetic analysis in the presence of *ILS*, arising from the theory of *coalescence* (i.e. the tracing back of two gene copies to a common ancestor) of population genetics (beginning with Kingman 1982). Doyle (1992), Brower et al. (1996), Maddison (1997), and Slowinski and Page (1999) provided early reviews of the question of gene trees vs. species trees from the perspective of systematics. The classic approach to phylogenetic inference for datasets consisting of multiple genes (the one preferred by Brower et al. 1996, often called the *concatenation* approach) combines data from all available genes in a single matrix to be analyzed under the same criteria as individual genes, with the assumption that the tree determined by the majority of the genes will be the correct one; *ILS* is expected to occur in only a minority of the genes. The alternative approach (derived from early work by Goodman et al. 1979; Page 1994) is using the individual gene trees as input for methods combining them into a species tree; this approach has become much more common in recent years, with many methods and computer programs now intended to assist phylogenetic inference under this situation (see Edwards 2009; Anderson et al. 2012; Roch and Warnow 2015; Mirarab and Warnow 2015). As noted by even the most ardent defenders of the gene tree approach (e.g. Edwards 2009), this approach is generally thought of as most applicable to studies closer to the population level, while studies at higher levels of divergence are usually thought to be tractable by the standard approach of combining all genes in a single matrix and finding the single tree that simultaneously fits all the data together. This is applicable even more forcefully to the case of morphology, where the characters used to infer phylogeny are rarely polymorphic (and can be presumed to have become fixed in the populations at the time of cladogenesis); misleading effects of *ILS* in morphological characters also seem unlikely if the morphological differences are caused by many genes instead of just one. This is particularly the case when the degree of divergence implied by differences in those characters is so high that it seems unlikely that individuals with and without the character could have (if in a single "population", however defined) been capable of interbreeding, thus precluding the conditions for *ILS* to occur. Therefore, it seems that the gene tree approaches now common in molecular studies can be safely ignored in the case of morphology. I am not aware of any serious proposal to use gene tree methods for sets of morphological characters (which would be very hard to properly define, even if one wanted to).

Even beyond the population realm, after clearly established clades have been produced, it is still possible (for example, if reproductive isolation is not being prevented by intrinsic factors) for subsequent hybridizations or horizontal gene transfers to alter the hierarchical scheme of relationships. This is expected to be unusual in animals, but it may occur more frequently in plants, viruses, or prokaryotes. Some authors (see review in Doolittle and Brunet 2016; Arnold 2016) have even suggested that there is not a single "Tree of Life", but instead a tangle of relationships, particularly in the Prokaryota. This view was contested by Brower (2020). The problem posed by reticulation has been discussed since the early days of cladistics; not surprisingly, botanists were among the first to call attention to this question (e.g. Bremer and Wanntorp 1979). Despite several papers with attempts to formalize some methods (from Nelson 1983 to more recent ones by Nakhleh et al. 2005; Huson et al. 2011; Yu et al. 2014; Wheeler 2015), the problem of reticulation continues to be rarely considered in empirical studies.

To play it safe, phylogenetic analyses based on morphology should be generally performed for groups at a level of divergence such that the problem can be safely assumed to have evolved beyond the population and phylogeographic situation and where reticulation can be assumed to be absent or at least infrequent. If there is no reason to anticipate a well-defined hierarchy, other methods or approaches are obviously called for. To some degree, it must be noted, the extent to which different sources of evidence (i.e. different subsets of characters) agree on a single tree also gives some idea as to how reasonable it is to assume that a particular problem is in the phylogenetic realm, instead of the populational or phylogeographic one—that is, the degree to which the distribution of characters in the data seems to fit the expectation of a general hierarchy. However, as clear as the definition of a phylogeny may be when it comes to relationships between groups with clear structural differences—and thus having diverged well beyond the realm of population or phylogeographic studies—it is important to note that the difference between the two is more one of degree than of kind. Thus, the limit between what constitutes tokogenetic and phylogenetic relationships is, necessarily, blurred and fuzzy at the boundary. This is not to say that the distinction between one and the other extreme is arbitrary; fully interbreeding populations vs. fully separated clades with no lineage sorting represent (in principle) two easily diagnosable situations. Perhaps even more interesting, from the perspective of phylogeneticists, is the fact that those blurry limits also mean that the notion of a single unique phylogeny is an idealization, an abstraction, not something that can be observed directly. The analogies with balls and urns, often used to justify model-based methods, are indeed very loose: it's not just that neither the balls nor the urns can be observed directly; they may not even exist or be definable in any strict way. This is not giving up realism: a phylogenetic analysis is based on finding the best explanation for observed patterns of character distributions (characters which are really present in the taxa under study and really have patterns of distribution). But a phylogenetic tree is not a "thing" lying out there to be seen or discovered; that can be a serious oversimplification.

1.8 OPTIMALITY CRITERION AND FOUNDATIONS OF PHYLOGENETIC ANALYSIS

The previous sections showed that, the larger the number of steps (=independent originations of similar features for a given character), the poorer the degree to which the tree allows explaining similarities observed in that character by common ancestry. As all the characters in a dataset are considered, the parsimony score is the total sum of steps for all the characters. If some characters are thought to provide stronger evidence of relationships—i.e. it is preferable to sacrifice our ability to explain some of the characters at the expense of providing better explanations for others—then the relative weights of each of the characters can be taken into account. Two characters are said to be *incongruent* when no tree can display both characters as free of homoplasy, which is to say, providing the best possible explanation for one of the characters requires inevitably postulating some homoplasy for the other. If there is a tree which can display both characters as free of homoplasy, the characters are *congruent*. For a set of perfectly congruent characters, determining the tree(s) of maximum parsimony is an easy task, but this quickly becomes complicated as the characters have more incongruence and we add more taxa.

In cladistics, the parsimony score is the criterion used to measure how good a hypothesis of relationships a tree is, regardless of where that tree came from. This is desirable: one would expect a scientific hypothesis to be amenable to evaluation based only on the available evidence and on how the hypothesis compares to other alternatives. All the possible trees for a dataset are then valid hypotheses of relationships, in the absence of evidence, and it is only the actual observations (characters) which filter out some trees, leaving those best supported by the data. A well-defined numerical criterion to compare among trees is called an *optimality criterion*, and having one is probably the most crucial component of phylogenetic inference. In this regard, parsimony is identical to several other criteria for phylogenetic inference (likelihood, distance fitting, compatibility); the degree of fit between trees and observations is measured differently by those alternative criteria, but a number of problems are shared, notably the one of how to select the tree which indeed maximizes the optimality criterion.

The explosive increase in numbers of possible trees with number of taxa places the problem of finding the optimal phylogeny in a family of problems known in computer science as NP-complete—the class of problems for which computing solutions requires a time that increases very rapidly with the size of the problem, much more than polynomially. This implies that solutions guaranteed to be correct—correct in terms of actually optimizing the numerical value that constitutes the objective criterion for the solution—cannot be produced except for very small problems. Analysis of larger datasets or problems must rely, by necessity, on approximate methods, which can with high probability find the truly optimal solutions but cannot offer mathematical guarantees of having done so. These methods are collectively called *heuristics*, and they rely on trial and error or deterministic algorithms expected to approximate the optimal solution in shorter times. A typical case of heuristic methods is global branch-swapping (see Chapter 5, Section 5.3.2), which requires an amount of time polynomial on the number T of taxa. Depending on details of implementation, completing branch-swapping has dependencies from $O(T^2)$ to $O(T^4)$, where the "big-O" notation expresses how the number of operations, or the time to obtain solutions, is expected to increase with the number of taxa. Understanding the behavior of these approximate methods is extremely important for a conscientious analysis, not only because it makes the phylogeneticist aware of the conditions under which the methods may perform better or worse but also because they prevent interpreting as real phenomena results that are in fact only an artifact of the method or algorithms used. An example of such misinterpretations is taking the fact that a dataset produces a single tree under a parsimony analysis as indication of the reliability of the results (which was commonly done in earlier days of cladistics; see cautionary note by Hovenkamp 1999). This is certainly a misinterpretation: when there are very large numbers of characters, exact ties in the parsimony score of alternative trees are very unlikely, and such cases typically produce just one tree—even when the dataset includes only random characters and thus cannot possibly produce reliable results. A more proper comparison, in that case, would consider the alternative scores of next-best trees, and for random data this will reveal that even if a single tree has better parsimony score than all the other trees, the score difference is minimal, and every possible tree is an almost equally acceptable hypothesis of relationships (how to evaluate this problem is discussed in depth in Chapter 8, Vol. 2). Much of this book, therefore, is devoted to covering details of the algorithms, their behavior, and their implementation in computer programs.

An optimality criterion does not establish directly any relationship with the truth of conclusions; it only measures the degree of conformity with evidence, or data fit, of different hypotheses. The probability of truth of the conclusions may be in direct relationship to the data fit, given certain specific assumptions (such as the models used in maximum likelihood methods, see Chapter 4, or the degree to which the assumptions implicit in a parsimony analysis are met). Deciding on the use of one of various possible criteria for selecting phylogenetic hypotheses may also depend on how well the data seem to conform to the assumptions implicit in the alternative methods, which is also a criterion of fit (at a different level of analysis) and conformity with evidence. But, after a given criterion to evaluate trees is accepted, once we are at the point in which we need to effectively establish conclusions from a given dataset, that problem ceases to be important: the only problem is now how to identify the tree, or hypothesis of relationships, that (in the light of the assumptions embodied in the criterion used to select trees) best fits the observations at hand. Except for trivially small or clean datasets, there is no way to derive the best tree automatically from the dataset. Then, for empirical datasets, the only way to establish phylogenetic conclusions is, in principle, evaluating all possible trees and retaining those that are optimal.

1.9 ON THE NEED FOR COMPUTER PROGRAMS

At this point, it should be clear why computer implementations of phylogenetic analysis are needed. Once a criterion of evaluation has been formalized and translated into appropriate quantitative methods, and the relevant observations have been summarized in a data matrix, the numerical solution to the problem just follows. However, in practice, deciding on the phylogeny for any non-trivial problem requires simultaneously considering the fit of many characters to trees with numerous taxa. For evaluating the fit of each of the characters onto a given tree, alternative reconstructions—pathways to the observed data—must be taken into account. And this must be done for many, many trees. Each of those evaluations requires numerous calculations, which could well be done by hand, but computers are error free and faster. Error free, of course, refers to errors in numerical calculations; it does not mean that the phylogeny can be inferred without errors; what a phylogenetic program is supposed to deliver is not the true tree but instead a tree with the maximum possible parsimony, or the maximum likelihood. In fact, a phylogeny program cannot even promise to produce the tree of highest parsimony or likelihood: it can only promise that the tree produced has the correct value of optimality (under the corresponding conditions or parameters) and that no tree in the vicinity of the tree produced has a better fit. While this may seem like not much, it is way more than one can do by hand. The value of a program is not, therefore, in its results being blindly trusted as a phylogeny. Some authors who are part of a sort of counterculture against computer programs (Mooi and Gill 2010; Mooi et al., 2011; Ebach et al. 2013; Williams and Ebach 2014: 175) complained about users believing that programs "recover real phylogenetic trees, actual schemes of ancestry and descent" (Ebach et al. 2013: 588). Leaving aside what some confused user may have occasionally interpreted, the utility of computer programs is instead in helping make the goals and criteria of phylogenetic analysis more explicit and in providing trees that are indeed more likely to fulfill those logical criteria. If, for a given dataset, a phylogeneticist happens to find, either by hand or divine revelation, a tree which is more parsimonious (or more likely) than the one found with a well-tested program, then that tree is indeed a better hypothesis of relationships for the data in question. The optimality criterion used (be it parsimony or likelihood) allows for that better, manually found tree to be evaluated and preferred over the tree found by the computer program.[5] Not that such a thing happening is impossible, but it would be a really astonishing feat—about the same as the chances of shaking all the parts of a car in a giant box and having all the pieces assemble together by themselves.

1.10 IMPLEMENTATION IN TNT: TREE ANALYSIS USING NEW TECHNOLOGY

Thus, the value of a good computer program for phylogenetics is in the program being a tool which makes it more likely that the researcher succeeds in finding the best tree (most parsimonious tree, *MPT*, most likely tree, *MLT*), and in facilitating a series of comparisons with other trees that also fit the data well. All these ideas have been incorporated into the general design of TNT (Goloboff et al. 2003). TNT is distributed in binary form (i.e. executable files only) for all three main operating systems (Windows©, Linux©, and MacIntosh©). For Windows, there are versions both with a menu system, or *graphical user interface* (*GUI*), and command-driven versions. After 2007, the program has been subsidized by the Willi Hennig Society and is provided free of charge thanks to this subsidy (users are expected to acknowledge the Willi Hennig Society and cite the papers describing the program, published in *Cladistics*; Goloboff et al. 2008; Goloboff and Catalano 2016). Copies can be obtained from www.lillo.org.ar/phylogeny/tnt.

As of this writing, TNT consists of about 160,000 lines of C-code, plus some lines borrowed from PVM and PicoC. It implements a large number of data types and optimality criteria based on parsimony, as well as facilities for performing most of the aspects of phylogenetic analysis covered in this book:

- powerful algorithms for finding optimal trees (including the possibility of using constraints, i.e. limiting the search to trees containing, or lacking, specific groups);
- diagnosing trees, producing either lists of character change (in studies of character evolution) or lists of synapomorphies;
- comparing and summarizing trees (in the case of ambiguous results, or when the results of different analytical criteria are to be compared, or when the results of different datasets are to be compared);
- evaluating degree of group support (i.e. measuring the strength of the evidence supporting a given group) by means of resampling and Bremer supports;

- character weighting, with a variety of methods that determine character weight (=influence) according to homoplasy;
- automating tasks, in the case of very specific or repetitive analyses, by means of a scripting language, which extends TNT's capabilities by a quantum leap.

A given analysis may or may not perform any of those tasks; the user decides the sequence in which TNT performs its work in a given run. Some programs, once execution has begun, have a fixed set of routines, perhaps modified by options at program startup but unmodifiable subsequent to this (as in RAxML, Stamatakis et al. 2005; Garli, Zwickl 2006; PhyML, Guindon and Gascuel 2003; or IQ-Tree, Nguyen et al. 2015). In contrast with those programs, and similarly to some of its predecessors (Nona and Pee-Wee, Goloboff 1993d, 1993e, as well as the illustrious Hennig86, Farris 1988 and PAUP*, Swofford 2001) and some of the contemporaneous versions of POY (Varón et al. 2010), TNT is fully interactive: once started, it will do nothing, just awaiting instructions from the user. The fact that TNT is so flexible gives it a lot of power but, admittedly, also makes it harder to use. To facilitate understanding the general behavior of the program, keep in mind that the general mode of operation of TNT results from the general logic of phylogenetic analysis outlined in this chapter. A graphic summary of the overall organization of TNT is in Figure 1.6. The data are first defined by the user, then read into the program, and only once the data have been stored in memory is it possible to define and evaluate trees; evaluating large numbers of trees and retaining those with the fewest (weighted) steps is—in a nutshell—what is needed to find trees of maximum explanatory power. Only once trees (calculated

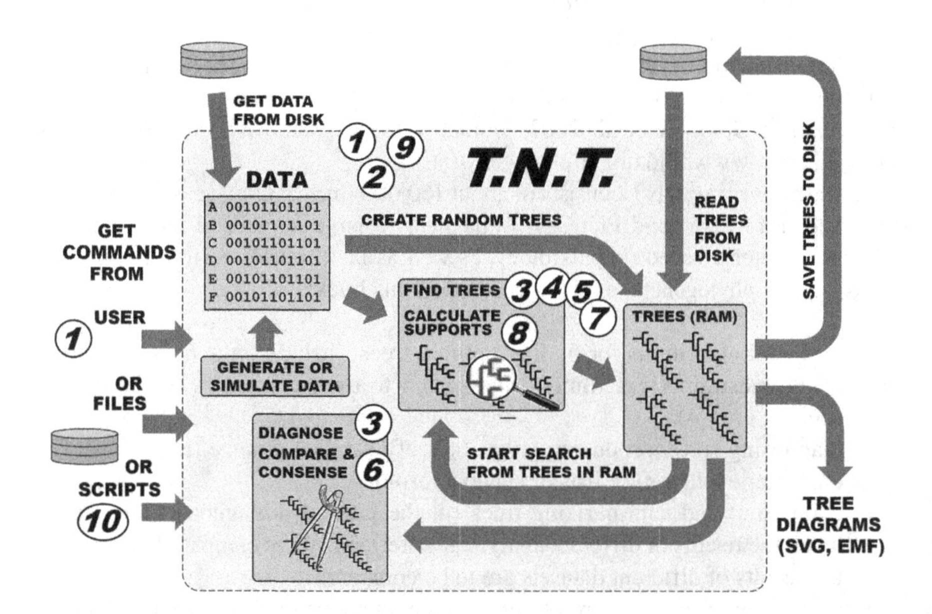

FIGURE 1.6 General organization of TNT. Circled numbers indicate the chapters that discuss the different problems in more detail.

in a search, found by a previous analysis and stored on disk, defined by the user, or created randomly) are held in memory can they be compared and used to examine their implications on the evolution of different characters or character systems.

Following that same logic, therefore, the user decides (by means of commands or menu choices) which of those options to use at every step. Once some trees are stored in memory, the user also must decide which of several actions to effect on those trees. It is possible to display the trees (in several available formats, some of them publication quality), diagnose them (e.g. producing lists of synapomorphies, matching groups against those of a pre-defined taxonomy), and summarize them (by means of consensus methods). The trees stored in memory can also be modified by the user (e.g. by "manually" moving branches to different positions) or used as starting point for new searches; they can also be saved to tree files for permanent storage. TNT can process only one dataset at a time; a new dataset can be read at any point, replacing the previous dataset (all the trees for the preceding dataset held in memory are cleared when the new dataset is read). Most chapters in this book include a table at the end (Table 1.1, in this case) summarizing the relevant commands.

TABLE 1.1
List of Basic Commands for TNT

Command	Minimum truncation	Action(s)
`clbuffer`	`clb`	Clear text buffer
`cls`	`cl`	Clear screen
`blocks`	`blo`	Set block limits, deactivate blocks
`cnames`	`cn`	Read character, state, and block names
`export`	`ex`	Save data in Nexus, Fasta, or TPS format
`help`	`h`	Display help, set string-matching options for commands
`log`	`l`	Open a text output file; in *GUI* versions, open a metafile
`mxram`	`mxr`	Set or report maximum amount of RAM to be used by TNT
`nstates`	`ns`	Set data size and type; convert types; rescale continuous characters
`procedure`	`p`	Read an input file; close current input file (`proc/`); mnemonic: proceed
`quit`	`qui`	Quit TNT (no questions asked!)
`quote`	`qu`	Copy from input to output (i.e. sends messages)
`report`	`rep`	Set progress reports for time-consuming operations
`run`	`run`	Read an input file, possibly passing arguments
`runc`	`runc`	Execute a C-style script
`screen`	`scr`	Change the size of the display (Windows character-mode only)
`shortread`	`sho`	Read a compact tree file from disk
`silent`	`si`	Set whether output is written to console, text buffer, or file
`svtxt`	`sv`	Write text buffer to log file
`taxcode`	`tax`	Deactivate or activate taxa; list active taxa
`taxlabels`	`taxl`	Report taxon names and numbers
`taxname`	`taxn`	Set whether lists of taxa are output with names or numbers; change taxon names
`tplot`	`tp`	Display trees; change how trees are displayed

(*Continued*)

TABLE 1.1 (Continued)

Command	Minimum truncation	Action(s)
`ttag`	`tt`	Handle branch labels, save tree diagrams (scalable vector graphics, *SVG*)
`txtsize`	`tx`	Change the size of the text buffer
`view`	`v`	View text buffer
`warn`	`wa`	Change warnings
`watch`	`wat`	Set whether execution is timed; report RAM usage
`xread`	`xr`	Read, edit, or save a dataset
`xwipe`	`xw`	Eliminate data from memory
`zzz`	`z`	Quit TNT; mnemonic: sends TNT to sleep, last command you run in TNT

1.11 INPUT, COMMANDS, FORMAT OF DATA MATRICES

As any computer program, TNT takes data or instructions (=input), runs the analyses or calculations, and then produces reports (=output). This section covers the organization and commands needed for input; the next section covers the basic formats of output produced by TNT. Other chapters cover details of the calculations.

1.11.1 Commands and Truncation

Input can be taken from commands typed at the TNT prompt, from a menu interface (in the *GUI* version), or from a file. In most cases, when taking input from a file, TNT behaves in exactly the same way as if the commands were being typed at the console.

In command-driven versions, commands are typed at the TNT command prompt, for example, `help`:

```
tnt*> help ;<enter>
```

In subsequent examples, the <enter> is normally assumed, not indicated explicitly. Several commands can be typed at the prompt, in which case they are executed in the same sequence. Most commands can take *arguments*—that is, numerical values or further options (the options can be keywords, names or numbers of elements, or, in many cases, one or more of the symbols =, –, +, /, \, [,], !, |, &, *, >, <). If additional options are possible for a command, the prompt changes to the name of the command and waits for the user to type the option or a semicolon to indicate that the default (remaining) options are to be used:

```
tnt*> help     [prompt changes to "help" here; user may
                want general help or help for a specific command]
help> ;        [general help actually displayed only after
                entering a semicolon]
```

A semicolon can be added at the end of commands that do not require (further) arguments, so it is always safe to end all commands with a semicolon.

Commands can be truncated; in that case, the first command in the list of commands matching the string provided will be executed. The sequence of commands is mostly alphabetical, with a few exceptions intended to give priority to commands used more commonly (e.g. the command `procedure` redirects input and is used very often, so it precedes the command `pcrprune`, which identifies wildcard taxa and is used much less frequently). All TNT commands are strictly alphabetical, so when the arguments are numerical values, TNT can identify the separation between command and argument automatically, without a need for the user to insert whitespaces between the command and the number. Likewise, numbers end when a whitespace or non-numerical symbol is encountered, so after a number you can just type the next command. For example, both options of this sequence of commands set the number of maximum trees to `hold` in RAM to 25, then run a search with five starting points with the `mult` command and calculate the strict consensus (with the `nelsen` command) ignoring the position of taxon number seven:

```
tnt*> hold 25 ; mult 5 ; nelsen / 7 ;
tnt*> ho25mu5;ne/7;
```

Note that a semicolon is not required between `hold` and `mult` (because `hold` does not take any alphabetical arguments), but a semicolon is required between `mult` and `nelsen` (because `mult` can take more arguments than the ones used in this example). Type as much as you like.

In *GUI* versions with a menu interface, the commands can be typed at a small command line at the bottom of the main window. A semicolon is automatically added at the end of every string of commands so that you do not need to type it explicitly (keep in mind that this prevents typing the command and its arguments with two separate <enter> keystrokes!). TNT may be expecting input either at the command line or the menu system; this is called the *focus*. After typing most commands, the focus is returned to the command line. If you wish to unfold a given submenu (e.g. the *File* submenu, by pressing <alt> + F), you can move focus between the command line and the menu by pressing <tab>, back and forth; if focus is on the menu, pressing <enter> always returns the focus to the command line.

In the Linux/MacIntosh versions, the last commands typed can be recalled (using the up or down arrow keys). Typing some letters and pressing <tab> will reinvoke the last sequence of commands entered at the prompt which matched the letters provided; pressing <tab> repeatedly cycles through all the command sequences matching the letters provided. The command history (up to 56 command sequences) is saved to a file `~/.tnt_history` on normal exit.

To help mnemonics, most of the commands that deal with definition, grouping, or comparison of *characters* start with the letter *x* (e.g. `xread`, `xgroup`, `xinact`, `xperm`, `xwipe`), and most commands for definition or comparison of *trees* start with the letter *t* (e.g. `tcomp`, `tchoose`, `tgroup`, `tread`); the command to *plot* trees is `tplot`. Similarly, many commands in TNT set options *on* and *off*; for most of those commands, the equal symbol (=) turns the corresponding option *on*, and the dash symbol (–) turns it *off*. For example, the `beep` command sets whether TNT makes a sound when an error or warning is encountered; `beep=` sets the sound to occur (this is the default), and `beep-` stops making sound (i.e. error and warning messages are displayed silently). Keep in mind that TNT is an interactive program, and all the settings you have changed during a run remain in effect if you read a new dataset or instructions file.

1.11.2 Input Files

The commands for TNT can be typed at the console or included in a file to be read by TNT. As TNT parses this file looking for commands, it behaves almost exactly as if you were typing the commands at the command prompt,[6] sequentially executing them. If an error is encountered during the parsing of an input file, the file is automatically closed and the rest of the commands in that file are ignored (in *GUI* versions, if `warning` is *on*, then you are queried on whether to just ignore that command and parse the rest of the file). Operations that are not errors but have potentially undesirable side effects (e.g. reading a new dataset, which discards all the trees held in memory) can issue warnings (this is changed with `warning=` and `warning-` or with *Settings > Report Levels*).

To have TNT read commands from a file instead of the console, you need to use the `procedure` command, followed by the `filename`. In *GUI* versions, the equivalent is *File > Open Input File*. Placing long and complex command sequences within a file (e.g. the commands defining a data matrix, see next section, or the list of character settings, see Chapter 2, Section 2.14) avoids having to type them at the console. The file itself should also contain, at the end, a `procedure/;` command. The slash indicates that the input file is to be closed—TNT parses nothing beyond that point. Nested input files are allowed (with a default maximum of 10 levels of nesting by default; should you need to allow deeper nesting, this can be changed with the `mxproc` command).

In addition to being able to understand commands read from the console or a file, TNT can also parse commands typed at the command line—i.e. when invoking the program from the console. In the case of Linux and MacIntosh, the usual Unix shell is bash, which uses semicolons to separate commands for bash itself. To facilitate using the command line in those operating systems, TNT therefore allows replacing (at the command line or command prompt) the semicolons with commas (this replacement is not allowed from within input files). Alternatively, in those operating systems, you can always use the (more annoying) option of passing special symbols (e.g. `|`, `&`, or `;`) from the command line to TNT as doubly quoted or escaped characters (i.e. with a backward slash). In the case of *GUI* versions, TNT parses as commands all the strings it finds *after* a semicolon; strings found *before* a semicolon are interpreted as names of files to parse. You can use this to pass several commands to TNT when invoking it from the console; e.g. to read an input file containing the dataset, open a log file with the `log` command (see Section 1.12.5), effect a run with the `mult` command, calculate the consensus, and exit:

```
C:\phylo> tnt ; p mydata.tnt; log results.txt; mul; nel; quit
```

Or, for Linux/MacIntosh (and assuming the executable file is in the path, e.g. /usr/local/bin in Linux):

```
john@mac:~/phylo$ tnt , p mydata.tnt, log results.txt,
mul, nel, quit
```

In addition to the `procedure` command, there are two other commands which instruct TNT to read from an input file, intended for use in more complex sets of instructions (see Chapter 10): `run` and `runc`. The `run` command is followed by a `filename` and a list of arguments to pass to the file. The list of arguments ends with a semicolon; those arguments can be referenced within `filename`; this is meant to be used in more complex scripting routines. The commands contained in `filename` are interpreted as regular TNT commands. The `runc` command (only in non-menued versions), followed by a `filename`, instructs TNT to parse `filename` as C-style code (Kernighan and Ritchie 1988), allowing calls to standard TNT commands.

1.11.3 Getting Help

TNT currently recognizes about 200 commands (the exact number depends on the version and how commands and options are counted). Remembering the details and options of every one of those commands is not only difficult, it is also pointless. A much more reasonable approach is learning the syntax of the most basic commands and getting used to work with the online `help` system for the rest of them. `Help`, if followed by the name of a command, displays a brief description of the command (or all the commands that match the string provided) and its (their) options. The same is achieved by typing the name of the command followed by a question mark (this displays help always for a single command at a time, i.e. the first matching command). `Help` without arguments (that is, with a semicolon) simply displays the full list of commands. Entering `help*` instead provides a brief description of each of the commands. These descriptions can be exported to a text output file (see Section 1.12.5).

TNT can apply string-matching algorithms to the commands typed at the console to detect typos (i.e. imperfect matches between strings). The Needleman and Wunsch (1970) algorithm is then applied to commands without an exact match and the closest alternative is suggested. The default is to have the string matching disconnected; to set string matching *on*, type `help>N` (where `N` is the similarity for strings to be considered potential matches, between 0 and 1). To disconnect the string-matching option, use `help<`.

1.11.4 Lists and Ranges; Numbering of Elements

Numbering for almost everything in TNT starts from 0. The last one of N elements is then numbered as $N - 1$. TNT adopts this convention (shared with several other phylogenetic programs, like Hennig86, Nona, Pee-Wee, and WinClada, Nixon 2002) because this is how arrays are referenced in standard programming. The only exception to this rule is the numbers for blocks of data because the number 0 is reserved in that case for the "block" composed of all data together.

Ranges can be used only for elements referred to by number; a range is indicated by separating two numbers with a period (this is because the dash is often used to specify options other than ranges). The two numbers must be contiguous to the period; a period immediately preceded by a non-number (whether a whitespace or another symbol) indicates that the range begins at the first possible element; a period immediately followed by a non-number indicates that the range ends at the last possible element. Therefore, to indicate "all possible elements", use just a period (with no need to remember what the actual values of the first and last element are). Examples:

```
15.21     from element number 15 to number 21, inclusive
  .21     from the first possible element to number 21,
          inclusive
15.       from element number 15 to the last possible one,
          inclusive
  .       from the first to the last possible element
```

Most commands that can take lists of elements use the default selection that is generally most appropriate. For example, the `nelsen` command calculates consensus trees; if no list of trees is specified, the command calculates the consensus of all trees; some taxa can be pruned from the consensus, but no taxon is pruned by default.

For commands that operate on lists of named elements, the elements to be included in a list can be specified by name (N.B. trees cannot be named; characters and states can; taxa must) or number. In any context in which a list of named elements is expected, the elements that contain a given string in their name can be indicated by preceding the string with a colon. This can greatly reduce the effort needed to produce the list. Consider as an example the following taxa, including in their names information about their taxonomy (taken from Wheeler et al.'s 2016 analysis of spider phylogeny):

```
Cyclocosmia_Mygalomorphae_Ctenizidae Aphonopelma_Mygalomorphae_Theraphosidae
Drymusa_Araneomorphae_Drymusidae     Poaka_Araneomorphae_Amaurobiidae
Ctenus_Araneomorphae_Ctenidae        Phoneutria_Araneomorphae_Ctenidae
Plexippus_Araneomorphae_Salticidae   Salticus_Araneomorphae_Salticidae
```

All the salticids and ctenids from that matrix (*Ctenus*, *Phoneutria*, *Plexippus*, and *Salticus*) can then be deactivated easily with the `taxcode-` command:

```
tnt*> taxcode - :Salticidae :Ctenidae;
```

or, equally easy, all the araneomorphs:

```
tnt*> taxcode - :Araneomorphae;
```

It is also possible to create *groups* of elements, which can subsequently be used to refer easily to sets. Groups can be defined for taxa (with the `agroup` command), trees (`tgroup`), and characters (`xgroup`). Groups can also be created from the menus in *GUI* versions (under the *Trees* or *Data* submenus). These groups can then be referred to by their name (or number), enclosing the name (or number) within braces, in any context in which a list of the corresponding elements is expected. Tree-calculating and tree-reading commands, by default, automatically place the trees in tree groups (named after the file from where the trees were read or the search algorithm used to find the trees). For example, the `mult` command places the trees produced in a group called `random_addseqs`, while the `randtrees` command (which creates random trees) places the trees produced in a group called `random_trees`, so that you can separately calculate the consensus (`nelsen`) for the trees resulting from each of those commands:

```
tnt*> p dataset.tnt; mult; randtrees 10;
tnt*> nelsen { random-adds }; nelsen { random-trees };
```

In any context in which a list of taxa is expected, the `@` symbol can be used to indicate all the terminal taxa which belong to a given node (*N*) of a tree (*T*), with `@T N`. In the case of characters, the `@` symbol indicates membership to a given block (using the numbering that corresponds to the reading frame within that block). Then, typing `@3 5` in a list of characters indicates the sixth character of the third block (instead of the sixth character of the entire matrix; remember that numbering for blocks starts at 1 and for characters at 0), and typing `@2 .` indicates all of the characters included in the second block (remember: the period indicates "all"). The numbering frame of the `@` symbol remains in effect within execution of the corresponding command; to return it to the frame of the entire matrix (instead of a particular block), use `@0`.

1.11.5 Reading Basic Data

To define a matrix of characters per taxa, the `xread` command is used. `Xread` is simply a command, one of the options of which is to define a matrix (other options are saving, generating, or modifying data, see Sections 1.11.13 and 1.11.14). As with most other commands, whether the `xread` command is included in a file or directly typed at the console makes no difference to TNT. An `xread` command with definition of data is almost always included in a file for practical reasons only. `Xread` allows a large number of formats and options; this section discusses only the basic settings, leaving more advanced options for subsequent sections.

To define a matrix, `xread` must be followed by the numbers of characters and taxa; if desired, a "title" (within single quotes, ASCII char 39; it can contain any character except a single quote; beware the possessive case and contractions) can be added between `xread` and the number of characters. The matrix follows the number of taxa:

```
xread ¶
'An example Mammalian matrix' ¶
4 8 ¶
Ornithorhynchus    0 0 0 0 ¶
Panthera_leo       0 1 0 0 ¶
Canis_familiaris 0 1 0 0 ¶
```

```
Mus_musculus      0 0 1 0 ¶
Pan_troglodytes   1 0 0 1 ¶
Gorilla_gorilla   1 0 0 1 ¶
Homo_habilis      1 0 0 1 ¶
Homo_clumsy       1 0 0 0 ¶
; ¶
proc/; ¶
```

The frame denotes that the `xread` command has been included in this case in a file; the symbol ¶ denotes carriage returns.

There is no pre-set limit for the number of characters that the program can handle; it is also possible to specify a "matrix" of 0 (zero) characters, consisting simply of a list of taxon names. Taxon names can contain neither blanks (use an underscore to separate genus and species names) nor the symbols used for arguments (as well as single or double quotes, numeral or dollar symbols, commas, and semicolons). By default, taxon names can contain up to 32 letters or symbols (this can be changed to any number N, with the `taxname+N` option prior to reading the data); letters beyond the maximum allowed for a name will simply be skipped. Duplicate taxon names are in principle allowed (at least in most contexts), but as they create ambiguity (e.g. when saving trees in parenthetical notation using taxon names), they are best avoided.

The character states for each taxon are specified after the taxon name. For basic matrices, numeric codes to indicate states from 0 to 9 can be used. For more than 10 character states, alphabetical characters are used, beginning with "a" to indicate state 10 and beyond. *Polymorphic terminal taxa* (e.g. when the taxon included in the matrix as a terminal is in fact a higher group variable for the character in question) are indicated by enclosing within square brackets all the corresponding states.

All the blanks in the file with the data are simply ignored; the only mandatory blanks in the example prior are those separating each taxon name and the first character state and the blank separating the numbers of characters and taxa. All carriage returns (¶) and spaces (whites, tabs) are interchangeable when the data are read as a single block (when reading multiple blocks, the carriage returns at the end of the characters for a taxon are mandatory; see Section 1.11.8).

The semicolon at the end is not mandatory when the data are being read in a single block, because TNT knows that after having read the state for the fourth character of the eighth taxon, the matrix is complete. However, it is advisable to add it anyway, because it facilitates troubleshooting errors in the matrix.

1.11.6 Different Data Formats and Sizes

If the data to be read have a format or number of states different from the defaults, this must be specified with the `nstates` command prior to `xread` (`nstates` affects the execution of `xread` itself). It is also possible to mix data formats, but this requires placing each format in a different block (see Section 1.11.8), specifying the format for each block individually; this section only discusses the case of a uniform format for the entire matrix.

The `nstates` command, in addition to specifying the format itself, is also used to specify the maximum numbers of states that are to be read in subsequent `xread` commands. TNT must know this before running `xread` to allocate memory for the matrix: when the data have few possible states (as in DNA characters), it is preferable to use fewer bits to store each character state, thus saving memory. Possible values for the numbers of states in `nstates` are 8, 16, and 32 (specifying a value in between these numbers sets the size to the next fixed number). The default value for maximum numbers of states is 16. You may well want to specify `nstates 8` and `nstates num` for numerical datasets where no character has more than 8 states. Note that `nstates dna` automatically sets maximum number of states to 8 and `nstates prot` automatically sets it to 32. Reading some continuous characters requires that `nstates` be set to 32.

In addition to specifying the *size* of the data, `nstates` indicates the *type* of data to be read:

- `nstates num`, for alphanumerical data, with states 0–9, and then from state 10 on, starting with "a". This is the default.
- `nstates dna`, for DNA (nucleotide) data, IUPAC codes. State correspondence is A = 0, G = 1, C = 2, and T = 3. Gaps are, by default, treated as a missing entry; if read as a fifth state, they are state 4. As is standard in DNA datasets, you can use a period to indicate (for all taxa but the first) that the state in the corresponding taxon is identical to that in the first taxon. Beware that the default setting of alphanumerical data also recognizes states A, C, G, and T (as states 10, 12, 16, and 29 respectively), but it does not recognize IUPAC codes for polymorphisms.
- `nstates prot`, for protein (amino acid) data. For single-residue codes, the sequence a, c, d, e, f, g, h, i, k, l, m, n, p, q, r, s, t, v, w, y, u corresponds to states 0–20. The period indicates matching as in the case of DNA data, and gaps are treated by default as missing entries (they can optionally be read as the 22nd state, i.e. state 21).
- `nstates cont`, for continuous (numerical) data. In this case, states are indicated as a real number, possibly with decimals; obviously, the successive number(s) for different characters must be separated by a blank in this case. See Chapter 9 for details on format and options.

The `nstates` command is also used to globally specify whether gaps (in `dna` or `protein` data formats) are read as an additional state (5th state in DNA, 22nd in amino acids) or as a missing entry (this can also be specified individually for each block, e.g. in matrices including multiple genes). To read gaps as an additional state, use `nstates gaps`, and to read them as a missing entry, use `nstates nogaps`. Note that this option affects how the dash (–) symbol is *interpreted* when reading in the matrix; this format cannot be changed once the data have been read (this is different from the way PAUP* works, where you can specify either `pset gapmode=missing` or `pset gapmode=newstate` subsequent to reading the data—TNT would require that the data are read again to change the `gapmode`).

1.11.7 Wiping the Dataset from Memory

Some command options (e.g. `nstates`, `mxram`, `taxname`, `thsrink`, `cost`, etc.) determine how the data are read and thus are forbidden once a dataset is already loaded in memory. You may well be in the situation where you have a dataset in memory but want to change some of those settings to read a new dataset—and you want to do this without exiting and reentering TNT. The `xwipe` command erases the dataset from memory and thus makes all those commands available again (note that, unlike the case of leaving and reentering the program, the options and settings changed remain in effect).

1.11.8 Multiple Blocks and Data Formats

TNT also allows reading the data in multiple blocks. The blocks themselves can be in different formats, so the multiple blocks are easily used to read combined datasets. Each block of data must be preceded by `&`, a special symbol used to indicate that a new block begins. The `&` symbol may be followed by a specification of the options and format for subsequent block, within square brackets (this may be omitted if the block has the same format as previously specified). Then, each name of a taxon included in the block must be followed by its character states; since TNT has no way to know ahead of time how many characters each block will contain, then every line of character states *must* end with a carriage return (¶) in this case. In principle, the same taxon must have an identical name in each block (but see the following); TNT establishes taxon identity by matching taxon names, and thus it is not necessary for the taxa to be listed in the same sequence in each block. A taxon can be included in some blocks but not others; for each character of a block where the taxon is not included, the taxon receives a missing entry. The taxa are numbered by TNT according to their first appearance in the file. In the case of multiple data blocks, the end of the matrix *must* be indicated with a semicolon.

To indicate a special data format for a given block, the strings `num`, `dna`, `prot`, `cont`, or `landmark` can be enclosed within square brackets right after the ampersand (`&`) which indicates the beginning of the block. Beware: if continuous characters are to be combined with discrete characters, then the continuous block(s) must be the first one(s), preceding all the discrete block(s). If you are going to use formats which require numerous states (such as `prot` or `cont`), you have to execute `nstates 32` before the `xread` so that TNT allocates the proper amount of memory before starting to actually read the data.

As example, the simple mammalian matrix shown prior could be expanded to include some DNA sequences and an additional taxon (giraffe):

```
xread ¶
'An example Mammalian matrix' ¶
14 9 ¶
& [num]
```

```
Ornithorhynchus      0 0 0 0 ¶
Panthera_leo         0 1 0 0 ¶
Canis_familiaris     0 1 0 0 ¶
Mus_musculus         0 0 1 0 ¶
Pan_troglodytes      1 0 0 1 ¶
Gorilla_gorilla      1 0 0 1 ¶
Homo_habilis         1 0 0 1 ¶
Homo_clumsy          1 0 0 0 ¶
¶
& [dna]
Ornithorhynchus     AGCGGCTTTG ¶
Gorilla_gorilla     GGAAG--GTC ¶
Homo_habilis        GGAAG--GTC ¶
Giraffa_camelop     AAAGAGGGGG ¶
Panthera_leo        AAAGAGGTTG ¶
Pan_troglodytes     AAAGA--TGG ¶
; ¶
proc/; ¶
```

In this case, the number of characters and taxa has been increased to their total values; however, when reading multiple blocks, specifying a number of characters or taxa *larger* than actually included in the matrix only causes a warning—TNT will reset the dimensions of the matrix to the proper value. This is very useful when mixing datasets and avoids having to make exact counts of characters and actually distinct taxa. Note, however, that the memory assigned to read the data cannot be freed after the data have been read so that some memory is wasted when specifying more characters and taxa than actually present in the matrix; if you had seriously overcounted characters or taxa, you may want to edit the data file and change the numbers of characters and taxa to their correct values.

In the example, note that the giraffe is *not* included in the first block, while *Homo clumsy*, the dog, and the mouse are *not* included in the second; this is perfectly valid. The reading of multiblock data in Nexus files (Maddison et al. 1997) is rather different, forcing the user to specify every taxon in every block and in the same sequence (Nexus files have the advantage that taxon names in all blocks but the first are effectively ignored and thus can be mistyped or altered without harm).

Within the square brackets, you can indicate (in addition to data formats `contin`, `num`, `dna`, `prot`, or `landmark`) other options specific for the block. These are:

- `gaps/nogaps`, to determine whether gaps are treated as an additional state (`gaps`) or as a missing entry (`nogaps`). Specifying one of these options for a block applies it to all subsequent blocks, unless the alternative is specified.
- `trimhead` and `trimtail`, to consider only the leading or trailing gaps as missing entry (obviously, useful only when the internal gaps are to be treated as an additional state). An obvious case to do so is when the sequence for

some of the taxa is incomplete, so the leading or trailing gaps indeed represent lack of information. This option is not remembered between blocks, and it must be specified for each block where the trimming is to be applied.

- `match N`, to compare names of taxa within that block with a string-matching (Needleman-Wunsch) algorithm. If the similarity in names with a taxon in a preceding block is equal to or greater than `N` (where $0 < N < 1$), TNT queries the user for whether the two taxon names are to be considered synonyms. Obviously, identical names are considered to belong to the same taxon without queries. This allows taking into account possible spelling errors (e.g. when mixing matrices from different authors). This option is not remembered between blocks, and it must be specified for each block where the matching is to be applied.
- `verbose`, to report the numbers of characters and taxa (for the block and the cumulative ones) as the matrix is read. It can be used to monitor reading of very large datasets and to detect errors in large and complex matrices. Like the two preceding options, this option is not remembered between blocks and must be set for each block where you want it to be applied.
- `/block_name`, to apply to the block the name following the slash. Block names can also be set with the `cnames` command (see next chapter), but naming them from within the same block may be more convenient when combining multiple files in a single dataset (see the following).

In addition to the `match` option for similar taxon names, TNT will not require an exact matching when a taxon name ends with an elipsis (. . .). If several already defined taxon names match the string provided (e.g. in the case of the mammalian matrix prior, `Homo...` matches both `Homo_habilis` and `Homo_clumsy`), then all of them are assigned the subsequently specified character states. This can be used to simplify matrices.

1.11.9 Combining Existing Datasets

The fact that TNT may read different blocks of data by simply preceding each block with the `&` symbol makes it easy to manually combine different datasets into one. All you need to do is cut-and-paste the data parts from each file into a new file; with the use of `&`, each dataset from a file will now represent a different block of data; you only have to make sure that the names used for the same taxon in different datasets are identical. In those cases when there are character specifications for each file, they can also be combined manually with ease, by using the `@` symbol to indicate that the reading frame for some character specification is within some specific block (as discussed in Section 1.11.4).

1.11.10 Datasets in Multiple Files

One of the alternative ways to input data into TNT is by using different files to store different parts of the data. This is especially useful when the data come from different genes: having each gene in a different file may make combining the datasets easier and more flexible. This was the approach taken by Goloboff et al. (2009).

Reading the data from different files requires that scripting (see Chapter 10) is enabled; this is accomplished with the `macro=` command. Once the scripting is enabled, whenever TNT finds a double `@` symbol, it will then read a filename and a `tag` to search in the file (arguments are possible although not used in this example), ending with a semicolon. TNT will then search in the file specified until it finds the string `label tag`; from that point onward, TNT will read as if the contents of the file were simply being copied into the input stream where the file was called with the double `@`. Input from the file will be interrupted when another double `@` is found, returning input to exactly where it was in the original, calling file. By placing the calls with double `@` in different data blocks (see prior), then all that is required to read data from multiple files is listing the files to combine within an `xread` with multiple blocks. Assume you have two files, for example `file_a.txt` and `file_b.txt`:

```
label data ¶
Ornithorhynchus   0 0 0 0 ¶
Panthera_leo      0 1 0 0 ¶
Canis_familiaris  0 1 0 0 ¶
Mus_musculus      0 0 1 0 ¶
Pan_troglodytes   1 0 0 1 ¶
Gorilla_gorilla   1 0 0 1 ¶
Homo_habilis      1 0 0 1 ¶
Homo_clumsy       1 0 0 0 ¶
@@ ¶
```

file_a.txt

```
label data ¶
Ornithorhynchus AGCGGCTTTG ¶
Gorilla_gorilla GGAAG--GTC ¶
Homo_habilis    GGAAG--GTC ¶
Giraffa_camelop AAAGAGGGGG ¶
Panthera_leo    AAAGAGGTTG ¶
Pan_troglodytes AAAGA--TGG ¶
@@ ¶
```

file_b.txt

Then, you can create a file (say, `filelist.tnt`, which is the one to be initially parsed by TNT) which contains calls to those files from within different data blocks of an `xread` command:

```
macro=; ¶
xread ¶
'Reading data from different files'¶
14 9  ¶
& [ num ] @@ file_a.txt data ; ¶
& [ dna ] @@ file_b.txt data ; ¶
; ¶
proc/; ¶
```

filelist.tnt

To read the data, you then parse `filelist.tnt`, with *File > Open Input File* (in the *GUI* version) or with the `procedure` command:

```
tnt*> p filelist.tnt;
```

By adding or removing blocks to the `xread` command in `filelist.tnt`, you can have different lists of files to produce different combinations of datasets. Beware that the semicolon after the label in `filelist.tnt` (`data` was used in the example, but you can choose any word you like) is required to indicate that no arguments are used when reading `file_a.txt` and `file_b.txt`.

1.11.11 Redefining Data Blocks

Sometimes, you may want to change the block definitions that result from reading the file. This can be done with the `block` command: a list of numbers following this command is interpreted to indicate the starting character for each of the blocks (the first one, always starting at character 0, is allowed but unnecessary; the end of a block needs not be indicated because every block ends where the next starts). This can also be done from the *Data* menu in the *GUI* version.

The `blocks=` command deactivates some data blocks. All the characters not in the specified blocks (a list of block numbers or names) are deactivated; all the taxa with only missing entries in the other blocks are deactivated. If you place different types of data (different genes, different organ systems) in different blocks, then you can easily run combinations of different datasets. Note that the `block=` command does not reactivate any character or taxon; before running it, you must make sure that all taxa and characters are active.

1.11.12 Other Data Formats

Data files in the Nexus and Fasta formats can also be read into the program. The Nexus format (Maddison et al. 1997) is very common in phylogenetic datasets; it is used by PAUP* (Swofford 2001), MacClade (Maddison and Maddison 2000), and Mesquite (Maddison and Maddison 2018), as well as other programs. The Nexus format can be quite complex, allowing the inclusion of different "blocks" of commands, each one starting with the string `begin`. This is so that the same Nexus file can be parsed with different programs; a Nexus-compliant program will simply skip those blocks in the file which are intended for other programs (e.g. PAUP* will skip those blocks starting with `begin Mesquite` and skip all the Mesquite commands until an `end` is found). Data blocks in Nexus files always begin with `begin data`, but the data can be given to the program in different formats.

TNT only understands Nexus data files when they have the simplest format. Other Nexus formats (e.g. when the taxon labels are defined first and then the data) are not understood by TNT. If you have a data file in such format, the easiest way to make it readable by TNT is reading it into PAUP* and then saving the data with the `export` command of PAUP* (`export file=newfilename format=nexus interleaved=no`). When saving the data, PAUP* uses a simple format, which

TNT can (normally) read without errors. Another alternative is reading the Nexus data file with Mesquite and then saving it in TNT format.

Nexus blocks for other programs (delimited by `begin` and `end`) are skipped by TNT, except for PAUP and TNT blocks. Within a PAUP block, the only commands recognized by TNT are `delete`, `undelete`, `usertype`, `ctype`, `exclude`, `include`, and `weights`; these commands are also recognized if outside a PAUP block. If you edit your own Nexus files by hand, keep in mind that numbering of elements in Nexus files starts with 1, not with 0 as in TNT. Step-matrices, defined with the `usertype` command in PAUP* and assigned to some character(s) with `ctype`, are accepted by TNT.[7]

In Nexus files, a block that starts with `begin tnt` can be used to embed TNT commands. This must be the last block in the file—subsequent blocks, if present, will be ignored by TNT. If you want the file with the `begin tnt` block to be Nexus compliant (e.g. to be run with PAUP*), then you have to include an `end` at the end of the TNT block (otherwise, PAUP* will complain). TNT simply ignores the string `end`—it may be omitted if you are going to read the Nexus file only with TNT.

Data files with Fasta format (used by many alignment programs) are also recognized by TNT. Fasta files use the > symbol preceding each taxon name, followed by the sequence for that taxon. To read a file as Fasta into TNT, you call it with `procedure`, but preceding the file name with the ampersand `&`. In TNT, taxon names in Fasta files must contain no blanks, and a carriage return after the taxon name is allowed but not required. The sequences must be aligned; TNT will read the first sequence in the file and then expect subsequent taxa to have sequences of the same length as the first one. If you want to read unaligned sequences, you can use the `=N` option (right after the ampersand) to tell TNT to add, at the end of any sequence shorter than N positions, enough gaps to make it of length N. This will obviously result in completely arbitrary "alignments" and can be used if you are interested only in the taxon names contained in the Fasta file or in other kinds of data processing.

1.11.13 Changing the Data

While in other phylogeny programs the data are unmodifiable once they have been read, TNT allows editing the dataset, both by changing taxon names and individual character entries. The taxon names are changed with the `taxname` command; the syntax is `taxname /NewName OldName`. Taxon names in GUI versions can also be changed from *Data > Edit Data > Taxon Names.*

Changing the matrix entries themselves is more involved. This is done with the equal option of the `xread` command. To change the entry for character `C`, taxon `T`, to be the state set `S`, use `xread=C T S`. As shown, both `C` and `T` must be numbers; no character names can be used in this context. To allow names, you can enclose the character or the taxon in square brackets, in which case you can give a list (using either numbers or names). Since this list can be given using the format for ranges specified prior, this allows for great flexibility in defining the data. For example, in the matrix with 73,060 taxa of Goloboff et al. (2009), the morphological characters were first read as all missing entries (as a block of data where only one taxon was included, with only missing entries; all taxa not included in that block had missing entries as well) and then redefined with `xread=`. The redefinition was done by taking advantage of the fact that taxon names contained all the taxonomic categories for each terminal in the matrix (separated by underscores). Thus, since character 575 in the matrix of Goloboff et al. (2009) was "mammary glands", it was possible to assign state 0 to all taxa, then change to state 1 all mammals, with just two lines:

```
xread = 575 [.] 0;
xread = 575 [:_Mammalia_] 1;
```

This will properly set the 2,401 species of mammal included in that matrix as having mammary glands and all the rest of the taxa as lacking them. Likewise, character 340 in that matrix corresponds to "opisthosomal silk glands" (exclusive of spiders), so that

```
xread = 340 [.] 0;
xread = 340 [:_Araneae_] 1;
```

will suffice to set all 655 species of spider included in Goloboff et al.'s (2009) matrix as having spinnerets and all other 72,405 taxa as lacking them. As a last example (and not that you would often want to do this in any real analysis!), you can make every single entry in the matrix to be a missing entry with `xread=[.][.]?`. As an example of how to take advantage of these options for editing large datasets, see Mirande (2017).

The list of character states for `xread=` can be any valid set of state symbols (including, of course, missing entries or gaps); for discrete characters, no square brackets are needed here to enclose multiple states in the case of polymorphisms because the end of the command is indicated by a semicolon; for continuous characters, two dash-separated values indicate a range. When the data have been read using different formats, you may want to define how the symbols are read in lists of states; for example, the letter `G` may mean state 1 if the data were read as nucleotides but state 5 if the data were read as amino acids or state 16 if the data were read as 32-states alphanumeric. By default, TNT will read the symbols according to the last format specification, but it is possible to explicitly indicate how the state symbols will be read using the `nstates/` command. For example, `nstates/dna` will read subsequent state specifications as nucleotides.

Once the data have been changed, TNT needs to recheck the maximum and minimum possible numbers of steps for all the changed characters (this is what determines parsimony informativeness and amounts of homoplasy). For a large matrix and many characters changed (e.g. by a script), this repeated checking is time consuming and unnecessary: it is strictly required only after the last change was made (since no searches or calculations are to be done before that). Using an exclamation mark right after the equal sign (`xread =!`) tells TNT to change the matrix but to not recheck the minima and maxima. Use this option cautiously; keep in mind that TNT will think that the minima and maxima have been checked and are those that correspond to the matrix *prior* to the changes—effecting a tree search or other tree calculations may produce the wrong results. To prevent this, you have to make sure you issue an xread == command (double equal) after having done the last change, which tells TNT to check minima and maxima for all the characters in the matrix, so that you are sure that character informativeness and homoplasy are properly assessed.

The data can also be changed using a graphic interface, in the *GUI* versions, accessed from *Data > Edit Data*. The data can be edited on a taxon-by-taxon basis or a character-by-character basis. The first (taxon based) provides lists of the alternative states to be assigned to a taxon, allowing moving from one character to other to easily. The second (character based) provides lists of the taxa with each of the available states for a given character. Both of these options allow editing a dataset using only the character and state names, without ever having to actually write the dataset as a numerically coded matrix.

1.11.14 Saving Edited Dataset in Different Formats

Changing a dataset in memory does not alter the dataset in disk. It is possible to save the dataset: the xread* option writes a matrix with all taxa and characters to output. Displaying a dataset just read (with xread* or with *Data > Show Matrix*) can also help confirm that the data have been read as expected. The xread- option is similar but displays active taxa and characters only. Adding an asterisk (i.e. xread** or xread-*), the character settings (weights, additivities, character names, etc.) are written to output as well. The output can be the screen or a file; you will normally need to place the modified dataset in a file, so you should first open a log file and (unless you want to see the modified matrix on screen) also issue a silent=console command (see next section). The dataset can also be saved, in the *GUI* version, from *Data > Save Data*.

To save a dataset in a format other than the native TNT format, you use the export command. The export command will save the dataset to the specified file, either in Nexus or in Fasta format in the case of discrete characters or in TPS format ("thin plate spline", used by programs such as MorphoJ, Klingenberg 2011) in the case of landmark characters.

Nexus is the default format for export; using just the name of the file saves the dataset. To save the trees in addition to the data, use an asterisk (export*filename.nex). If step-matrices are defined, you have to explicitly ask TNT to save those step-matrix definitions, with a plus sign (+) preceding the file name. To save just the trees, with no matrix, use export-filename. The saved matrices or trees should be directly readable by PAUP* or any Nexus-compliant program. To save as Fasta, precede the name of the file with an exclamation mark (!). This will include the gaps if present in the matrix. If saving the data file in Fasta format with the intention to realign the sequences, then you can get rid of the gaps (i.e. saving only the original sequences) with a double exclamation mark: export!!orig_seqs.fas. To save as TPS format, use export|filename; in this case you need to specify, after filename, a character number (the TPS format specifies only a single configuration) and, optionally, a tree number (if specified, the landmark coordinates for the ancestor, as reconstructed on the tree, are also saved to the file). See online help of the export command for additional options.

1.12 OUTPUT

As TNT does different kinds of analyses, it produces reports of the results. These reports can be shown on screen only (the default) or written to a file as text. Tree diagrams are shown as line characters in this case and cannot be inputted back into TNT from these files. It is also possible to save the trees in files which can be subsequently input into TNT and as graphics files (publication-quality diagrams).

1.12.1 Text Buffer

All the output produced by TNT is stored in an internal text or display buffer. In *GUI* versions the text buffer is automatically displayed in the main window after execution of each command. In command-driven versions, the text buffer can be viewed with `view` (arrows move display, the <esc> or Z keys quit the viewer and go back to the command prompt). The text buffer works cyclically: as additional output is produced, it is added at the end; as the text buffer is filled, lines of text written earlier are erased. The size of the text buffer can be changed with `txtsize N` (`N` is the size in KB); this operation clears existing contents. The display fits the entire terminal screen in Linux and MacIntosh; in command-driven Windows versions, the size of the display can be changed to `C` columns and `L` lines with `screen CxL`; pressing the > or < keys while viewing the buffer changes the number of lines to display, and \ or / changes the number of columns.

The text buffer can be searched for a string using `view:string;`. The string itself can contain whitespaces, and what indicates the end of the string to search is a semicolon, which is here mandatory; if you leave a blank between *string* and the semicolon, the program will search for "string " (with a blank space at the end), which may or may not exist in the text buffer. If the string to be searched exists, the text viewer is positioned at the point where that string starts (pressing `&` searches for the next occurrence of the string).

If a file is opened for output (see Section 1.12.5) after having produced and stored some output in the internal text buffer, it is possible to transfer the contents of the text buffer to the file with the `svtxt` command. `Svtxt` without arguments (i.e. followed by a semicolon) saves the entire contents to the file; using instead `svtxt` L_1 L_2 saves only the contents from line L_1 to L_2.

1.12.2 *GUI* Screens (Windows Only)

In *GUI* versions, the text buffer is displayed automatically in the main window. When a tree is to be plotted (with any command that produces tree diagrams), the tree goes first to a separate screen, the *pre-viewing* screen. In this screen you view the tree (possibly coloring branches if mapping characters), then decide whether to save the tree diagram to the text buffer (pressing S), show next tree diagram (<enter>), or stop displaying trees (<esc>). Pressing H while in the pre-viewing screen displays other possible options. Once all the trees have been displayed or saved to the text buffer, the pre-viewing screen is closed and the text buffer is displayed again in the main window. The pre-viewing screen can be deactivated (so that trees go directly to the text buffer, no questions asked) or activated again with *Format > Preview Trees* (or with the commands `tplot]` and `tplot[`). The pre-viewing screen does not function well in machines with touch-screen control, so to use TNT properly you may have to deactivate the touch-screen option (or disable the pre-viewing screen).

In addition to the main display and the pre-viewing screen, a third type of window exists, the *tree-viewing* window, accessed with *Trees > View*, where TNT displays the tree(s) in memory, allowing the user to modify them or make mouse-based taxon and shrinking selections by clicking on tree nodes. Editing a tree requires that the tree is "unlocked" (with the padlock tool or with *Settings > Lock Trees*). In non-*GUI* versions, the same is done with the `edit` command (with no need to unlock trees in that case).

A *shrink* is defined as a group of taxa to be shown as collapsed if present in the tree, displaying only the name of the group. Defining shrinks for all families, for example, may facilitate visualizing the relationships *among* the families. Shrinks are defined either by double left-clicking on a node or with the `tshrink` command. Invoking any taxon-selecting option from the menus, with some nodes selected in the tree-viewing window, will display as initial taxon selection all the taxa belonging to nodes marked in red (with a single left click), which are not in turn descended from nodes marked in green (with two single left clicks; don't do them too fast or Windows registers them as a single double click!). This can be used to easily define taxon selections corresponding to paraphyletic groups.

1.12.3 Outputting Numbers or Names

When TNT outputs lists of taxa or characters, it uses the numbers by default. Alternatively, it is possible to use the names. In the case of taxa, using taxon names is set *on* and *off* with `taxname=` and `taxname-` respectively. The use of taxon names instead of numbers applies to trees saved in parenthetical notation as well. For characters and states (obviously, this requires that characters have been named, which is optional; see Chapter 2), using character names is set *on* and *off* with `cnames=` and `cnames-` respectively. Using names instead of numbers applies also to mapping character states onto trees, in which case (to save space) only the state sets differing from those of the ancestor are shown on the tree. In *GUI* versions, the use of names or numbers is controlled from the *Format* submenu (as well as from the dialogs for character mapping, under *Optimize > Characters*).

1.12.4 Table Formats

Whenever TNT outputs a list in the form of a table, it can use two alternative formats. In the first format (set with `table-` or with *Format > Optional Table Formats*), all the values occupy a space in the table (with a dash or blank if not calculated or applicable), with 10 values per line. Assume 20 trees are present, and you ask TNT to calculate the lengths of trees 8 to 15; the default table format will display this as:

	+0	+1	+2	+3	+4	+5	+6	+7	+8	+9
0	--	--	--	--	--	--	--	--	583	559
10	567	549	546	541	547	525	--	--	--	--

You can also display the values for those trees in an alternative format (set with `table=`), which is more compact when only a few values are to be displayed:

```
       8        9       10       11       12       13       14       15
     583      559      567      549      546      541      547      525
```

The alternative table format prints up to 10 values per row when using numbers. If the list is a taxon list and the `taxname=` option is in effect, or if the list is a character list and the `cnames=` option is in effect, then the second table format will use the names instead of just the number (printing in this case only five values per row, as the columns in this case are expected to be wider).

When tables print out values with decimals (e.g. under implied weights or when landmark data are present), then the default number of decimals is five; this can be changed, so the table prints fewer or more decimals, with `table/N` (where `N` is the number of decimals to use).

1.12.5 Output Files

Log files are opened with the `log` command, followed by the name of the file to open. In *GUI* versions, log files are opened from *File > Output*. After the file is opened, any output produced by TNT will go to the file (unless silenced, see Section 1.12.7). `Log/` closes the open log file. A single output file can be open at any time. Output files can be opened in append mode, with `log + filename`; this adds output to the end of the file (instead of overwriting contents, as in the default mode). Normally, a `log` file can be opened at the beginning of a TNT session so that all the work done is recorded. Alternatively, the contents of the text buffer can be saved after opening a `log` file, with `svtxt` (in *GUI* versions, with *File > Output > Save Display Buffer*).

1.12.6 Quote Command

If you need to show a message, the `quote` command copies from input to output until a semicolon is found. If you need to display a semicolon in the output, use a period immediately followed by a comma. With `lquote` the literal quotes are activated (`lquote=`, not using an enter at the end of each `quote` command) or deactivated (`lquote-`, always add an enter at the end). With `lquote[` the conversion of ASCII character numbers is enabled; in this case, the expression `&N` within the `quote` command is converted into ASCII character number N on output. For example:

```
Quote This is an &34example&34  of ASCII conversion.;
```

(note double space after the second `&34`) produces on output:

```
This is an "example" of ASCII conversion.
```

1.12.7 Silencing Output

Depending on the type of analysis, some of the output produced by TNT may not be of interest. In that case, it is possible to tell TNT to not write output to the console, the text buffer, or the output file. The command to control this is `silent`; `silent=` silences the subsequent list of devices (`console`, `buffer`, `file`, or `all`; any of these can be truncated), and `silent-` makes them non-silent. Note that *GUI* versions display on screen the contents of the text buffer so that the options `console` and `buffer` become synonymous. Keep in mind that when output is silenced and an error occurs (e.g. a mistyped command, attempt to open a non-existent file, etc.), the output to the `console` is reactivated (so that you can see error messages); the output to text `buffer` (in command-driven versions) or `file` remains silenced.

1.12.8 Progress Reports, Warnings

For potentially time-consuming operations, a progress report is done. The reporting is turned *on* and *off* with `report=` and `report-`. Keep in mind that, in *GUI* versions, calculations can only be interrupted when a report is being made; if the reports have been disconnected, the program may appear to have frozen when in fact it is doing some demanding calculations. Operations that have potentially undesirable side effects can also issue warnings; warnings are connected and disconnected with `warn=` and `warn-` respectively. In *GUI* versions, both the reports and the warnings are controlled with *Settings > Report Levels* (with settings remembered between sessions).

1.12.9 Graphics Trees

Commands that display trees (trees, consensi, etc.) produce diagrams as text, using ASCII symbols. The diagrams are part of the text buffer and so can be saved to files. The tree diagrams can also be produced in two other formats: *windows metafiles* (with the EMF extension) and *scalable vector graphics* (*SVG* extension).

Metafiles are available only in the *GUI* versions. Pressing M while in the tree pre-viewing screen, TNT queries for the name of a file. Trees are saved one at a time. Alternatively, it is possible to have a metafile opened before the commands that display the tree(s), either from *File > Output* or with `log& filename` (`log/&` closes the metafile). Once the metafile is opened in this way, any subsequent tree-displaying command automatically saves the diagrams to the metafile, no questions asked.

Scalable vector graphics (*SVG*) are available for all versions of TNT, but their use is a little more involved. For this, it is necessary to use the *tree-tag collecting* facility of TNT, activated with `ttag=`. Once tree-tags are activated, the first command displaying a tree will store in a separate buffer both the tree topology and all the branch labels (=tags). Subsequent commands displaying trees will concatenate the branch labels at the corresponding branches (labels for groups not present in the first tree are lost, as those groups do not exist in the multitagged tree). The tags can be displayed, with `ttag` (no arguments), or saved as an *SVG* with `ttag& filename.svg`.

To discard existing tags, use the `ttag-` command. Thus, the sequence of commands for a simple black-and-white tree diagram of the first tree in memory is:

```
tnt*> ttag=; tplot 0; ttag& mytree.svg; ttag-;
```

This can also be used to produce trees with colored branches (e.g. from character mapping); see details in Chapter 3, Section 3.13.3, and Chapter 10, Section 10.4.1. The *SVG* files can be opened with any web browser and are easily converted into publication-quality figures (there are a number of free web-based converters; I regularly use the one at https://image.online-convert.com/convert-to-tiff).

1.12.10 Saving Trees to Files

A different kind of output is the saving of trees to files so that they can be read back into TNT (or other program) without repeating calculations.

TNT can save (and read) the trees in two different formats, the *compact* and the *parenthetical* notation. The compact notation is recognized only by TNT; it produces trees as binary files, which occupy much less disk space than parenthetical trees. This format can be used only to save trees containing all taxa (any taxon missing from the tree is added at the base of the tree). These compact files are read back into TNT by parsing the file with `shortread compact-file-name`. The parenthetical notation uses the same format as Hennig86, with a `tread` command followed by the parenthetical trees (separated by asterisks; a semicolon indicates the last tree in the `tread` command); terminal taxa are indicated with either names or numbers and without commas separating tree nodes. For either of these two formats, you first need to open the file with `tsave filename` (the default is the compact format; use an asterisk between `tsave` and `filename` to indicate parenthetical). Once the file is opened, the `save` command (with a list of trees to save) effects the saving. Trees can continue being added to the file until it is closed. Adding a + symbol after `filename` opens it in "append" mode. These same options, in the *GUI* versions, are available via *File > Tree Save File*. When tree-tags have been defined with `ttag`, they can be saved (in parenthetical format only) with the `save*` option. The tags are saved, preceded by an equal sign, after the node to which the label should be attached.

In addition to the native TNT formats, TNT can export trees in other formats (Nexus, simple Newick) with the `export` command. If tree-tags have been defined, they can be saved in Nexus format as either branch lengths or labels, with `export> filename` or `export< filename`, respectively. The trees thus produced can be read into programs like Dendroscope (Huson et al. 2011), FigTree (Rambaut 2009), or TreeView (Page 1996).

1.13 OUTLINE OF THE REMAINING CHAPTERS

While all that is needed for the basics of cladistic analysis are the principles established in the preceding sections, in practice a series of secondary problems become important. The graphic summary of Figure 1.6 shows the chapter where each of the different operations of TNT is discussed. Those aspects, after additional preliminaries in Chapter 2, are discussed starting at Chapter 3:

- For a given tree, the optimal ancestral reconstructions could be found by enumerating all possible reconstructions and retaining the best, but this would be very time consuming. Specific algorithms are required to find optimal ancestral reconstructions efficiently, and these constitute the subject of Chapter 3. A more detailed consideration of the properties of optimal ancestral reconstructions also illuminates several aspects of phylogenetic analysis, such as questions of rooting and polarity, recognition of ancestral taxa, and the use of polytomies as expressions of ambiguity instead of multiple simultaneous speciation. The problem of calculating the score of a single tree also occurs with likelihood methods in a manner almost completely analogous to that under parsimony; these problems are discussed briefly in Chapter 4, with a general overview of model-based methods.
- Having methods for calculating the score of a given tree in practice, the problem is now how to increase the chances of having found the tree that truly maximizes the chosen criterion. Evaluating all trees is prohibitive in practice for all but the smallest datasets, as the number of possible trees increases very rapidly with numbers of taxa. It is then necessary to use methods that make calculations possible. This is the topic of Chapter 5, "Tree Searches".
- With any optimality criterion chosen, when the data are insufficient or conflictive, it may well be that more than a single tree optimizes the tree score. In such case, the data determine a series of alternative phylogenies as possibilities. Or, alternatively, several different datasets may exist for a set of taxa, each of which supports slightly different trees. This is the topic of Chapter 6 on summarizing and comparing trees. How best to proceed in such a case? How is the information common to a number of trees best summarized? How similar are the trees obtained?
- Chapter 7 deals with character weighting. This is especially important in the case of morphology, where datasets typically involve small numbers of characters with very diverse characteristics, some of them probably providing stronger evidence of relationships than the others. How can those stronger characters be identified without requiring that the investigator assumes a priori which are the good and the bad characters?
- Another problem is that even when a phylogenetic analysis chooses a single tree as the best hypothesis, alternative trees may differ in their fit by only a marginal difference or a large one. This is the concept of support: how strong the preference of the best tree is over the alternatives. Related problems and methods pertaining to this aspect of the problem are discussed in Chapter 8.
- Most phylogenetic analyses deal with discrete, qualitative characters, but quantitative characters (measurements, shape characters with subtle differences best represented by means of geometric morphometrics) are also

an important source of information, especially at lower taxonomic levels. These are discussed in Chapter 9.
- Many of the elements in Chapters 3–9 need to be integrated in any phylogenetic analysis that goes beyond the basic aspects. The integration of these elements, together with general methods for further automation of complex tasks, is discussed in Chapter 10, which closes this book.

NOTES

1 The discussion concerns the dismissal of evidence without any grounds to do so other than protecting the conclusions against rejection—one of the meanings of the term *ad hoc*, if not the only one (Votsis 2016). That the term *ad hoc* is also used in other meanings is not relevant here.
2 Some authors (e.g. Kluge 2005) prefer to use the term *retrodiction* for this kind of reasoning, on the grounds that it concerns future observations of *past* events. Such distinction between retrodiction and prediction continues cropping up with some regularity in the phylogenetic literature. While there is nothing wrong with the distinction, I do not find it particularly useful and continue using the more common *prediction*—it is clear that the term is being used to refer to future observations, not to the future evolution of taxa or characters.
3 It is, in general, difficult to judge how much of the dismissal of the pattern notion stems from loyalties to specific research groups. In fact, A. Quinn, a young philosopher, exemplifies that, because she is a frequent collaborator of K. De Queiroz, well known anti-pattern cladist. Her opposition to that approach could then have been predicted without having read her paper.
4 For the record: this is a type of example which I first showed to N. Platnick in the early 90s, with the hopes of convincing him that the three-taxon approach is fraught with difficulties. Platnick, instead of being convinced, reinterpreted it and proceeded to publish it (Platnick 1993), crediting me. Although I gave him permission to use the example, I also made it very clear that I entirely disagreed with his reinterpretation, but he persisted in using it to attempt a justification of three-taxon statements. This is the main reason why I then participated in the Farris et al. (1995) paper discussing this example that so distressed Platnick (i.e. see the "Open Letter to the Gang of Thirty" in Platnick et al. 1996: 248).
5 Modern-day Bayesian programs based on *MCMC* (such as MrBayes or Beast) are perhaps the only exception to this; depending on the settings chosen for an analysis, the trees produced cannot be judged on an optimality criterion, but just accepted as good estimates. See Chapter 4 for details.
6 The only exceptions to this rule are the reading of tree files in compact notation with the shortread command, the execution of a C-script with the runc command, the reading of a Fasta file with the `&` option of the `procedure` command, and the parsing of some configuration files by the `dcomp` command.
7 Note that characters with asymmetric transformation costs may be treated somewhat differently during searches by both programs; PAUP* considers all possible tree rootings when asymmetric characters are included, not just those where the outgroup is the sister group to all the remaining taxa in TNT. PAUP* can be made to work just like TNT if a positive constraint is defined for all the taxa but the outgroup. TNT can be made to work just like PAUP* if an all-missing outgroup is added to the dataset.

2 Characters, Homology, and Datasets

2.1 THE GREAT CHAIN OF CHARACTERS

What first calls the attention when observing differences among living organisms is how these differences are organized in particular ways. While some points of resemblance are maintained, others change, so the differences can be decomposed in particular units. In taxonomy, the individual points of difference and resemblance among species are called *characters*. The different, alternative conditions are called *states*. Different states can consist of the origination of a structure de novo; the characters where this occurs are usually called "presence/absence characters" (occasionally, *neomorphic* characters; Sereno 2007). Alternative states can also consist of the differentiation of an existing structure (sometimes called *transformational* characters; Sereno 2007). This distinction between neomorphic and transformational characters is, of course, contextual—a character may well include three alternative states, such as absent, small, and big, so that some of the differences between states are neomorphic and some transformational.

Different states of a character are considered to be corresponding alternatives of "the same" condition—thus, homologs. Character states arise during evolution, as modification of preexisting conditions, in a great chain that does not originate in God and results in more variation and diversity being added to the story of life as it evolves. Thus, paired limbs originated with Gnathostome vertebrates, first as fins; they subsequently evolved into legs in Tetrapoda and then into wings in Aves. This can be considered as a single character with four states (limbs absent, fins, legs, wings); in terms of a formal numerical analysis, the decision to code these four conditions in a single character implies that, whatever the results, every one of the conditions observed transforms from a previous one, into a subsequent one, or both. That is, in every possible tree, a given state will occur at the root, and any other condition will be a transformation that can be traced back to that original, primitive state. Note that if we took the Tree of Life and rooted it in one of the branches connecting a sparrow with a canary, that rerooted tree would imply unambiguously that having wings is the primitive state, then transforming into legs and fins, and that all "invertebrates" form a monophyletic group. But, even on that obviously incorrect tree, the conditions are all linked: part of the same character.

On the "correct" tree (with invertebrates at the base), the primitive condition is absence of limbs, but this is implied by the tree, not by any intrinsic property of the character states. That the primitive condition, historically, is absence of limbs could also be decided, of course, on the basis of other considerations (e.g. fossil record), but the important point is that there is nothing in the observed conditions themselves which can tell the direction of the transformation. Evolution proceeds by reducing

DOI: 10.1201/9781003220084-2

unnecessary structures (e.g. loss of wings in many insects, leglessness in snakes) just as often as it proceeds by adding complexity, making it impossible to establish general rules as to a general progression from the simple to the complex.

It is important to note that the observed conditions of a character are mutually exclusive; an individual anterior limb cannot be both in the form of a wing and a leg at the same time; one precludes the other. On the vertebrate tree, having paired members (as fins, or some transformation thereof) is a synapomorphy of Vertebrata (Gnathostomata, actually); walking legs (or some transformation thereof) is a synapomorphy of Tetrapoda; and wings is a synapomorphy of birds. Each of those modifications entails successively closer degrees of homology: fins and wings are homologous, at some level, because both are transformations from the original (undifferentiated, absent) condition. Wings and legs are homologous at that same level but also at the less inclusive level of legs (of which wings are a modification)—i.e. wings and legs share a closer level of homology than wings and fins. The example did not consider bat wings—obviously different from bird wings, in their own structure, so that (in this overly simplistic example) they would have to be considered as a fifth state, also derived from walking legs:

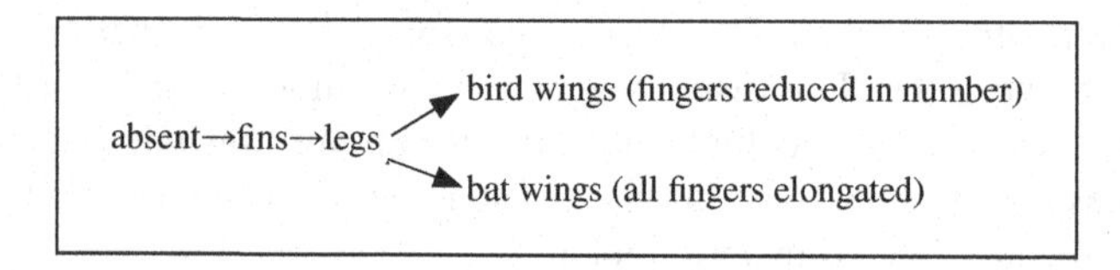

Thus, bird and bat wings are not homologous at the level of wings, but they are homologous at the level of tetrapod limbs.

When the distribution of all conditions of a given character is examined on a tree, all the conditions are homologous at the most inclusive level—the character itself. A phylogenetic analysis—that is, comparing character state distributions on different trees and selecting the most explanatory tree(s)—cannot test such a level of homology. The homology between the character states at the most inclusive level is given before the analysis and should be based on careful anatomical considerations. The phylogenetic analysis, in contrast, determines whether the alternative *states*, as individual conditions, can be considered homologous or not. A tree which places all birds in a monophyletic group can account for the observed presence of wings in all birds by virtue of homology. It is precisely the differences in the degree to which the individual states of the character can be considered homologs that is used to select from among different trees. The great chain of characters is, then, a conclusion of phylogenetic analyses—it is not used as evidence, as it cannot be observed directly. It is what one will arrive at after establishing phylogenetic conclusions, not the starting point. The evidence is in the plain similarities among states, i.e. what can be observed in the present time, and this leads to further considerations on homology.

2.2 HOMOLOGY

One of the key concepts in phylogenetic analysis is that of *homology*. Homology is also a relevant concept in other fields of biology outside of phylogenetics and systematics. Unfortunately, the same term indicates several different concepts, and the ensuing confusion is among the most pervasive in biology. There can be little doubt that communication and understanding would be greatly facilitated if the different concepts were denoted with different terms. Using a single term with different meanings (and endlessly arguing over what homology truly "is") is bound to create problems.

2.2.1 Two Main Meanings of Homology

The term *homology* was introduced in biology by Owen (1843), who borrowed the term from geometry. Owen (1843) considered that homologous organs, or parts, were those that could be considered essentially equivalent, despite possible differences in form or function. The sameness, or equivalence, was one based on similarity and organization. In terms of the alternative states of a character, this applies to the conditions considered to be "the same state" (no two individuals are exactly the same, so there is always the need to separate differences into relevant and irrelevant), as well as the decision to consider all alternative conditions part of the same character. As this preceded Darwin (1859), there was naturally no explicit reference to common ancestry in this early notion of homology; the "transformation" was ideal and figurative, just like the transformation of images in the successive frames of a motion picture; no biological evolution is needed or implied. Darwin (1859), in fact, considered that observations about homology—i.e. that homology was consistently observable in characters of closely related species—were one of the important pieces of evidence supporting his theory of descent with modification. Darwin considered that, once his theory of descent was accepted, the figurative references to transformations between parts, often made before 1859 and meant to be abstract, idealized transformations, could be interpreted *literally*, as actual, physical transformations having occurred along the course of descent. In this sense, evolution provides an *explanation* of the pattern of similarities observed among related taxa; the pattern itself is (among other observations) the *evidence* supporting the theory.

Note how this matches the formulation of cladistic analysis proposed in Chapter 1. The evidence used to establish phylogenetic relationships consists of the observations on similarities and differences; a tree is preferred over others precisely because it better lets us understand those similarities and differences in terms of common ancestry. On this view, homology *is* the essential similarity or correspondence; it is an abstraction. As Darwin's theory of descent with modification became widely accepted, and the so-called "modern evolutionary synthesis" (Huxley 1942) was forged in the first half of the 20th century, the general emphasis shifted from considering that homology is essential similarity to considering that homology *is* similarity shared by common ancestry. The definition was changed from a specification of the observed condition to a definition based on accepting (prima facie) the *explanation* of that observed condition. This second notion of homology is the one more widely used today—under this view, the essential similarity is only evidence for common ancestry, not homology itself. This is the notion of homology that was used in Section 2.1 on "The Great Chain of Characters"—there, we called two conditions homologs if due to common ancestry.

Although not the only ones, these two definitions—homology is the general correspondence of organization or homology is that same correspondence only when effectively due to common ancestry—have been the main alternative usages of the term, at least since the late 1800s. Thus, Lankester (1870) proposed to use the term *homology* for the observable correspondence and *homogeny* for those similarities caused indeed by common ancestry. Most of the authors who discussed homology have taken one or the other point of view, many siding with the idea that the term *homology* should be reserved for the essential similarity and not for the genealogy-based one (Boyden 1935, 1947; Zangerl 1948; Remane 1952; de Beer 1971; Sattler 1984; Brady 1985; Panchen 1992; Rieppel and Kearney 2002), and many siding with the idea that *homology* should be used only for those cases where common ancestry is the hypothesized cause of similarity (Hubbs 1944; Haas and Simpson 1946; Hennig 1966; Mayr 1982; Wenzel 1992; Kluge 2003; Wiley and Lieberman 2011; Farris 2014c).

Most of those who defend the non-evolutionary definition of *homology* accept that evolution is the best explanation for the observed pattern of homologies, and most of those who defend the common ancestry definition of *homology* accept that similarity is the evidence with which to test common ancestry. For most authors, however, the value and meaning of both concepts the term *homology* refers to is clear and acceptable. The differences among authors then refer not so much to principles but more to the terminological problem: to which of the two concepts should the name *homology* be applied. There are exceptions, of course: an extreme example is Kluge (2003), who found the notion of homology as similarity "repugnant". Grant and Kluge (2004) were equally disdainful of using similarity as evidence for phylogenetic conclusions (see Farris 2008, for a rebuttal of these extreme views; Farris 2008 used the term *homology* for the evolutionary concept but made it clear that the *evidence* for phylogenetic conclusions comes from similarities).

That the term *homology* can be used for both the evidence (similarity) and the conclusion (similarity caused by common ancestry) certainly does not aid communication, and confusion arises often. In an attempt to clarify matters, de Pinna (1991) proposed the terms *primary homology* (for the notion of observed, essential similarity) and *secondary homology* (for the notion of similarity due to common ancestry). This terminology has been embraced, to some degree, and it helps but is perhaps not the most felicitous, as noted by Nixon and Carpenter (2012a). It is not the most felicitous because it suggests that both are the same thing, in different degrees, with "primary" somehow being in the way to "secondary", when in fact both are different concepts—the similarity observed in a primary homology is not diminished, rejected, or altered in any way if the similarity is concluded to not be a secondary homology. An adenine changing to a cytosine at the Nth position of a gene in two independent, distantly related lineages is not due to homology in any historical sense, but the two cytosines at the Nth position continue being identical. However, Nixon and Carpenter's (2012a) alternative ("use *hypothesis of homology* and *homology* in their places") suffers from the same problem: the proposal that two features are primary homologs does not constitute a preliminary assessment that they will be secondary homologs after a phylogenetic analysis—it is just a statement on observable comparability. The correction suggests that once a hypothesis of homology has been rejected, we are left with nothing, as if the earlier assessment of similarity was just erroneous. Yet the similarity does not vanish; the observations on which the "hypothesis of homology" had been based correspond to a real phenomenon; it is that real phenomenon the one for which we need a name, and the amended version does not really provide one. Of course, one could just use *similarity*, but in the case of primary homology, several specific criteria of similarity (intrinsic similarity, topological correspondence, developmental origins, etc.) must be separately satisfied, not just a plain superficial resemblance; the mere word *similarity* seems to fall short of capturing this deeper concept of resemblance. But a statement about similarity, however specific or profound, is about similarity, not ancestry; a statement about ancestry (probably, post-phylogenetic analysis) is intended to explain the similarity, similarity which is observable independently of ancestry.

My intent here is neither to propose any new terminology nor to restrict *homology* to one of these two specific meanings. Instead, the intent is only to make clear (a) that the term *homology* can be interpreted to mean more than one thing, (b) that although the same word is used for different concepts, the context usually serves to distinguish the intended meaning, and (c) that the concept used for homology in this book is important in systematics and phylogeny. That one restricts *homology* to a certain set of concepts does not mean, either, that other concepts the term has been used to refer to are not useful or important within their own contexts (e.g. Van Valen 1982: continuity of information; Roth 1984; Wagner 1989: developmental); they are simply not too relevant for systematists.

2.2.2 Types of Homology

The comparability implicit in the notion of homology can be meaningful in different contexts. This leads to the frequent distinction among serial, special, and general homology. *Serial homology* is the repetition of similar structures within a single individual: the first pair of "legs" in an insect is serially homologous to the second and third pairs. Serial homology does not (and cannot) make any reference to common ancestry and is usually of no direct relevance for establishing taxonomic relationships, but given the obvious connection between those parts, one needs to be careful not to multiply the counts of similarities among different individuals or species by coding them as different characters. For example, if a group includes some insects with long legs and some with short, the number of characters potentially separating a taxonomic group is clearly just one: the length of the legs. Counting three separate characters (one for the elongation of each leg!) would amount to counting derivations that necessarily occur together as a single one. If the character describing the length of the legs is in conflict with other characters which would suggest a different, incompatible split, then that would amount to improper counts of the number of truly unexplained similarities. This does not mean, of course, that all serial homologs must be exactly identical—consider the raptorial legs of mantises (modified first legs). The raptorial leg in a mantis is still a serial homolog of the second and third leg, even if not identical, and using the character in a phylogenetic analysis would require that the proper comparison is established—that is, the first leg of species A must be compared to the first leg of species B.

Another type of homology is *general homology*. General homology is concerned with the question of the primordial form that may have given rise to a structure. A good example is the origin of abdominal spinnerets and spider silk. Spiders are the only animals ever to have developed abdominal spinnerets, yet the origin of the spinnerets (and their concomitant associated glands and extruding apparatus) is not clear. Some authors (e.g. Decae 1984; Shultz 1987) have considered the question of how those structures arose (e.g. modified abdominal appendages?), and these investigations belong to the realm of general homology. General homology is, like serial homology, of no immediate interest for the systematists.

The homology of most interest for systematists is that of comparable structures found in different species or individuals—this is known as *special homology*. It is the comparable structures that correspond to special homology that can, in some cases, be meaningfully attributed to common ancestry (i.e. one of the main meanings of homology). The interest in defining homology as similarity due to common ancestry stems, no doubt, from a desire to define concepts in a concrete and material way. A desire, that is, to break away from idealistic approaches to biology. This desire is understandable in general, but it is of dubious literal application in the case of homology. Even if the ancestor-descendant relationship were to be known with certainty, the correspondence between *parts* of organisms is still not direct. As put by Wagner (1989: 54), "morphological characters are not directly inherited but are built anew in each generation". There is no physical continuity between your hand and the hand of your father; your hand did not directly bud from your father's hand, so there is no actual connection. Embryogenesis in the early stages of your life produced a structure in the exact same location as your father's hand and with an almost identical structure; the conclusion that your hand and your father's are both "hands" corresponds to a classification, the detection of a pattern. An abstraction. The fact that there is no one-to-one correspondence between single genes and complex features also points in this same direction: even if a few genes were mostly responsible for hand creation and configuration, that presupposes the existence of a genetic environment, and different genetic contexts can make the crucial genes act differently (de Beer 1971; Minelli 2009). The same point about the difficulties with too literal an interpretation of homology as similarity due to common ancestry had been made earlier by Sattler (1984). No direct observation of homology *qua* common ancestry is possible, except perhaps in extremely simple organisms or those that reproduce by budding. There is phylogenetic continuity, of course, in the genetic makeup determining wings in all birds (both in the genes themselves that determine wings and the rest of genes that create the framework—bones, for example—without which wings could not exist), but the wings themselves are not what is continuous. Even in an individual farmyard and among siblings, the wings of every chicken have formed independently, via identical but separate developmental processes, without physical continuity in the wings themselves.

2.3 CRITERIA FOR HOMOLOGY

It is widely understood that the main criteria to recognize homology must rest on similarity. If homology is defined to be essential similarity, regardless of common ancestry, then these criteria *are* homology—they refer to the same thing (Brady 1985). If homology is understood instead as due to common ancestry, these criteria are only what is used as *evidence* to detect the shared ancestry, but the criteria themselves are the same. The "sameness" refers to similarities in the whole body plans of organisms so that they are generally listed as independent phenomena, at different levels.

1) *Intrinsic similarity.* To the extent that the organs or parts share a significantly similar structure, they can be considered as homologs. Evidently, more complex organs (with more parts, layers) can be homologized more confidently; they provide ampler opportunities for detection of differences. Only very simple statements can be made regarding qualities such as length or size, but the principle is the same.
2) *Connections to other parts.* The other parts of the organism with which a structure is connected provide an additional criterion of homology. This is sometimes called topological correspondence. Finding that a possibly homologous muscle attaches to the head of the ulna in one species but to the head of the radius in another weakens the idea that the two muscles are homologous. Again, the complexity of the connections of some parts is greater than that of others. E.g. an "ulna" stands in a specific relation with the radius, the humerus, and carpals; in contrast, a "dorsal fin" (in a fish) just connects to the dorsum. The criterion of interconnection, therefore, may provide a weaker or stronger argument for homology, depending on the specific part being considered.
3) *Developmental similarity.* Just as with the criteria of intrinsic similarity and topology, this criterion does the same on a temporal sequence. This criterion asks about the embryonic layers from which a structure is derived or, alternatively, in deciding the homology of an organ present only in earlier stages of development, whether it transforms into the same final product—or at least a product that can be considered as homologous in the subsequent stage. Just as the preceding criteria are not absolute, neither is this one. Sometimes similar structures—which are present in related organisms and, as far as can be ascertained, must have been present in their common ancestor—originate from different embryonic layers, but they are still considered as homologous. Whether that decision is sensible will, of course, depend also on the degree to which the other criteria are fulfilled.
4) *Conjunction test.* A criterion often cited is that homologous structures (structures with special homology, that is) cannot coexist in an individual. Finding a vertebrate with both wings and arms would make us doubt of the usual homology wings ≈ arms. That would mean, evidently, that there are two characters involved (i.e., armed ≈ armless, and winged ≈ wingless). Finding two structures coexisting is a strong reason to reject their homology, but of course not finding them is a relatively weak argument in favor of their homology; this is a test which can provide strong rejection but only weak corroboration.

A separate criterion is often added, and this is what comes after performing a parsimony analysis:

5) *Congruence with other characters.* After a tree has been established, on the basis of the majority of the characters, some of the similarities (i.e. those that passed criteria 1–4 prior) may nonetheless be found to occur in different groups—when the similarity of the organ or part in question does not vary coordinately with other characters. This can be thought of (particularly, if one is a pattern cladist) as elevating the notion of "common body plan", inherent in criteria 1–4, to a subsequent level of abstraction. For example, echolocation emitting sound via tongue-clicking occurs both in some bats and swifts—very distantly related groups. The only reasonable conclusion seems that such type of echolocation evolved independently in those two groups. In practice, once a tree is established, hypotheses of homology of the different states within a character are interrelated; e.g. a feature found in two distantly related taxa but not in the intervening groups (as in Figure 1.1d) can be nonetheless considered as a "homolog", but this necessarily implies that the absence of the feature in all the intervening groups is not due to homology (as in Figure 1.1e). The realization that, given tree topologies that require homoplasy, homologizing one condition necessarily means that the other condition is non-homologous was one of the most important contributions of cladistics over traditional taxonomy. In discussions about character evolution before the advent of cladistics, it was common to see features shared as separate occurrences by distantly related taxa attributed to some form of "primitive retention", without realizing what this would have meant for the alternative condition in all the intervening taxa—endless numbers of parallel derivations of the alternative. This criterion of congruence with other characters, at least as postulated here, can only take effect together with some form of character mapping, so even if it is often referred to as the criterion of "congruence", it is more than just that. A criterion truly based on congruence and nothing more would simply look at whether a feature co-occurs with particular states in other characters, as if homology—secondary homology, that is—could be established without reference to a tree. We retain the term *congruence test* because this is the name most often associated with a parsimony analysis, but keep in mind that the criterion is more than just the "congruence" between pairs of characters. Consider as example the following case (Figure 2.1): the apparently apomorphic state of character 0 is shared by taxa A–C; the apparently apomorphic states of characters 1 and 2 are shared instead by C–F. A naïve check of congruence would seem to indicate that characters 1 and 2 define a group and character 0 is not a homology. But the full dataset includes also many other characters (3–8), as shown in Figure 2.1, and the characters *other* than 0–2 determine the tree shown on the right, where the coincidence in what seemed originally to be a plesiomorphic state of characters 1–2 for taxa A–B is best interpreted instead as a reversal (i.e. secondary apomorphy). Since A–B is a subgroup of A–C, at the level of the group circled in

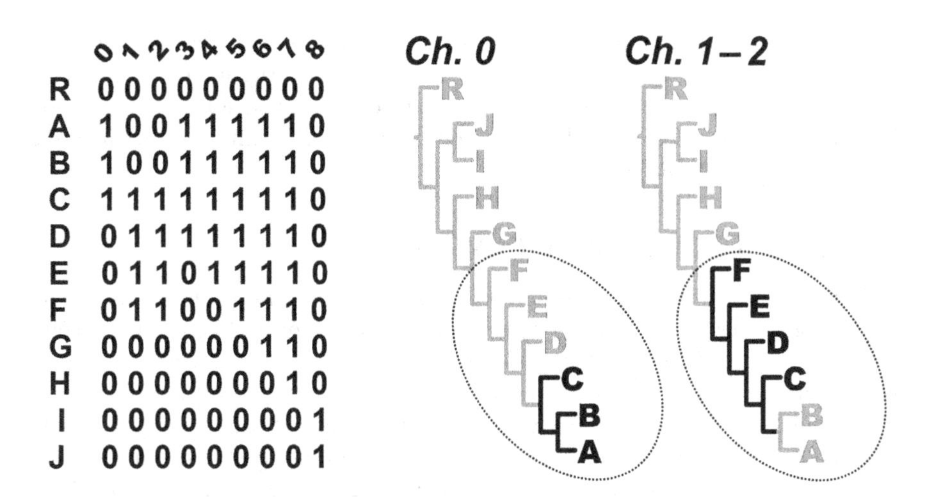

FIGURE 2.1 Example showing that the so-called congruence test for homology cannot operate at the level of simple pairwise comparisons between characters but instead on the tree simultaneously determined by all characters together.

> the figure (a group determined by the rest of the characters in the dataset), there is no conflict between character 0 on one hand and characters 1–2 on the other. The example shows that the congruence test is not so much a test between characters but rather a test of the congruence between the character and a *tree*.

Since the advent of evo-devo in recent decades, the genetic basis of some aspects of the development and morphogenesis of many animals and plants has become much better understood. It is often the case that particular features, or teratologies, can now be associated with specific genes or gene complexes. This, of course, can be used to provide an additional criterion to test hypotheses of homology. One could subsume considerations of this nature under criterion 3, developmental similarity, as the genetic basis of features refers, ultimately, to how the features originate (during morphogenesis and growth). Thus, any similar condition in distantly related taxa provides evidence against placing them in separate groups; finding that the genetic basis of the otherwise identical condition is different in both taxa certainly weakens the evidence, strengthening the idea that the feature in question is not homologous on grounds separate from the tree itself. An identical condition but a strongly different genetic basis is unlikely to be observed within a group of closely related organisms; in addition, this criterion will be limited only to those few cases where the genetic basis of different morphologies is known in most of the taxa under study. The use of this criterion in practical cases is therefore very uncommon, but in theory it could be used whenever applicable.

Behavioral characters are routinely used for phylogenetic analysis. Establishing homology in the case of behavioral characters (Wenzel 1992) uses criteria similar to those of anatomical characters. Behaviors can often be decomposed into units, and the relationships between these units (e.g. their sequence, timing, etc.) as well as the context in which the behavior is expressed are akin to the criterion of connections. The study of motor patterns (or the anatomy related to the behavior, e.g. innervation or musculature) provides an additional criterion of intrinsic similarity between the behaviors (Lorenz 1958; Wenzel 1992).

2.4 HOMOLOGY BY SPECIAL KNOWLEDGE?

With the term *homology* being used in so many different senses, it is not surprising that homology lies at the heart of some controversies on how phylogenetic conclusions should be established. This involves homology both in the sense of what it means and how it is recognized and has led some systematists (mainly morphologists) to hold quite unfavorable views of quantitative methods, the use of parsimony, and most modern developments in phylogenetics. Among other problems, those views conflate the two concepts of homology—the abstract and the genealogical. Those systematists would like to return to days past, when the trees were drawn by hand:

> Increasingly, cladistics is viewed in terms of algorithmic implementation via one or another computer program (e.g. TNT Goloboff 1999), which aim to convert any binary data matrix into a branching diagram (or diagrams) purportedly summarising the original data economically. In doing so a crucial connection has been lost with the overall methodological goals, namely to find homologous relationships. That connection can be rediscovered if solutions (branching diagrams) were arrived at by hand (non-algorithmic implementation), with no other tool than pencil and paper.
>
> **(Ebach et al. 2013: 588)**

They think that the proper study of homology is the best way to produce phylogenetic hypotheses and that this requires neither parsimony nor computer programs. Note Ebach et al. use *homologous* as a qualifier of taxonomic relationships, not the usual qualifier of conditions observed on different taxa. This is because Ebach et al. (2013)—following Patterson (1982b) and Nelson (1989a, 2004)—think that homology and monophyly are inextricably tied. In their view, a homology is that which defines a monophyletic group—discovering real homology is discovering natural groups. They often decry that quantitative parsimony analysis as presented in this book is based on undefendable principles, which they characterize as "phenetic". Thus, in the view of these authors, parsimony analysis is

> a pseudoscience based on quantitative manipulations of questionable phenetic data [that] has focused on cladograms at the exclusion of character analysis. Reversing this trend will require a return to the refinement of Hennig's theories and methods and the integration of diverse available data.
>
> **(Wheeler 2008: 2)**

They accused parsimony analysis (which they sometimes incorrectly call "Wagner parsimony"; Williams and Ebach 2012a, 2012b; Ebach et al. 2013) of being phenetic because it uses matrices with 0 and 1 states, grouping the states by similarity, without distinguishing apomorphic or plesiomorphic states a priori, and potentially grouping by 0s when they appear to be reversals; thus, they argue, the plesiomorphic state is being considered informative—"just as in phenetics". In their view, a proper classification is done by finding true homologies, which are exclusively those that characterize taxa, because homology = synapomorphy. Plesiomorphies or convergences are not homologies, so when finding true homologs—which they think characterize natural groups—the true groups can be identified easily, with no computer or quantitative analysis needed: "The difference between numerical cladists and Hennig is their interpretation of relationship, that is, homology. Since the computer algorithms can only implement a phenetic approach, they rely solely on some form of similarity (homology = similarity)" (Ebach et al. 2013: 592). This depends on how *phenetics* is defined, of course, but the meaning I use (and the one originally used by Sneath and Sokal, as well as other prominent representatives of that school) is that a method is "phenetic" when it groups based on overall similarity. Parsimony certainly does not operate in that way, as explicitly shown (Figs. 1.4d–f) in the previous chapter—species which share more states in common may well not be placed in the same group; parsimony creates groups so that the maximum possible number of similarities can be attributed to common ancestry, which is *not* equivalent to grouping most similar things. The criticism that parsimony is a "phenetic" method is especially surprising coming from authors (like D. Williams and M. Ebach) who have held the view that classification is not about phylogeny—the notion that classification is not based on evolution had often been considered (also incorrectly) as "phenetic" on the grounds that it seeks classifications reflecting what is observable instead of hypothetical genealogical relationships (e.g. Ridley 1986: 91–92; Scott-Ram 1990: 145). The layers of confusion continue piling up, however, since these critics of quantitative cladistics maintain that (true) cladistics differs from phenetics in using only characters that truly define groups: "Real relationships, in the sense of occurring in the real world, are not abstract entities. Relationships are observed when, for example, the same forearm manifests itself in different individuals of the same taxon or across different taxa" (Williams et al. 2010: 186). So, it seems, relationships can be observed directly, and "sameness" is something very different from "similarity":

> The concept of homology concerns sameness rather than similarity. . . . By sameness we mean different manifestations of the same thing: a bat's wing and a whale's fin are both manifestations of fore-arms. Similarities are simply likenesses or measurements (that is they are phenetic).
>
> **(Ebach and Williams 2013: 588)**

> Cladistics . . . is about discovery, about finding repeating patterns, finding the same relationships, finding relationships that are not method dependent, finding relationships that are reflections of the world as it is.
>
> **(Williams and Ebach 2018: 3)**

One can certainly concede to these authors that, if homology designates those characters that truly define natural groups, then if we had the means to discover which ones those characters are, we would be automatically discovering real, natural groups (see also Williams et al. 2010: 186–187, for a twisted discussion on "relative relationships" and the use of "similarities in order to quantify sameness"). There is no need for computers, according to those who wish to return to the bucolic days of hand-drawn trees, because the systematist can magically sort out the true from the false characters:

> Present practices focus on solutions to matrices rather than on character homology, and rely on algorithms and statistics rather than biology to determine relationships. . . . Characters that can be hypothesized to be homologous provide the evidence for phylogenies that we recognize as synapomorphies. If the homology of characters is identified correctly, the insertion of a fourth taxon into Nelson's (2004) set of three . . . should not alter the relative positions of the initial taxa—elegant, logical in an evolutionary sense, and stable.
>
> **(Mooi and Gill 2010: 26–27)**

"Homology identified correctly" would have the additional benefit that once you discovered truth, classification would become stable (after all, "what is true today, cannot be false tomorrow", and yes, they really say this: Mooi and Gill 2010: 31). The problem with this approach is that neither they, nor any of several other advocates of homology-by-revelation (Nelson 1994; Borkent 2018; Ebach et al. 2011; Williams et al. 2010; Williams and Ebach 2012a, 2012b; Mooi et al. 2011), provided any clues as to what kind of reasoning or method one could use to correctly identify homologies, other than carefully studying the organisms to classify and engaging in meditation until the truth is revealed.

2.5 NO SPECIAL KNOWLEDGE OF HOMOLOGY IS POSSIBLE . . . OR NECESSARY

The previous section shows that some authors indeed believe that determining homology is the crucial aspect of any phylogenetic study. But the notion that determining the correct homologies automatically allows identifying monophyletic groups, without the need of quantitative analyses, is mostly a play on words. With homology defined as those character states that identify monophyletic groups (i.e. Patterson's 1982b equation of homology = synapomorphy), then one is left with no independent source of evidence with which to test monophyly, or homology, other than some nebulous unspecified source of wisdom. What appears plesiomorphic cannot be a homology (only synapomorphies are), in this view, but some "plesiomorphies" result from reversals, and these are, in fact, synapomorphies (of the less inclusive group, characterized by the reversal: legs characterize tetrapods, but lacking legs is a synapomorphy of snakes!). Much the same can be said of parallelisms.

Of course, identifying that the "reversed" state is different from the "original" state, or that the same state is derived independently in different parts of the tree, one could code the different instances differently, and then every homology would really define a group—the reversed and the original state, as well as the parallel states, are simply not homologous, even if they look exactly identical. But how could one identify such non-homology prior to having a tree, other than finding physical, observable differences, which can be non-existent in the case of reversals and parallelisms? The authors just cited provide no clues as to how that would be done. True, there is really no replacement for careful study of the morphology, development, and function, as stated by Borkent (2018), but why is that supposed to be mutually exclusive with a quantitative parsimony analysis? Why should one replace the other? Both are needed; parsimony alone is no good, but neither is careful morphological examination. Borkent (2018: 108) complains of the ease for "plugging in any set of features into a matrix and generating a phylogeny", but what prevents anyone from plugging *a set of well-studied features* into a matrix? Any proponent of quantitative phylogenetics will insist on a thorough study of morphology, which is indeed the way all studies (computer assisted or not!) should be done.[1] Despite careful anatomical study, sometimes no differences between the independently derived states can be detected—they can only be distinguished on a tree, and this requires considering how characters fit different trees. That many papers (e.g. those fly taxonomy papers criticized by Borkent 2018) present only superficial morphological data (e.g. taken from the literature rather than from specimens) is hardly a reason to eschew modern cladistic analysis; it should only be a reason to demand deeper and firsthand anatomical studies. Similarly superficial analyses occur as well in molecular studies (e.g. see discussion of problems of determination of homology in molecular studies, by Springer and Gatesy 2018), but this does not mean that the whole discipline needs to be discarded—only that standards need to improve.

In the real world, not all characters—regardless of the anatomical thoroughness of the study—define mutually compatible groups, and from this follows that some characters *must* be homoplasies or reversals and thus indicate incorrect groups (mutually incompatible groups cannot all be correct at the same time!). Contrary to what the authors cited in the previous section imply, we do *not* have the means to discover which ones are the good characters, other than collectively examining all of them together. A computer program does the collective examination of characters in exactly the same way as the human mind would—if only (alas!) a human life could last enough eons. As (contrary to the assertions of extreme pattern cladists Nelson, Ebach, Platnick, and Williams; see Chapter 1) common ancestors of a hypothesized group must have had *some* character state, either the phylogeneticist in his mind, or the computer program used to reconstruct character evolution, must consider alternative explanations of how the character evolved—and this is done under the same logic in manual or computer analyses. When two or more characters conflict, either one or the other character (or both) indicates an incorrect group; this applies even conceding the ability to assign (based on whatever arguments might be reasonable) prior differential weights to the characters. Even in that case, the groups supported by the interactions among characters cannot be visualized mentally, and an automated consideration of the implication of alternative trees is required. The only way to avoid this is to focus on some characters and resolutely ignore the others. Any attempt to simultaneously consider all the evidence (both that which supports preferred conclusions and that which contradicts them) quickly runs, in the real world, into solutions of compromise.

The solution for that compromise cannot come from a special prior knowledge of which structures are more likely to correspond to real homologies, because—recall—this situation may occur even when the characters have been assigned different prior weights (e.g. if one is willing to assume specific rules of evolution, for example, that simpler characters are more likely to originate in parallel), so the information presumably available as special knowledge to the phylogeneticist, of which characters are more likely to indicate incorrect groups, has already been used in the form of prior weights. Even with these differential character weights, in the presence of significant character conflict, a tree will normally not be uniquely determined by a set of perfectly congruent characters (unless many characters are given zero weight, i.e. ignored completely), and detecting these ties in aggregates of characters is simply beyond the possibilities of the human mind. A formal, quantitative analysis of some kind is therefore always required. The conflict between characters may exist as well when the aggregate of the (possibly weighted) evidence determines a unique tree but where a number of groups are contradicted by some character(s)—that is, in *all* real datasets with more than a handful of taxa and characters. What the best-fitting tree suggests is that some of the characters delimiting alternative groups, apparently being the same in all the members of that alternative group, are not really "the same". This is one of the senses in which the principle of "reciprocal illumination" (Hennig 1966; see Farris 1983) can work: the tree suggests that some observations may be incorrect, which leads us to reexamine specimens. For the characters that define conflicting groups, a more meticulous examination of the morphology—the part for which these authors rightly yearn—may indicate that some of the conditions that on a more superficial examination seemed identical actually have differences (e.g. originating from different embryonic layers, having differences in fine structure, not connected to other parts in the same way, etc.). This would be the only way to eliminate that conflict: this resolution can only come from more detailed comparisons in the form of empirical observations about previously undetected differences. If there is an empirical basis to consider that the observations were indeed incorrect, then the data are corrected and reanalyzed. This may well favor the same tree or trees as in the first cycle or different ones (depending on which characters had to be rescored after reexamination).

What certainly should *not* be done is coding as separate states otherwise identical conditions in different taxa, just because of an inner conviction that the condition is not homologous in all the taxa having it or with the aim of improving the congruence between the characters scored in the matrix; this is a very dangerous practice. Separation of states should be done *only* on the basis of actual observations. By suitably recoding characters that do not agree with the groups we happen to believe are monophyletic, we can easily end up with perfectly congruent "data" sets that have very little connection to actual observations but instead describe what *we believe* the relationships are. Perfect fit . . . to our beliefs. The amount of homoplasy is used to measure the degree to which the tree agrees with the empirical observations for a set of characters—not the conformity of the tree with our own conviction—and this number *must* duly reflect the lack of conformity with observable evidence. Honesty requires that both the favorable and the *contradictory* evidence are presented in the form of data;[2] if no difference has actually been detected among taxa, they must be considered as having the same state. Unless we can boast of knowing so much about evolution that we can give zero weight to all the characters that contradict the true tree, a quantitative analysis continues to be needed. Simultaneous consideration of all characters is not preferable because it produces truer trees than speculations about which characters define correct groups; it is preferable as a method of analysis, as a more rational way to establish conclusions. As recognized even by some rather traditionally oriented phylogeneticists, "there is no contradiction between Hennig's ideas and the use of computers" and "the implementation of the cladistic method by computer is welcomed" (Wägele 2004: 107); Wiley et al. (2011) maintained a similar position.

While the view that phylogenetic analysis does not need quantification is extreme, the notion that figuring out homologies is a crucial and very difficult step in a phylogenetic analysis is not uncommon (e.g. Mishler 2005; Winther 2009). It is for that reason that this topic needs to be discussed. Taxonomists often anguish over whether some observed similarity must be really coded as homology. Don't! Realizing that that the only evidence needed prior to a phylogenetic analysis is the observable similarities, and that the function of a phylogenetic tree is precisely to explain as homology (i.e. secondary homology) as many similarities as possible, is liberating. It lets taxonomists rely on the special knowledge they indeed have—details of fine anatomy—instead of expecting them to rely on a special knowledge of homology that cannot possibly be obtained. The use of computers facilitates quantitative analysis of all the features taken together and, rather than being a reason to neglect careful morphological studies, provides the speed needed to quickly test alternative scorings and makes it clear that what determines differences in results is mostly the data, so that more time and energy can be spent at the lab, the microscope, and the field. The final phylogenetic hypothesis, if proper methods exist, will automatically follow from the dataset. The biggest value in a phylogenetic study is in the form of the new information (morphological, molecular) accrued for the group in question, properly discriminating among the conditions for all the taxa, indicating the individual points of resemblance and difference. A matrix should contain this information, as precisely as possible, but merely this information.

2.6 LIFE STAGES, COMPARABILITY, ONTOGENY

The comparison of structures and organs in different taxa can (or should) also involve considerations of the life stages. Most organisms undergo metamorphosis or at least some kind of change during their development. The correspondence, or lack thereof, between different structures can be established not just by careful morphological observation but also by considering whether the potentially equivalent conditions occur in the same stage of the life cycle. This is roughly captured by Hennig's (1966) concept of *semaphoront* (the "character bearer"). Evidently, in sexually dimorphic characters, the comparisons should be done between the corresponding sexes. This does not mean that comparable structures occurring in different life stages can never be homologized. It is in regard to such situations that the notion of a phylogenetic tree being the means to explain character distributions attributing similarities to common ancestry is liberating, as discussed in the preceding paragraph. To the extent that the species sharing similar structures in different life stages can be grouped without doing violence to other characters, the tree which places them together will allow explaining that common occurrence by inheritance (and, of course, a shift in the life stage where the structure occurs). Homology is often considered to be an all-or-none concept, and this is true when talking about homology as similarity inherited from a common ancestor. Sattler (1984; see also Sattler 1994; Minelli 2009: 111–112), however, argued that structures can be homologized (as primary homology, i.e. without reference to common ancestry) to different degrees—some organs or parts can be "more" homologous than others, given that homology is a composite of criteria (intrinsic similarity, topographical correspondence, etc., all of which can vary in degrees in the case of complex structures). Even if homology itself is considered to be an all-or-none concept, the *evidence with which to recognize it may come in degrees*, so that in cases of conflict, the benefit of homologizing some structures may overweigh that of being able to homologize others. Example: Some peculiar type of seta occurs only in the early stages of a species, and a similar seta occurs only in adults of potentially related species; grouping those allows considering the origin of funny setae as explained by common ancestry; if this is in conflict with another character of less dubious homology (say, an unambiguous character shared only with the species with setae occurring in early stages), then the characters can be given relative prior weights (so that non-homology of the setae costs less than non-homology of the unambiguous character). The unusual setae would be stronger evidence for grouping if occurring in the same life stage.

The study of ontogeny deserves a brief comment from another perspective. In the early days of cladistics, when phylogenetic analysis was presented as requiring a prior assessment of "polarity", it was proposed that ontogeny could provide a way to polarize characters. Hennig (1966) already referred to the ontogenetic criterion as "an important aid in phylogenetic systematics". Nelson (1978; see also 1985) argued that ontogeny allows establishing that some character states are more (occurring in the life stages of both species) or less (in fewer) "general" and that this would allow establishing primitive (occurring in the corresponding stage of all species) and derived (occurring in the corresponding stage of only some species) states. With this, Nelson (1978) attempted to reestablish Von Baer's "law" (i.e. the old maxim that embryos of related groups resemble each other in their initial stages, and diverge at later stages) in such a way that it would not depend on an evolutionary notion (the notion that evolution often works by adding new steps to preexisting developmental programs; see discussion in Rieppel 1985). That would avoid the "infinite regression" perceived (see, e.g. Weston 1988) for outgroup comparison, which always relied on higher-level hypotheses but never established an "absolute" basis for polarity. Nelson's (1978) approach was criticized by Kluge (1985: 13), who noted that "dedifferentiation, paedomorphosis, and the insertion and deletion of developmental stages make it impossible to deduce the genealogical hierarchy from only ontogenetic transformations series". More interestingly, Fink (1982), Wheeler (1990),[3] and Mabee (1989, 2000) proposed that ontogenetic changes should be interpreted in the context of parsimony. That is, alternative trees would differ in the way in which they let us visualize and understand the different developmental programs, and it is this fact which makes some of those trees more explanatory—and this is measured by parsimony. Ontogenies can be modified during evolution by addition of terminal stages, which is perhaps a more common form of alteration of developmental programs, and when this happens, the rule that the more "general" state is primitive will be fulfilled. But ontogenies can also be modified by eliminating later stages of development (i.e. in neotenic or paedogenetic organisms) or by deleting non-terminal stages (e.g. wasp nest construction, Wenzel 1993, or direct development in *Eleutherodactylus* frogs). Consider a character (Figure 2.2) in a hypothetical case of five species, A–E, with development passing through five stages in species C–E (e.g. metamorphic taxa, with final stage as a full circle). Species A and B differ from the others in never reaching the stage of full circle. Applying the ontogenetic law would imply that the circle is a derived state, but given the tree in Figure 2.2, the most parsimonious conclusion is that the developmental program of species A and B diverged from that in the other species by becoming truncated and never developing a circle. Lacking a circle is thus a synapomorphy of species A and B. More properly, ancestral ontogenies could be reconstructed (so that as much similarity as possible in developmental programs can be attributed to common ancestry; Mabee 2000), as shown in lighter color in Figure 2.2 for the *HTU*s. The study of development and ontogeny, and inferences about changes in development along evolution, is then based on the same principles as other characters; ontogeny is an additional source of differences among taxa that can be explained by reference to common ancestry, and it may provide evidence on the homology, or lack thereof, of different parts or organs (via the same criteria used to initially establish homology). Minelli (2009) provided an excellent review of the enormous array of modifications of modes of development, giving many examples showing how much of an oversimplification there is in the classic notion of "ontogeny recapitulates phylogeny". The difficulty of establishing general "rules" for the evolution of development highlights the need to study and understand development in a genealogical context, just as it is done with normal characters. In general, it seems

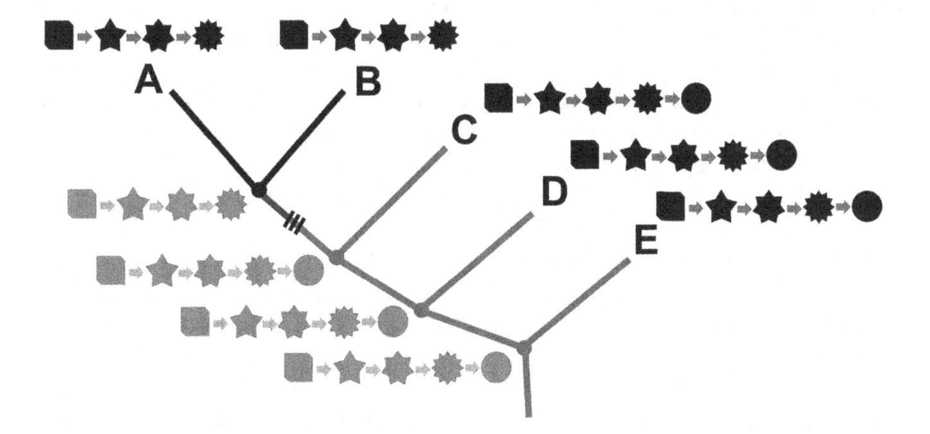

FIGURE 2.2 Whether characteristics appearing later in ontogeny represent apomorphies or plesiomorphies can only be determined in the context of a tree, where the developmental programs are mapped. The circle appears at the end of ontogeny in some species and remains as a star in others. Given the tree shown, the developmental program where a star transforms into a circle can only be interpreted as plesiomorphic at the node shown.

likely that consideration of development and ontogenetic characters will be most relevant in the case of distantly related organisms, with widely different body plans—for example, in different phyla or classes, where the development of only some representatives of each taxon is known. In the case of more closely related organisms, it is unusual that the development of a significant enough number of taxa is known and that they present differences which may be phylogenetically relevant. There are exceptions (after all, this is biology!), such as the salamander genus *Ambystoma* (only some species of which are neotenic), or shape characters which are sometimes subtly influenced by heterochrony in closely related species (Catalano et al. 2019).

2.7 GATHERING MORPHOLOGICAL DATA

As should by now be clear, two main steps can be distinguished in a morphological phylogenetic study. The first step has a strong observational basis and involves translating observed morphologies into specific scorings, e.g. matrix cell entries, conforming to a specific set of characters and states. The second step involves obtaining a phylogenetic hypothesis from the matrix generated during the character analysis, taking those data as given. This second step—after the data have been taken as a fixed set of observations—is much more amenable to formalization, and that formalization constitutes the bulk of this book (and most books on phylogenetic analysis). The first step is much harder to formalize, for several reasons. One of those reasons is that this is the part that provides the empirical link, the connection between the analysis and the real world—the analysis itself is only interpretation. The gathering of morphological data—i.e. building the matrix itself—is, then, in a very definite sense, the most important part of the cladistic analysis: all the rest of the results are interpretation and (given accepted and well-defined phylogenetic methods) should logically follow from it; the creation of the matrix is the only step which contains empirical statements about the taxa included. As this step involves many practical aspects, establishing strict guidelines becomes very difficult; only very general notions can be mentioned. Another reason for the difficulty in attempting a cookbook approach for the gathering of morphological observations is the enormous diversity in structures and organ systems found in different taxonomic groups and the fact that morphology behaves very differently from molecules in this regard. For example, major differences between arthropods lie in tegumentary structures, but major differences between vertebrates are in the endoskeleton. There is thus no easy way to predict which organ systems one should study—it depends on the taxon being examined. In particular, focusing on single organ systems—e.g. only cranial characters in a group of vertebrates or only leg characters in a group of arthropods—is bound to miss important differences that could otherwise contribute crucial information about phylogeny. Few general recommendations can be made beyond this. One is that a detailed knowledge of previous work on anatomy is imperative and that centuries of systematic work show that key features of clades are rarely concentrated on a single character system. Within any given taxonomic group, a large corpus of knowledge has been established long ago, and accessing this requires both careful consideration of treatises on the respective groups and initially working under the guidance of experts. Another general recommendation is that, for clarifying the relationships between a given set of taxa, it is impossible to focus on a predefined set of characters; rather, as many characters as possible should be identified which allow some subset of species to be distinguished from the rest, and these characters constitute the raw material of phylogenetic analysis. Focusing on a predefined set of characters (i.e. markers, genes) often works well with molecular characters; after all, as long as gene X is shared by all the species under study, a mutation could occur at any position of the gene and then potentially identify a clade. That is clearly *not* the way morphology evolves, as it is much more conserved and hierarchically organized—the modification of some characters presupposes certain conditions for other characters or organ systems. One might score the presence or absence of thoracic wings, wing venation, and presence of halteres in a dataset to resolve the phylogeny of fish, but doing so would be a completely futile exercise—three entirely invariant characters. Resolving fish phylogeny with morphology requires consideration of characters relevant for fish relationships, not characters for insect relationships. Centuries of systematic work have showed that detailed knowledge of the characters and anatomy of a group cannot be extrapolated easily to another—someone who has worked for decades and gained a deep knowledge of spiders will have to start learning a whole new set of anatomical characters and observational abilities to begin serious work on mammals. And within a given clade—say, a class or order—the taxonomist first studying in detail a family not previously revised will have to discover new characters and synapomorphies of the subgroups within that family—some of the characters relevant for distinguishing those subgroups will never vary even in related families.

The most useful recommendations that can be made, in terms of data gathering, are those that help avoid common malpractice. While this advice may seem superfluous to experienced phylogeneticists,[4] it is still common to see papers that would have benefited from these considerations.

a) *Avoid relying on preestablished lists of characters.* Although systematists have identified the most salient characters distinguishing major clades in most groups, many other more subtle features often contribute enormously to diagnosis of different clades. These differences may have gone unnoticed to previous researchers, e.g. unique modifications characterizing smaller groups of species. Discovering that subtle, hidden, previously unnoticed anatomical differences can help diagnose important taxonomic groups is one of the most thrilling experiences of systematics. Think outside the box, scan the specimens without assuming that all the anatomy is known, and when finding some feature which does not ring any bells, go back to other specimens for reexamination.
b) *Be aware that different characters define groups at different taxonomic levels.* This is connected to the previous point. To restate the obvious: morphological characters cannot be "sampled" in the same way as molecular characters. Morphological characters are typically applicable to particular taxonomic groups. If one were to make an exhaustive list of the characters used in spider taxonomy and then score those same characters (and only those characters) to try to resolve the relationship between classes of arthropods, one would have no resolution of the major groups—a completely

frustrating but totally expected outcome. Perhaps the tree would have some resolution among spiders, but certainly not among arthropod classes: most characters would be invariant for any non-spiders. Arthropod classes differ in many features (antennae, mandibles, leg number, segmentation, etc.), but those features are constant in spiders. Unless those characters varying indeed among major groups are included in the matrix, there can be no hope of resolving the relationships between those classes or the classes themselves.

c) *Incorporate as much information as possible from previous work.* One of the consequences of the previous point: as much information as possible that is relevant for the level (or levels) of the analysis at hand should be included in the matrix. This should need no comments: taxonomic work advances through the accumulation of data, not by discarding what we knew before and starting from scratch. Taxonomy is one of the sciences where giving credit to predecessors is already established most strongly (e.g. in the form of authorship of taxonomic names, regulated by the codes of nomenclature). Then, as long as the characters can be properly defined and seem to be heritable, follow this tradition and make every effort to include all the characters used to diagnose any clade by earlier workers in the phylogeny of the group. This extends to cases where the characters were used in informal analyses and mentioned only in discussions (i.e. never included in a matrix); scoring for all the taxa in the matrix those characters previously mentioned in descriptions or just anatomical research for a few taxa may greatly increase the quality of the data. A phylogeny gains support as it can explain greater quantities of evidence; the more, the better.

d) *Examine firsthand as many specimens as possible, without relying only on literature.* With phylogenetic analysis being now almost mandatory in systematics, it has become common to describe new species and give them a phylogenetic placement by plugging them into an existing matrix. When that is done, the previous matrix is often taken just as published by some earlier author, simply scoring the same characters for the new species and rerunning an analysis with the new terminal taxon. Without firsthand knowledge of the rest of the taxa in the matrix, this practice may easily lead to propagation of egregious errors in scoring or even to improper scoring of many characters. For example, a character may consist of the states "short", "intermediate", or "long", but unless you have actually seen the other taxa and how they differ, the assignment of a given state to an isolated species may be a difficult decision. Matrices provide information about features but in an extremely simplified way. More reliable information can be derived from good illustrations or photographs, but the ideal is always the direct, firsthand examination of specimens. A similar practice is that of creating a new matrix from characters described in the literature; whenever possible, direct examination of specimens is much preferred.

e) *Use taxonomic distinctiveness, not geographical or political boundaries, to decide which taxa to include in your matrix.* As exemplified in subsequent chapters of this book, some taxa can present character state combinations that affect not only the placement of that taxon but also the relationships of *the other* taxa in the matrix. The matrix should ideally contain all the taxa known for the group, but any analysis is always a tradeoff between the ideal and the practical; the difficulty of handling specimens, loans from museums, and the resulting matrix increases exponentially with the number of taxa, so it is often necessary to limit the number of taxa to consider. As the decision to include additional taxa is made, bear in mind that those taxa that help determine plesiomorphic conditions of characters in some groups (i.e. "groundplans") and those that have mixtures of apparent synapomorphies of different groups are often critical taxa in phylogenetic analysis. Of course, those groups whose monophyly is in doubt should ideally be represented in an analysis by at least two taxa (the most distinct)—this is the only way to test the monophyly of the group. If the monophyly of the entire group is in doubt, testing it requires that the outgroup (i.e. putative non-members) includes two or more taxa, preferably including several of the most closely related species (or suspected members of the group). Adding taxa that (for the characters in question) are identical (or almost identical) to other taxa in the matrix is of no help; they sometimes can even be safely removed from the matrix without otherwise altering results (e.g. as in procedures for "safe taxonomic reduction", Wilkinson 1995; Siu-Ting et al. 2015). Taxonomists with a general knowledge of the diversity within the group under study can thus make the best choices of taxa to include in a matrix. It is often the case that the most relevant taxa for a study do not live in the study region; even in revisionary work of groups limited to some region, the study of related groups from other biogeographic regions can increase the quality of the analysis more than a complete sampling of the same country or continent.

f) *Avoid wasting effort on characters with so much variability that homology decisions cannot be made.* In many groups, there are often key characters to identify species. An example of this type of characters are genitalic characters. Genitalia in some groups have an extraordinary diversity of forms, very rarely being identical in two separate species. This makes those characters extremely useful for distinguishing species. But, by the same token, it makes those characters with a myriad subtle variations in shape or size very hard to use in phylogenetic analysis. Depending on the case and the character definition, it could either end up having a different state in every terminal taxon—thus being, for practical purposes, autapomorphic in every single species—or having variations found in some taxa which are very difficult to homologize with variations in other taxa. Characters with such extreme variability are useful only when it is possible to identify specific components of the modifications (e.g. presence or absence of entire sclerites, more than subtle variations of sclerite shape) shared by groups including several species.

g) *Do not exclude characters based on a priori considerations of their reliabilty.* The goal of simultaneously including all the relevant information in a matrix, and analyzing it *in toto*, is that the groups determined by the bulk of characters provide, in themselves, a guide as to which characters are reliable and which are not. The characters which, prima facie, seem poorly correlated with phylogenetic groups must be included in a matrix as well, for only by doing so can the degree to which the matrix fails to provide an acceptable genealogical explanation for those characters be measured. The only grounds for exclusion of characters are that the characters are not heritable or that the researcher has no way to confidently establish proper scorings (e.g. absence of specimens, inappropriate preservation).

h) *Keep in mind that a character being variable in one species does not imply that it is variable in all of them.* Even when characters are fully heritable, some of the populations may present individuals with both alternative states (i.e. genetic polymorphism). A character that can be present or absent in every one of the species in the matrix is uninformative—no tree for that set of taxa will require any homoplasy, thus making choice of trees impossible. But a character being variable in one or a few species of the matrix does not invalidate the use of that character to distinguish the *other* species, the ones in which the character is fixed; the character continues being useful to discriminate among *some* trees. In other words, not all trees are the same with respect to the character; those trees in which the species with one of the states fixed form a clade relative to the species where the character is fixed at the other state provide a better account of those differences. The species where the character is variable can be placed in any part of the tree without making any difference in terms of fit for that character, so many trees are possible, but this is not the same thing as being unable to discriminate between any trees.

i) *Try to obtain scores for as many taxa in the matrix as possible.* With matrices often being based on characters extracted only from literature, it is also common to see that characters are assigned to taxa based only on extrapolation from their membership to certain groups. Except for the most conspicuous characters (those which a taxonomist would never have missed even in a cursory examination of material), this is a dangerous practice. Finding that a few species from a group share a character does indeed suggest the conclusion that the most recent common ancestor of all the species had the character, but exceptions abound, and scoring the rest of the (putative!) members of the group as having the character amounts to considering that our notion that the taxa form a group is part of the "evidence". Birds, mammals, and some tuna fish are capable of thermoregulation; in the absence of further observations, the conclusion that the common ancestor of Vertebrates could thermoregulate is a logical one. The problem is that there are countless instances of non-thermoregulating species interspersed between any two of those taxa. Thus, avoid attributing certain character states to terminal taxa, based on their putative membership to groups where only a few members have been observed to have the character.

All those considerations pertain to the treatment of morphological characters as a category of evidence with intrinsic differences from molecular sequences. A clear example of a case where these two categories are treated as if they were the same can be found in the paper by Bond et al. (2012) on relationships of mygalomorph spider families. Mygalomorphs are a group of primitive spiders (with 15 families in 2012); previous morphology-based analyses (e.g. Raven 1985; Goloboff 1993c) had considered the characters classically used (since the late 1800s; e.g. Simon 1892) to separate main groups of Mygalomorphae, as well as some new characters (for a total of over half a dozen characters changing at the base of the mygalomorph tree). Bond et al. (2012) used three nuclear genes and morphology in their analysis. But the list of morphological characters was not designed to address the problem of relationships between mygalomorph families; it was instead taken verbatim from a previous study by Bond and Hedin (2006), a study designed to address only the monophyly and relationships of a single subfamily of mygalomorphs (Euctenizinae, Cyrtaucheniidae). The list of characters of Bond and Hedin (2006) did not include, naturally, any of the characters long used to recognize major groups in mygalomorphs, as those characters would have been uninformative (i.e. invariant) regarding relationships of Euctenizinae. But things change dramatically when the problem to solve is the relationships of groups at much higher levels. Bond et al. (2012) then proceeded as if a sample of morphological characters was equivalent to a sample of genes, where every one of the positions might conceivably change in any point of the tree. The utility of those characters that define higher-level groups is precisely that they have very little homoplasy—changing early, only once or a few times in the entire history of spider evolution, and therefore having a good association with major groups. It is this same condition—the fact that changes in those characters happened only once or a few times in evolution—which makes them useless for resolving the relationships within a small group like Euctenizinae; Bond and Hedin (2006) had no need to include those characters in their analysis of euctenizines. But, by the same token, if those characters are not included in an analysis of relationships between major groups of Mygalomorphae, the evidence known already to be most relevant for that problem is not being considered. Bond et al. (2012) were unable to obtain much resolution from the reduced set of morphological characters they had included and concluded from that that morphology offers a limited prospect of helping settle phylogenetic problems. Ironically, their conclusion is mostly a consequence of their having, in fact, ignored the most relevant known characters bearing on the relationships of major groups of mygalomorphs. And this belief that morphology provides very little assistance in resolving relationships finally led, in subsequent efforts on mygalomorph phylogeny by workers in the same lab, to molecule-only analyses (Opatova et al. 2020) where the authors proposed several new families for which no diagnosis can be provided—those new groups simply showed up on their preferred trees, but no obvious characters (not even molecular ones) can be summoned to support the group.

2.8 CHARACTER INDEPENDENCE

It has long been recognized that the characters used for evaluating a phylogenetic hypothesis should be independent, under both parsimony and model-based methods. That is required to properly evaluate the fit between the data and different trees, and it does not depend on whether the fit is considered to be a measure of abstract fit, homoplasy, ad hoc hypotheses, or probabilities. If two characters support a group (say, AB), and another character supports an alternative grouping (AC), this would be evidence for favoring AB. However, if the two characters supporting AB are completely interdependent, then a change in one necessarily implies a change in the other, and then explaining one (by common ancestry) amounts to explaining the other (as a necessary consequence of the first). The number of independent observations supporting the alternative groupings AB and AC is in such case equivalent, and no decision between the two groupings can be made without invoking additional criteria. Farris (1983) put this in reference to the number of ad hoc hypotheses needed to defend a genealogy:

> If two characters were logically or functionally related so that homoplasy in one would imply homoplasy in the other, then homoplasy in both would be implied by a single ad hoc hypothesis. The "other" homoplasy does not require a further hypothesis, as it is subsumed by the relationship between the characters. This is the principle underlying such common observations as that only independent lines of evidence should be used in evaluating genealogies, and that there is no point to using both number of tarsal segments and twice that number as characters.
>
> **(Farris 1983: 20)**

A practical consequence of the characters being independent is that characters can be examined one at a time under parsimony, summing up their scores. Every character, that is, can be optimized on the tree without considering any of the other characters; this is one of the reasons for the speed of parsimony methods. Farris (1983) emphasized *logical correlation*, or different ways to express the same observation; this is generally much easier to identify and can be eliminated from the data without big difficulties. It still needs to be considered, however, as it may occur in hidden or more subtle forms. A case would be to establish a ratio between measurements X and Y, X and Z, and Y and Z. There is no logical dependence between any two of those ratios, yet the three ratios are not collectively independent—knowing the value of two lets you calculate the value of the third. And in the case of serial homologs, with strong repetition and little or no variation, using separate characters for otherwise identical structures also amounts to the same obvious artificial multiplication of characters—they cannot vary independently. The correlation that is much harder to evaluate than logical correlation is *functional correlation* (see the following).

As clarified by several authors (e.g. Kluge 1989; Goloboff 1995, among others), the very existence of a process of descent with modification creates a co-dependence between characters—they have all evolved on the same tree. In other words, when referring to character independence, the expectation is that the characters have no correlation for reasons *other* than phylogeny, so that when numerous characters agree on delineating compatible groups, the idea that those groups are indeed present in the phylogeny gains support. This will work as expected when there is no logical or functional correlation. For example, there is no strong logical or functional dependence between the presence of a mineralized skeleton (as in Gnathostomes) and maternal secretions (as in Mammalia), yet the two characters are strongly correlated: no animal with mammary glands lacks a skeleton. That correlation is due to phylogeny (mammary glands originated only once, in a group which had already acquired a mineralized skeleton); the correlation is not due to a logical or functional constraint.[5]

Phylogenetic methods almost universally assume character independence; parsimony is not the only method to do so. In the case of likelihood, the standard model assumed is that of the probability of change for all the characters (or all the characters in each partition) varying in concert but the characters themselves being independent. Under this *common model*, there is a common parameter for all the characters along a branch (rate of change and time), and all the characters must be used to estimate the probability of change along the branch. This means that how a character, X, is evaluated will in practice depend on the *other* characters: a change in character X along the branch is quite probable if many other characters seem to change along the branch but very improbable otherwise. Characters in maximum likelihood, that is, cannot be evaluated independently; they have to be evaluated collectively. This could seem on first thought to be a violation of the requirement of independence. It is not; all the characters depend on a common parameter, branch length, and they are thus all used to estimate that parameter. Once the branch lengths have a set of fixed values, the likelihood for each character is calculated by examining alternative character state reconstructions independently, one character at a time, and the final likelihood is the product of the character likelihoods—as always done for independent probabilities, with a multiplication. There is thus no violation of the assumption of independence but instead co-dependence from a common parameter.

The assumption of full independence is very hard to guarantee in practice; it is one thing to assume that the characters evolve independently of each other and quite another to have the means to test that. The functional correlation can be strong and obvious in some cases, and there have been some attempts to model such dependence, particularly with molecular sequence data. An example is Fitch and Markowitz's (1970) "covarion" model for codon positions. Another is in methods for cases where secondary structure is known (as in tRNA or ribosomal DNA), leading to the expectation that a change at a given position will probably have to be matched by a specific change in another (e.g. Schöniger and Haeseler 1994; Tillier et al. 1995). But correlations could also occur under more general circumstances, and they can be hard to identify. As a rule of thumb, model-oriented phylogeneticists tend to be much more optimistic about the properties of the data that can be estimated, but even Felsenstein (the champion of model-based phylogenetics) noted that functional character correlation is very difficult to identify with certainty. He first noted (Felsenstein 2004: 343; this in relation to the bootstrap, but his discussion applies in general) that a partial correlation between characters is to be expected and that it will cause "the appearance of too much evidence for [some] groups on the tree". Then he went on to state,

> Unfortunately, there is usually no easy way to know how much correlation there is between characters. . . . In certain cases, such as molecular sequences, one may be able to assume that the correlation of characters occurs mostly between nearby sites in the sequence. For example, we might have correlations that are mostly between sites that are within five nucleotides of each other.
>
> **(Felsenstein 2004: 343)**

So, not even for molecular sequences could one easily come up with methods to identify character dependence in general. In the case of morphology, the difficulties are much more obvious. A first problem is that the characters themselves are organized hierarchically—the differentiation in some of the characters presupposes a particular state for others. One would think that the presence of molar teeth is entirely independent of a kneecap; after all, animals with knees may either have or lack molars. Yet both characters depend on a bone structure providing support, so they are not truly independent (as soon as you know an animal lacks bone, you know it has neither molar teeth nor kneecaps). Although they both depend on a third character, within the domain of a more restricted phylogenetic analysis, the characters "molar" and "kneecap" can be considered, for practical purposes, as being independent, or not correlated. More subtle examples can exist where there is no such obvious hierarchical dependence; consider the existence of a sagittal crest (on top of the skull, where temporal muscles attach in animals with very strong chewing), a sagittal keel (a longitudinal thickening of the suture along the parietals but without attachment of muscles; found e.g. in some, but not all, humans), or a simple sagittal suture (with neither thickening nor muscle attachment). That character is obviously correlated with the size (and attachment points) of the *temporalis* muscle; there has to be a correspondence that is close to one-on-one. But the size and strength of the teeth cannot be wholly independent of the strength of the *temporalis* muscle. In turn, the size and strength of the teeth cannot be independent of the diet and thus of the morphology of the hindgut. Transitively, structure of sagittal suture is thus not wholly independent of the morphology of the hindgut. An animal, in other words, is a whole unit. The direct correlation between any two neighboring members of the transition sagittal suture→temporalis muscle→teeth→diet→hindgut may be strong enough to avoid using them together in an analysis, but more distant components of that sequence (e.g. form of sagittal suture and hindgut) will be correlated much more weakly and thus amenable to inclusion in a single data matrix. The approach of considering characters as independent units is a simplification needed for a phylogenetic analysis (whether model based or not!). The sensible requirement is therefore not that the units are really, completely, and totally independent but instead that their dependence can be considered, for the practical purpose of a phylogenetic analysis, weak enough to not strongly affect the results.

2.9 CHARACTER "CHOICE"

A suspicion of strong dependence between several characters may be grounds to select only one of those to be represented in a matrix. Likewise, variation that is not heritable (or has very weak heritability) should not be included in the matrix (just like extrinsic attributes, such as biogeographic or stratigraphic distribution; cf. Chapter 1). This type of variation is not considered to fall under the rubric of "characters". For example, many plants produce different toxic chemical compounds, and some insect larvae can sequester and accumulate those—the absence or presence of the compound in an individual larva will depend on it having grown in an environment where the appropriate plants were present. Thus, the presence of the compound itself is clearly circumstantial and not an intrinsic quality of these larvae. The ability to sequester compounds may itself be heritable (thus, a useful "character", if present in only some of the species under study), although verifying whether each of the species under study is capable of sequestration requires carefully designed experiments going beyond simple checks of the presence or absence of the compound.

How can it be decided which (independent, heritable) characters are included in a matrix? The theoretical answer is quite simple: given appropriate methods of analysis, *there cannot be any logical reason to exclude from the data any character that shows heritable variation within the study group.* Of course, some characters may provide better evidence of relationships than others (see Chapter 7), but an appropriate, ideal method of analysis should take into account that some characters may be more reliable than others. The better and worse characters, therefore, should be discriminated as part of the analysis—the differences in character reliability should be a conclusion of the subsequent numerical analysis, instead of a premise. Therefore, just the suspicion that a character is very prone to homoplasy does not provide any grounds for excluding it from a data matrix. This practice was more common in the past, but even today, one can still encounter authors who believe that the standard practice for morphology-based datasets is one based on "choosing a selection of characters that undergo few changes" (Puttick et al. 2017: 8) or who believe that "the act of choosing which characters to use, and what the states should be is typically performed by an expert examining populations of samples, and deciding which facets of organismal form vary" (Wright 2019: 5). The practice of selecting some characters for inclusion in an analysis and discarding the rest is neither well-justified nor common—most authors understand that there is no obvious way to distinguish between good and bad characters before an analysis and that even bad characters may help resolve some parts of the tree (e.g. Källersjö et al. 1999), with the implication that all characters must be simultaneously considered in deciding a hypothesis of relationships.

An alternative and more defensible argument to ignore some characters is those cases where the intraspecific (or within-terminal) variation approaches or exceeds the among-taxon variation. This criterion was first proposed by Farris (1966). It is the same rationale invoked to ignore characters where most of the taxa are polymorphic (but see Poe and Wiens 2000); just one or a few taxa being variable does not pose big difficulties to a phylogenetic analysis, but the situation becomes worse as more taxa present polymorphisms, and thus the character carries less and less information (i.e. allowing more and more equally parsimonious locations of those taxa in the tree).

Evidently, taxonomists being humans, every dataset is a compromise between the ideal and what is possible. There can be practical reasons to focus on some characters when building a data matrix for a particular phylogenetic problem. Note that these reasons do not generally belong in the realm of "choice"—all characters one can appropriately study, evaluate, and understand must be included, and there is really no "choice" in that case. Some discussions have considered the problem of character inclusion, but even if they do so under the rubric of character "choice" (e.g. Simmons 1993) or "selection" (Poe and Wiens 2000), they clearly point out that there is generally no reason to intentionally exclude characters from an analysis. Of course, expertise in certain character systems will determine that a particular phylogeneticist focuses on some characters at the expense of others. Someone well trained in the anatomical study of muscles will probably not be able to score firsthand many behavioral characters and vice versa. Availability of specimens and their modes of preservation are also determinant in the types of characters that can be studied (as paleontologists are well aware). Any author attempting a synthetic study to summarize several previous analyses is under the obligation to consider all the evidence that has been published. Only poor descriptions and documentation are admissible reasons for ignoring some previously proposed characters. A character dominated by missing entries, where only a minimum fraction of the taxa could be effectively studied for the character in question, is also likely to provide very little help in resolving the tree. This is particularly the case when the few taxa that have been scored are expected to show up in distant positions of the tree; the few shared observations might be more useful in resolving part of the tree if they all correspond to a subset of taxa that are (or are suspected to be) closely related.

2.10 CHARACTER CODING AND CHARACTER TYPES

Most of the preceding discussion in this chapter centered on a very abstract, idealized conception of characters and variability. The key to a cladistic analysis is connecting that theory into a formal set of rules allowing practical application of the criteria. This consists of translating the observations in the form of a data matrix, representing the data symbolically so they can be analyzed and understood by computer programs and appropriate algorithms. In a sense, the matrix is the only aspect which may indeed contribute new information in a phylogenetic study; the tree itself is merely an interpretation. We said "in a sense" because even the most basic observation—does this animal have three or four pairs of legs?—connotes some type of interpretation (is the short stump a fourth leg, or is it just a cuticular extension?). But the matrix itself is considered as a given when deciding on a tree; the hypothesis of a tree is built on top of those observations (accepting, at least temporarily, all of them as correct), adding a new separate layer of interpretation.

The variation observed in the different characters and states can be represented in the matrix in different ways, and this is also part of the information contained in the matrix. The choice of how to represent the observed variation in the matrix is known as *character coding*. This determines how the transformations between states are evaluated and must also incorporate information on the observed variation and the degree of differences between the states themselves. Depending on how the variation manifests, the best way to represent differences among taxa may by means of discrete, continuous, or landmark characters.

2.10.1 Discrete Characters

Discrete characters are the ones most commonly used in phylogenetic analysis. Since early phylogeny programs only implemented algorithms for discrete characters, these became almost a synonym of cladistic analysis until recently. Discrete characters are those where the states can be clearly delimited, the states being well-defined conditions and separated by obvious gaps. Of course, no two organisms are actually identical, so the separation into discrete states is a simplification needed for a phylogenetic analysis (simplification begins with the decision to consider variation in atomized ways, i.e. as separate characters).

When only two states are present in a character, the character is called *binary*. The most common way to include these characters in a matrix is as symmetric characters, i.e. characters can equally well transform in any direction. The *transformation cost* between states is the value of a step between those two states; these are equal in a symmetric character (i.e. $Cost_{0 \rightarrow 1} = Cost_{1 \rightarrow 0}$). When there are more than two states, the character is called *multistate*. The costs of transformations between different states can be all equal or different. In the same way as the goal of a phylogenetic analysis is being able to attribute to common ancestry as many similarities as possible (Chapter 1), when some of the states within a character are more similar to each other than they are to a third state, that degree of difference can be used to establish degrees of homology (primary homology, that is) among the states of the character (Maslin 1952; Lipscomb 1992). That degree of difference should be reflected in the transformation costs. A larger transformation cost between less similar states has the consequence that, when the tree requires that some states are independently derived, independent derivations from more dissimilar states produce a larger penalty for the reconstruction of the character evolution. That is, a step occurring between more dissimilar states is weighted more, in keeping with the idea that a phylogenetic tree is a theory by means of which we explain as much similarity as possible in terms of common ancestry. Changing the transformation costs for a character may well lead, on a given tree, to different optimal reconstructions of the evolution of the character. Optimality—i.e. the reconstruction which implies the fewest weighted steps and its value—can only be defined for a given set of transformation costs. Note that the transformation costs are not changed arbitrarily: they should reflect, as best as possible, the differences between states. These costs will depend on whether there is a state that can be considered as closer to some state(s) than to other(s).

When the states are equally distant from each other (Figure 2.3a), all the transformation costs can be considered equal; this is called a *nonadditive* (nomenclature used in the documentation of Hennig86; Farris 1988) or *unordered* character (nomenclature used in the documentation of early versions of PAUP; Swofford 1985). In the case of Figure 2.3a, illustrating the shape of a hypothetical spot, there is no obvious relationship between any two states; the chevron does not seem closer to the star than to the circle, and the circle does not seem closer to the chevron than to the star. Thus, all three transformation costs must be considered equivalent. This also means that the structure of the tree is less determined by the character; a tree where the chevron appears as intermediate between the star and circle has the same fit as a tree where the circle is instead intermediate between the chevron and the star. Other typical cases of nonadditive characters are alternative different shapes of structures or colors (e.g. when biochemical pathways determining the different colors are not known; otherwise, the coding where some color transformations are less costly may be preferable).

When the similarity between some of the state pairs is greater, then the appropriate course of action is to reflect that situation in their transformation costs. This can be thought as identifying—even if intuitively—the various components of the "similarity" between the states, in the same atomistic way as the organism is decomposed into characters. Consider the example of Figure 2.3b, where a structure can be smooth (state 0) or have a small lateral point (1), a longer hook-like projection (2), or a two-point projection (3). Examining the relationships between states (as drawn in a simplified manner in Figure 2.3b), it is evident that state 3 (long, two-pointed hook) is morphologically more different from the smooth state (0) than it is from a long, single-pointed hook (2). A diagram indicating the relationships between the states is often called a *transformation series* of the character (less commonly, this was called a *morphocline*, following Maslin 1952). A reconstruction for the character where state 3 originates as deeply nested within a group of taxa having state 0, in other words, implies that the tree is deviating from the morphological relations among the states; another tree where state 3 originates instead from the (much more similar) state 2 deviates less from the observed relationships. Thus, it is clear that, in terms of closeness to the observed morphological relations between the states, the

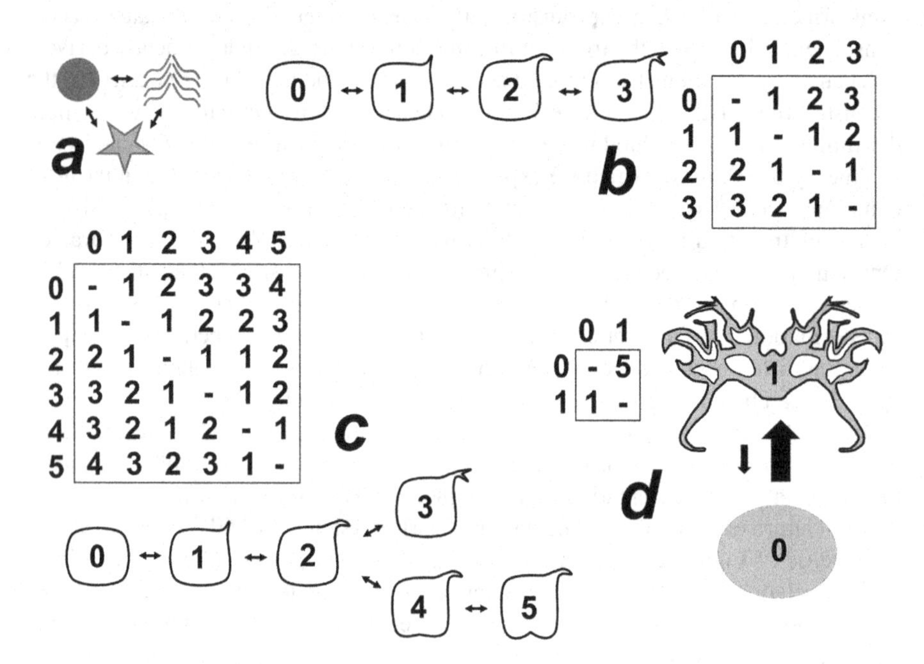

FIGURE 2.3 **(a)** A nonadditive character can be used to represent cases where relative degrees of similarity are the same among all states. **(b)** When a state (e.g. state 3) is more similar to some states (2) than to others (0), an additive character can be used to penalize cases where a state independently originates from a more different condition (e.g. multiple derivations of state 3 from state 2 are penalized less strongly than from state 0). This can also be expressed in the form of a matrix that specifies the costs of transformation between states. **(c)** When the similarity relationships between states are more complex, it may be possible to represent them by means of a branched additive character, where the cost of transforming between two states in the character state tree is the sum of the costs of transformation between all intermediaries. **(d)** Cases where a complex structure is strongly simplified are sometimes represented with asymmetric transformation costs because the evidence to homologize the more complex state is stronger than the evidence to homologize the simpler.

first tree (implying 0→3) should be penalized more strongly than the second (implying 2→3). The way to achieve that is to give 0→3 transformations a higher cost than 2→3. This is sometimes called the *hierarchy of the character* (i.e. it refers to more and less inclusive statements about homology of the states). Given the hierarchical relations among the states, with some states intermediate between the others in a lineal sequence, this is best accomplished by assigning the state at one end of the sequence state 0, increasing state number toward the other end, and then considering that the cost of transforming between any two states equals their numerical difference. This is indicated in the small matrix of character state transformation costs (see Figure 2.3b) next to the diagram of the states; these matrices are often referred to as *step-matrices*. In the particular case indicated in Figure 2.3b, the transformation cost between any two states equals the sum of costs to transform between all the intermediate states; they are hence named *additive* characters.

It is essential to understand that the morphocline, character state trees, or matrix of transformation costs do not indicate that some states are believed to have evolved from others during the course of evolution. That is, they do *not* represent a statement of the historical course of transformations or an idea that every state is uniquely derived from some other, only once. Interpreting them as such (as many authors do) leads to logical errors. They can logically be seen as representing, preferably, (a) a statement about the relative degrees of similarity among the states or, less desirably, (b) a statement on the probability (i.e. frequency) with which one state can transform into another during the course of evolution.

Lineal additive characters can be easily incorporated into existing phylogenetic software, for which the increasing levels of homology (number of independent qualities shared by the states) are indicated with sums of steps as the transformation is to be effected traversing more states in the sequence. This can also be visualized as if the transformation between two states needs to effectively pass through the intermediate states (hence the name "ordered"); the idea of actually "passing" through intermediate states is an assumption about the evolution of characters that is not required to justify the use of differential transformation costs between the states. Instead of the idea that evolution passes through intermediate states, the notion that during the course of evolution, certain transformations between states are more likely to have happened, is more sensible. However, it is usually a less defensible argument than relative degrees of homology for giving some character state transformations a higher cost than others. Even if their consequence is similar (an additive character), the two justifications are of a different nature. Rather than imposing on the analysis ideas on how evolution works, it seems preferable to let those ideas be a conclusion of the analysis, instead of a premise. A justification for additivity based on degrees of homology does not appeal to evolutionary mechanisms—only to observations or, more properly, to the degree to which a tree explains the different degrees of similarity among the states.

In the past, nonadditive characters were often preferred over additive characters; it was often argued that additive characters could only be justified on the grounds that, during evolution, transformation between some pairs of states necessarily had to pass through other intermediate states. Thus, additive characters were considered by some to require the idea

> that the states that are most similar morphologically share a closer evolutionary relationship to each other than to some less similar state (i.e. phenotypic intermediacy equals phylogenetic intermediacy). While this concept makes some intuitive sense, it operates under the assumption that the process responsible for the character state change is phyletic gradualism . . . —a punctuated change in character state evolution is ruled out. The validity of assuming a purely gradual mode of evolution, however, has been questioned. While both gradual and punctuated modes of evolutionary change have undoubtedly occurred, it seems unwarranted to restrict the interpretation of data to one or the other processes in the absence of phylogenetic information. . . . The *a priori* ordering of a transformation series requires reference to a particular evolutionary process.
>
> **(Hauser and Presch 1993: 257)**

In the view of Hauser and Presch (1993: 244), a hypothesis of additivity "is a restrictive statement that *excludes* all other possible hypotheses of character state order" (italics added). Hauser and Presch's (1993) argument was often used to justify routine treatment of all characters as nonadditive (e.g. Scott 2005; Hormiga 1994), but the notion that an additive character forbids some transformations is incorrect. Sometimes the additive coding was rejected on the grounds that it could produce "bias"; for example, Gauthier et al. (1988: 110), in their classic paper on fossils, stated that "all multistate characters were treated as unordered to avoid biasing our conclusions by the traditional, theory-laden interpretations of how such transformation series are supposed to have evolved". Gauthier et al. (1988) were correct in that it is desirable to not use arguments on evolutionary pathways for deciding character additivity, but their statement glossed over the fact that additivity can be also decided by considering relative degrees of homology, when observable. It is also true that coding a character as additive will lead to a preference for some results over others, but why does that necessarily amount to a "bias"? Conversely, coding the character as nonadditive does not make bias impossible—and it will *also* lead to preference for some results over others. If just leading to prefer some results over others is "bias", then evidence itself is "biased"—evidence is what leads to preferring some conclusions over others! Gauthier et al.'s (1988) reasoning for preferring nonadditive characters was backward.

The crucial justification for considering a character as additive is one of principle, not an argument about the correctness of results: the arrow diagram in Figure 2.3b does not indicate which states can transform into which or how the character is thought to have actually evolved; instead, it indicates the degrees of (primary) homology among the states. On an actual tree, any state can transform into any other; what the scheme of costs determines is the penalty to impose on additional transformations between some states, and this penalty is what determines that some reconstructions (possibly the ones implying less costly transformations) are chosen over others. The penalty is ideally not imposed on the grounds that a transformation is more or less likely, but instead on the basis of homology: "I can think of no justification for discarding observed similarity as evidence of homology in favor of a transformation series that is more descriptive of the tree" (Lipscomb 1992: 61). A similar recommendation was made by Slowinski (1993).[6] Thus, transformation costs refer to the empirical evidence that serves as the basis for homology decisions, not to a belief that some transformations can occur with a lower or higher probability. These costs serve to evaluate different scenarios of how the character has evolved on a tree; the costs refer therefore to how the evaluation of homology is done, not to a belief of the evolutionary steps through which the character has evolved. The final conclusion of how the character has evolved can be obtained only when a tree is accepted as the best hypothesis of relationships and the character is mapped onto the tree. This hinges, of course, on the general assumption that the best way to explain observed similarities is by means of phylogenetic hypotheses; that notion applies both at the level of characters and states. Without the notion that explaining similarity by hypotheses of common ancestry is the goal of cladistic analysis, there can hardly be a morphology-based phylogenetics.

Admittedly, the exact ratios between costs within a multistate character cannot be determined with precision. One reason is that the costs are not intended to reflect actual probabilities of transformation between states (if they existed, if they were constant throughout evolution, and if they were measurable, those probabilities would have exact values, but that is three big *ifs*). The costs are instead intended to reflect the desirability of homologizing different states. Phylogenetic analysis can hardly be an exact and precise science, but lack of precision should not be confused with arbitrariness. There is no way to know the exact cost ratios for the different transformations in Figure 2.3b, but not every set of costs is possible. For example, $\mathrm{Cost}_{0\rightarrow1}$ must be less than $\mathrm{Cost}_{0\rightarrow2}$, and this, in turn, must be less than $\mathrm{Cost}_{0\rightarrow3}$. The simplest course of action, and the one most commonly used, is to simply assign a cost of one step between states that are adjacent in the graph.

Another myth that was occasionally encountered (e.g. Scott 1996) was that a character could be coded as additive only when the polarity of the character was known; on this reasoning, characters with unknown polarity had to be coded always as non-additive. First, this confuses the notion of "polarity" (i.e. direction of the transformations) with that of "ordering" (i.e. the distances between the states, regardless of direction). Second, as discussed in the previous chapter, the very notion of "polarity" is flawed, but even if it were applicable, it would have nothing to do with the costs of transformation between different states. For example, for the character hierarchy of Figure 2.3b, the ancestral state (i.e. in the true course of evolution) might well be state 1, as shown in Figure 2.4. In Figure 2.4 the distances between states are exactly the same as in the character state tree of Figure 2.3b, and so the tree has a perfect fit to the character. There is thus no requirement that, in a lineal transformation series, evolution must have begun from one of the tips; it may well have started from a state intermediate between two (or more) others. Therefore, the costs of transformation between states are decided without reference to polarity, and the outgroup, if known, is simply coded as having the corresponding state—the one it indeed has.

The example of Figure 2.3b is one where the relationships between states can be expressed in a lineal sequence. In more complex cases, with modifications to different parts of a structure, the relationships between states may have a branching pattern, as shown in Figure 2.3c. This is called a *character state tree.* States 4 and 5 have a ventral cleft (incipient in 4, well developed in 5). The character coding in Figure 2.3c is also additive (it still fulfills the requirement that the steps to transform

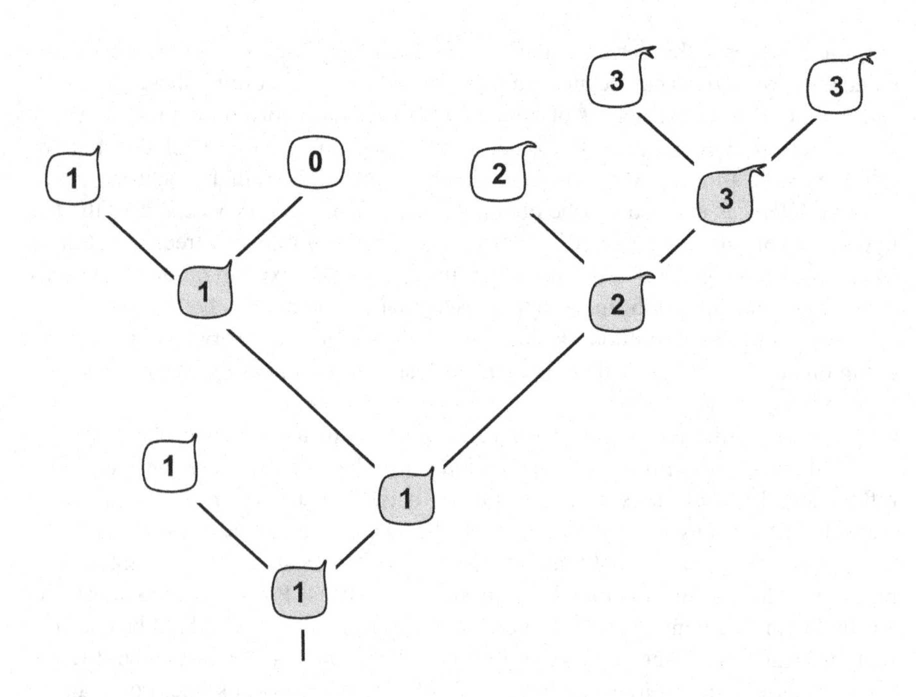

FIGURE 2.4 The notions of "polarity" and "additivity" are separate. In the character state tree of Figure 2.3b, even if state 0 is one of the "extreme" states, the plesiomorphic state (given the tree shown) must be state 1. The tree perfectly fits the character state tree (i.e. does not separate adjacent states, and every state originates from one of its most similar states), even if the plesiomorphic state is intermediate in the character state tree.

between any two states is the sum of steps to transform into all intermediate states), but it is not lineal. In computer programs, those characters either have to be recoded (as it is possible to express the same relationships between states with several binary variables, see Chapter 3, Section 3.4; for tree calculations, TNT does this internally) or coded as a step-matrix character (where the distances between the states are expressed in a step-matrix, as in Figure 2.3c).

All those types of coding schemes are used to represent different ideas of the homology relationships between states of a character. All preceding cases correspond to undirected graphs, where the links are symmetrical and there is a single step between the states connected via a direct link. There are cases where the state transformation costs cannot be expressed with such a type of graph. An example is a character where we want to set $Cost_{0\leftrightarrow1} = Cost_{0\leftrightarrow2} = 2$ and $Cost_{1\leftrightarrow2} = 3$. These costs are symmetrical but neither additive nor nonadditive; finding the optimal state assignments for such a character requires methods different from those that can be used for the preceding types of characters. No recoding of the character in several variables is possible either (unlike the case of Figure 2.3c). The same need for special algorithms to find optimal reconstructions arises when the transformation costs are asymmetrical: that is, $Cost_{0\rightarrow1} \neq Cost_{1\rightarrow0}$. In general, it seems difficult to justify the use of such a type of transformation cost. The most common argument is the idea that one of the transformations is less likely to occur than the other. The quintessential example is insect wings; there is ample consensus that maintaining such a complex structure is metabolically very expensive and that, as a consequence, insect wings very easily get lost (in dozens of independent events), yet there seems to be a single origination of wings (see, however, Whiting et al. 2003, and subsequent controversy, Telford and Budd 2003; Goldberg and Igic 2008). This agrees with the belief that a structure as complex as insect wings is unlikely to originate twice independently. This is a justification based on a theory of how evolution proceeds. This type of argument always needs to be used with caution because it means that (to some extent at least) conclusions will be determined by prior assumptions instead of just the available evidence. Nonetheless, assuming asymmetric transformation costs for a case such as the insect wings may be the most reasonable course of action if no other argument is available. A problem with asymmetries is that the similarity relationships between the states are, by definition, symmetrical: any state A is as similar to another state B as state B is to A. Thus, the argument of morphological closeness between states cannot be used to justify asymmetries in transformation costs (Goloboff 1997). An argument that does not directly resort to evolutionary mechanisms could, in some cases, be based on the relative complexity of the different states and the certainty with which homology can be established for each of the states. Complex structures, particularly as they have more parts or details, can be homologized with greater confidence; the opposite is true of very simple structures. Consider a case like the one shown in Figure 2.3d, where a complex structure is either present (state 1) or absent altogether (state 0). Given the choice of homologizing one or another state—e.g. in a tree with the structure . . . 0(0(0(1(10)))), where the option of two gains or one gain plus one loss would be equally parsimonious under symmetric transformation costs—a preference for homologizing the more complex structure can be expressed in a higher cost for the transformation from the simple to the complex (so that a single gain with a subsequent loss is preferred, i.e. "cheaper" than two gains). Whether this argument can be reasonably applied will depend on the details of the structures being compared.

2.11 TRANSFORMATION SERIES ANALYSIS

From the mid 80s to the early 90s, following the proposal by Mickevich (1982), some authors (Mickevich 1982; Mickevich and Weller 1992; Schuh 1991; Lipscomb 1992) recommended that the scheme of transformation costs be tested by means of congruence and subsequently modified to better fit the trees obtained in the analysis. This was called *transformation series analysis*, or *TSA*. As with characters (cf. the discussion of the congruence test and the example of Figure 2.1), that can only be done on the final trees. However, just as with homology, the empirical observations about lesser or higher degrees of homology between states continue being as valid as before (just as in the case of primary and secondary homology for whole characters), regardless of the resulting trees. Being able to gauge the agreement, or lack thereof, between the tree and those original observations requires that the costs be maintained as part of the data. The argument to weight characters or state transformations on the basis of congruence is of a different nature than the argument to assign transformations costs on the basis of state similarities; one is part of the data, i.e. given prior to the analysis, and remains unmodified regardless of the results, while the other can be modified subsequently (Goloboff 1997; see Chapter 7 for additional discussion). Much of the discussion on *TSA* conflates the problem of how the character truly evolved during the course of phylogeny with that of how to ponder the evidence provided by multistate characters; those two problems are best analyzed separately (Goloboff 1997).

2.12 CONTINUOUS AND LANDMARK DATA

Morphological variation can sometimes be adequately captured by a variable representing a simple magnitude, as is the case of ratios, lengths, areas, and volumes of biological structures. This information can be incorporated into the phylogenetic matrices in different ways. One possibility is to discretize the variation into different states that represent the full range of variation—for instance, defining states as "short", "intermediate", and "long". Every state thus accommodates all the observations that fall within a certain interval, the upper and lower limit for the category. This can be done when the different taxa have their values mostly grouped around a few values, with no or few taxa occupying intermediate positions. In this case, the specimens can be assigned a state by visual inspection. Alternatively, the limits between states may have to be expressed numerically when the differences are harder to detect visually. For instance, the state "short" may include every value below 0.05 mm, the state "intermediate" those between 0.06 and 0.1 mm, and the state "long" values higher than 0.1 mm. Although the limits are still somewhat arbitrary, once those limits have been defined, the observations can be assigned to each state without ambiguity. Although in these cases the original variable is quantitative (continuous), the information is introduced in the phylogenetic matrix as a discrete character (typically additive, as these variables form obvious morphoclines). This process of dissecting continuous variation into discrete states is called discretization. Given that discrete characters have historically been the most commonly used characters in phylogenetic analysis by far, discretization is one of the most common methods of coding continuous characters. Discretization, however, has many limitations. Particularly when the original observations represent direct measurements on several specimens (so that terminal taxa are actually assigned a distribution of values, instead of a unique values) and the variation covers a finer gradation of all alternative values (without obvious gaps between most taxa), the most acute problems are:

a) the subjectivity in the placement of the limits between states;
b) the loss of information regarding differences among values for taxa that are included in the same state;
c) the loss of information regarding the magnitude of the differences among states;
d) observations that represent the same difference in terms of the original values can be included either in a single state or in different states, depending on how close they are to the values used as limits for the intervals according to the presence or not of intermediate values.

Several approaches to discretization have attempted to alleviate these problems (e.g. Thiele 1993; Archie 1985; Strait et al. 1996), but most of these suffer from serious drawbacks. One of those drawbacks is the possibility of assigning taxa with significantly different values to the same state or assigning taxa without significant differences to different states (as discussed by Simon 1983; Farris 1990).

In view of the problems caused by discretization, if the original data are quantitative, a wiser approach is to avoid the discretization altogether and treat the character directly as a continuous variable. In fact, the earliest algorithms for finding most parsimonious reconstructions (Farris 1970) were described to work with quantitative variables; nothing in cladistic analysis precludes ancestors to be assigned a real value for a continuous variable, and these ancestral values then have to be assigned in such a way that the sum of differences between every ancestor/descendant pair is minimum. Despite this, most of the early phylogenetic programs (e.g. Hennig86, Farris 1988; PAUP, Swofford 1985) only implemented analysis of discrete characters. It was not until the implementation in TNT (Goloboff et al. 2006) that continuous characters started being incorporated in phylogenetic matrices more commonly (e.g. Granara and Szumik 2007; Bertelli 2017; Randle and Sansom 2017; Groh et al. 2019). The logic behind Farris's algorithms for optimization is simple: every value represents a state (=condition), and the cost of transformation between two values is the numerical difference between states (or directly proportional to that difference, if the character is weighted or rescaled). The main cause that delayed the inclusion of continuous variables in cladistic studies is probably the generalized preconception that quantitative characters are somehow phenetic in themselves. Perhaps this preconception stems from associating real-value differences with distances. In fact, distances are not intrinsically phenetic, but even if they were, distances are not the same thing as real-valued characters. A more profitable distinction is instead that methods for grouping taxa (Rae 1998; Catalano et al. 2010) can be either cladistic (if grouping so that similarities can be maximally attributed to common ancestry) or phenetic (if grouping so that most similar taxa cluster together). Other details of continuous characters, and the algorithms used by TNT to find optimal ancestral reconstructions, are presented in Chapter 9.

Sometimes the characters to be used for the phylogenetic analysis represent variation in shape, which by its very nature is gradual but cannot be easily captured by a single one-dimensional value and cannot be easily ordered in a sequence. Sometimes the shape differences can be described by means of ratios between two continuous values (for example, the ratio length/width), then considered as a single character. However, in most cases ratios of linear measurements are poor shape descriptors (Rohlf 2000); to make matters worse, ratios have additional problems of interpretation (Mongiardino et al. 2015) and quite peculiar statistical properties (e.g. see Marsaglia 1965, 2006; Broda and Raymond 2014). The field of quantitative analysis of shapes, or *geometric morphometrics*, *GM*, provides numerous techniques for the formal study of shape variation. Most *GM* methods are based on landmark configurations: groups of anatomical loci that can be recognized in all the specimens, the coordinates of which (in 2D or 3D space) represent the shape of the structure to analyze. All *GM* methods define *shape* as the geometrical information that remains after differences in size, rotation, and position have been sorted out (Kendall 1977).

For landmark data, the original information is the coordinates for each landmark (2D or 3D, according to the sort of shape considered), transformed after removal of non-shape differences via the superimposition of landmark configurations (Gower 1975). Once the configurations are superimposed, the quantification of shape change is based on differences in positions of comparable landmarks. The use of landmark data in phylogenetics, although resisted in the past (Bookstein 1994; Monteiro 2000), is becoming more common, with several methods proposed in the last 20 years (e.g. Rohlf 2002; Catalano et al. 2010; Klingenberg and Gidaszewski 2010). A detailed description of the use of *GM* in phylogenetics is presented in Chapter 9.

2.13 IMPLEMENTATION

The different considerations about homology and contents of the dataset can be incorporated in several ways into a numerical analysis with TNT. The most obvious way is with options to determine how to treat character state transformations. Characters can be treated as either additive, nonadditive, or step-matrix. This is done with the commands `ccode`, `cost`, `cstree`, `smatrix`, and a few others. In addition to those settings for the character themselves, to facilitate handling and interpretation of results in morphological datasets, it is advisable to include full names for the characters and states with the `cnames` command. The character and state names can subsequently be used both for inputting lists of characters in commands to manipulate or map characters and for producing more readable output. Many studies are based on modifying or combining existing datasets; if character and state names are the same (or sufficiently similar) in two separate datasets, most of the work needed for comparison and combination of the two datasets can be automated with the `dcomp` command. Another type of information that can be included as part of the data is taxonomic information, which can subsequently be used as a series of reference groups, against which a tree (e.g. one resulting from the analysis of the matrix) can be tested with the `taxonomy` command. All these commands are explained in more detail in the following and summarized in Table 2.1.

TABLE 2.1
TNT Commands That Handle Specific Settings for Datasets (e.g. character types, names, reference taxonomic groups)

Command	Minimum Truncation	Action(s)
`agroup`	`ag`	Create a group of taxa (use in lists, enclosing in braces)
`ancstate`	`an`	Force assignment of state in outgroup to root (for asymmetric characters)
`ccode`	`c`	Change character settings, weights, and activities
`ckeep`	`ck`	Store current character settings (then restored with `ccode!`)
`cnames`	`cn`	Define names for characters and states; define names for data blocks
`copytree`	-	Various copying functions, including transfer of a tagged tree to taxonomy
`cost`	`cos`	Set state transformation costs

Command	Minimum Truncation	Action(s)
cstree	cst	Define state transformation costs as a diagram
dcomp	dc	Compare and merge datasets
fillsank	fill	In taxa polymorphic for a step-matrix character, add intermediate states to state set
nstates	ns	Determine how states are read; convert data formats
smatrix	sm	Read and apply step-matrices to characters
xgroup	xg	Create a group of taxa (use in lists, enclosing in braces)
xinact	xi	Deactivate uninformative characters
taxonomy	taxo	Handle taxonomic comparisons
usminmax	us	In step-matrix characters, set value of minimum steps (TNT uses by default an approximate algorithm)

2.14 CHARACTER SETTINGS

2.14.1 Basic Character Settings: ccode Command

The basic character settings are determined with the `ccode` command. The `ccode` command of TNT is inherited from Hennig86; it is used for activating or deactivating characters; making characters additive, nonadditive, or step-matrix; giving characters prior weights; and setting internal steps in the case of implied weights analyses. The `ccode` command can be executed at any point in time as long as the dataset has been read into memory; it is perfectly possible to read the matrix into the program and then change some character settings by typing commands at the command line or (in Windows) by selecting the *Data > Character Settings* menu option. It is, however, highly advisable to set basic character options in the file so that the commands to set options do not have to be repeated every time you start a new TNT session and read the data file into the program. In the data file, the character settings must be included after the matrix itself, following the definition of the data (with `xread`, which should end with a semicolon).

The `ccode` command uses a series of specifiers, which tell TNT the action to be performed on the subsequently listed character(s). Specifiers are cumulative within the execution of a single `ccode` command (but not for different instances of the command). The asterisk (*) disables all previous specifiers so that no specific action is taken on subsequently listed characters unless a new specifier is given.

```
Specifier     Make following character(s)
+             additive
-             nonadditive
[             active
]             inactive
(             Sankoff
)             non-Sankoff
/N            of weight N (if N not a number,weight is 1)
=N            with N additional steps of homoplasy
              (if N not a number, 0)
*             disable all previous specifiers
```

By default, TNT treats all characters as active, nonadditive, with weight unity, and no internal steps. To make characters 1, 6, and 12 additive, you would type at the TNT prompt:

```
tnt*> ccode + 1 6 12;
```

If you want to also give a weight of 5 to character 15, you have to tell TNT to turn off the additivity specifier (+) by using an asterisk:

```
tnt*> ccode + 1 6 12 * /5 15;
```

otherwise (if you don't use the asterisk), character 15 would receive a weight of 5 *and* be made additive (of course, you can issue two `ccode` commands, one for making characters 1, 6, and 12 additive and another for giving character 16 a weight of 5, but this will often be impractical). Note that the nonadditive specifier (–) used to switch off the additive one:

```
tnt*> ccode + 1 6 12 - /5 15;
```

would also have the effect of making character 15 nonadditive; this may or may not be what you intend (e.g. you may have previously set character 15 to be additive and you want it to *stay* that way, only changing its weight, not its additivity). Only the asterisk (or repeating the `ccode` command) disables all previous specifiers and ensures that the additivity status of character 15 is unchanged when giving it a weight of 5.

In TNT, characters can be given different weights between 1 and 1,000. Zero weights are not allowed (this amounts to deactivation, achieved with the `]` specifier). Character weights can be decided prior to the analysis (e.g. on the basis of the certainty of homology decisions, if relevant and possible) or during the analysis (i.e. under implied weighting, see Chapter 7). The final weight is the product of both; prior weights are part of the data, fixed during calculation of trees. Prior weights simply multiply the number of steps in any report of tree scores, without affecting character mapping itself.

Of course, just like you type the `ccode` on screen, you might as well include it in the data file after having defined the dataset with `xread`:

```
xread ¶
  . . .
 (data definition here)
  . . .
; ¶
ccode + 1 6 12 * /5 15; ¶
proc/; ¶
```

As discussed prior, including character specifications in the file is highly advisable. When TNT finishes reading a file containing an `xread`, it stores in a separate buffer all the character settings when the file is closed. If you change those settings when running the program, you can always go back to the settings originally contained in the file by typing:

```
tnt*> ccode!;
```

The equivalent is accomplished in the Windows GUI versions with *Data > Character Settings*, clicking the *Restore settings* button.

If you want the current setting to be stored (so that subsequent executions of `ccode!` will return to current settings, instead of those at the end of the data file), you type:

```
tnt*> ckeep;
```

In Windows *GUI* versions, the same is done with *Data > Character Settings*, clicking the *Store current settings* button. In addition to the settings defined with `ccode` itself, `ckeep` and `ccode!` store and retrieve the settings of state-transformation costs in Sankoff (`cost` and `smatrix` commands) and ancestral characters (`anc-states` command), as well as the block definitions for all characters.

When you type a `ccode` command but make no changes to the current settings (i.e. `ccode`;), the current settings are displayed in a way that is readable by the program. You can use this to save the current settings to a log file, to be later read into the program. Lists of characters with a specific setting can be obtained by typing `ccode` and the corresponding specifier, without providing a list of characters. For example, `ccode+;` will output a list with all the additive characters and `ccode];` a list with all the inactive characters (using the corresponding table formats, see Chapter 1).

2.14.2 Step-Matrix Characters (and Ancestral States)

Characters where the user determines any set of transformation costs between states are called *step-matrix characters* (so called because, in other programs, the transformation costs are specified by using a matrix). This is often used to reflect the relative similarities between the different states when these relations are too complex to be expressed in additive or nonadditive characters—for example, three states may be more similar to each other than they are to a fourth. These characters are sometimes called Sankoff characters because the number of steps is calculated with algorithms first proposed by David Sankoff (Sankoff and Rousseau 1975).

The costs can be symmetrical or asymmetrical, i.e. changing from one state to the other does not cost the same as transforming back. This allows taking into account special cases, such as those characters which are believed to be easily lost but very rarely gained. One of the most common examples where such an assumption may hold is the case of wings in insects, believed to have originated only once (in the Pterygota) and been lost dozens of times (in many different insect orders). Thus, the user may choose to penalize wing gains more heavily than losses. This is equivalent to preferring, given the alternative, the homologization of insect wings,

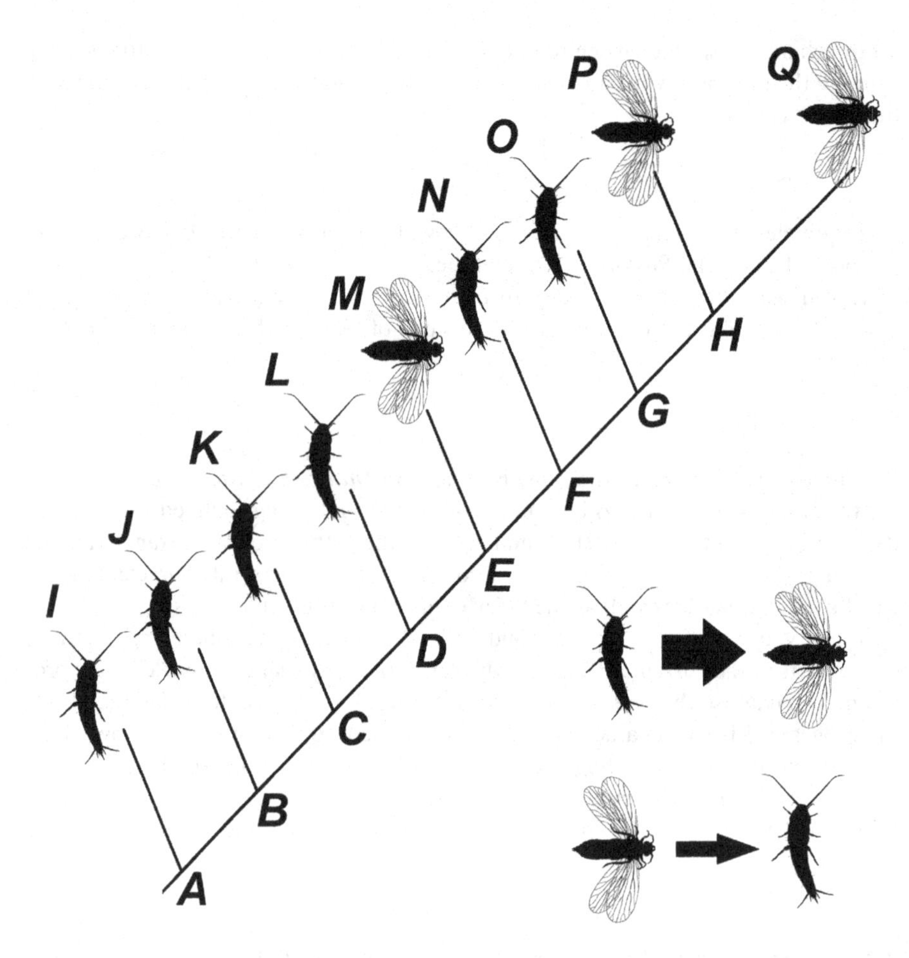

FIGURE 2.5 The choice between reconstructing the character as two independent wing gains or as a single wing with two successive losses depends on the degree of asymmetry in the transformation costs. See text for additional discussion.

not of their absence. For example, Figure 2.5 shows a tree where nodes E–G might be reconstructed as either having wings (in which case wing presence in the three winged species is considered homologous) or as not having them (in which case the presence of wings in the most basal winged species would be considered a parallelism). If the character is assigned symmetrical transformation costs, with cost of *wingless→winged* being the same as the cost *winged→wingless*, then nodes E–G are best reconstructed as lacking wings (thus, wings are a parallelism). But if the cost of *wingless→winged* is 3 and the cost of *winged→wingless* is 1, then the reconstruction of wings as homologs costs 5 (one wing gain at node E, with cost 3, plus two parallel wing losses in terminals N and O, summing another 2 steps) and is preferred over the more costly reconstruction of wings as a parallelism (of cost 6, two wing gains, at nodes M and H). Thus, giving wing gain a higher cost implies that, given the choice, it is preferable to homologize wing presences, not absences. Note that, for the case shown in Figure 2.5, increasing the asymmetry in costs even more, such as cost ratios 1:5 for wing loss and gain, the character may be best reconstructed (given the information in the matrix itself) as winged at the root node A (with 4 parallel losses for terminals I–L, with a summed cost of 4), thus avoiding any wing gains on the tree (which would cost as much as 5 losses). This may or may not be reasonable, depending on what is known to happen outside the group of taxa included in the matrix. If the *wingless* state is known to also occur in several successive sister groups not included in the dataset, then assigning state *winged* to the root of the tree (node A) will in fact imply either a gain (relative to taxa further down in the phylogeny) or many more losses and considering state *winged* as primitive for the whole group (i.e. further down node A). To prevent this from happening, the `ancstates` command can be used to force the assignment of states to the root of the tree so that the states to be assigned to the root are selected from the set of states present in the terminal taxon designated as outgroup (effectively making these "ancestral" states). Using `ancstates+` followed by a list of characters sets subsequently listed characters to be ancestral; `ancstates-` does the opposite. The `ancstates` command has no effect on characters with symmetric transformation costs because, in that case, the root states will always contain at least one of the states in the outgroup (with no need to force any state assignment). But, when transformation costs are asymmetric, the `ancstates` option can make a difference in tree costs and potentially lead to different trees being selected. The same is true when character state transformation costs are being weighted dynamically (see Chapter 7, Sections 7.8.3 and 7.8.4) because the implied transformation costs may be asymmetric, even when the prior transformation costs are symmetric. Applying the `ancstates` option to a set of characters will depend on information extraneous to the dataset at hand and should be considered carefully.

The definition of the transformation costs by means of commands can be done in three different ways, with the commands `cost`, `smatrix`, and `cstree`. These naturally apply only to discrete characters; any costs defined for continuous or landmark characters are just ignored. The `cost` command defines costs and applies it to some characters (individually). You first specify a list `L` of characters to which the costs are to be applied, then an equal sign, and then the costs `C` of transforming between states `X`, `Y`, and `Z`:

```
cost L = X>Y C_{X→Y} X>Z C_{X→Z} Y>Z C_{Y→Z};
```

The list `L` can be any valid set of characters. The arrow symbol (>) defines the cost $C_{X \to Y}$ but leaves the cost $C_{Y \to X}$ unmodified. If you want to define cost $C_{X \to Y}$ identical to the cost of $C_{Y \to X}$, use a slash (/) instead of the arrow. Any unspecified cost (e.g. Y→X in the example prior) is left as unity (or as whatever value was defined in previous instances of the `cost` command for the character(s) in question). You can specify as many state pairs and costs as you want; all of them will be applied to the characters in the list `L`; if more than a specification is done for a given transformation, the last one takes effect; the `cost` command finishes when a semicolon (or a syntax error) is found. The states `X` and `Y` are specified using the same symbols as in the data matrix (i.e. alphanumeric codes for morphological data, IUPAC codes for DNA or amino acid data). When the data have been read in multiple blocks, with different formats, the `nstates/` command determines how the `cost` command will read the states; use `nstates/num` for morphological, `nstates/dna` for DNA, and `nstates/prot` for amino acid sequences. Note that the `nstates/` option determines how to subsequently read state specifications; it does not alter the dataset itself. When reading state specifications with the `cost` command, use square brackets to enclose polymorphisms (in which case, character state combinations between both state sets are given the subsequently specified cost). The question mark (?) indicates, as in the case of data matrices, all possible states. You can take advantage of this to define all costs to have some value, then specify only the exceptions. For example, for a DNA dataset, the command

```
tnt*> cost . =?/? 3 R/Y 4;
```

will first set, for every character ("`cost . =`"), all possible transformations to have a cost of 3 steps ("`?/? 3`"), then it will change the cost of transversions to 4 ("`R/Y 4`"). Any change that is not a transversion is a transition, so this example sets the ts:tv ratio to be 3:4.

It is important to keep in mind that the `cost` command defines the costs that will be in effect for the characters listed *in case that they are to be optimized as step-matrix*, but they have to be explicitly made step-matrix with the `ccode` command if you want the costs to be applied during optimization:

```
tnt*> cost . =?/? 3 R/Y 4; ccode(.;
```

If you make all characters step-matrix without having specified any transformation costs, they will all be unity, i.e. they will be optimized as nonadditive. The transformation costs for each character, in other words, are saved and kept, regardless of whether the character is additive, nonadditive, or step-matrix; only when making a character step-matrix with `ccode(` do the differential transformation costs take place.

The `cost` command also allows outputting the current set of transformation costs. With no arguments (i.e. `cost;`), the costs are displayed for all the characters that have been made step-matrix with `ccode(`. The output produced in this case is readable by humans but not by the program itself; to make the output TNT readable (e.g. to save the current set of transformation costs to the log file for subsequent reading into TNT), use `cost*;`.

The second possible way to define character state transformation costs is using the `smatrix` command. This defines sets of transformation costs which are stored and can be subsequently applied *in toto* to different sets of characters; you can use this to facilitate trying several alternative schemes of transformation costs. The syntax for the `smatrix` command is:

```
tnt*> smatrix =N (name) [costs]
```

where `N` is the number of step-matrix (up to 32 different sets of costs can be defined) and `name` is the name (optional) for the set of transformation costs, delimited by parentheses. The number (and name, if given) are followed by the costs, in the same format as for the `cost` command. Once a matrix of costs with name `name` has been defined, it can be applied to any list `L` of characters:

```
tnt*> smatrix +name L; ccode (L;
```

Optionally, step-matrices can be referred to by their number, instead of their name. As with the `cost` command, this defines only the costs for the characters; to actually optimize the characters with those costs, a `ccode(` command is needed (as in the example prior). The step-matrices defined can be listed (with `smatrix;`) or saved to output file in a format that is TNT readable (with `smatrix*`).

The third command with which special transformation costs can be defined is `cstree`. This is primarily used to define transformation costs in the case where the relationships between states can be specified hierarchically, as a tree (hence the name of the command: character state tree). An example is the character state tree shown previously in Section 2.1, "The Great Chain of Characters":

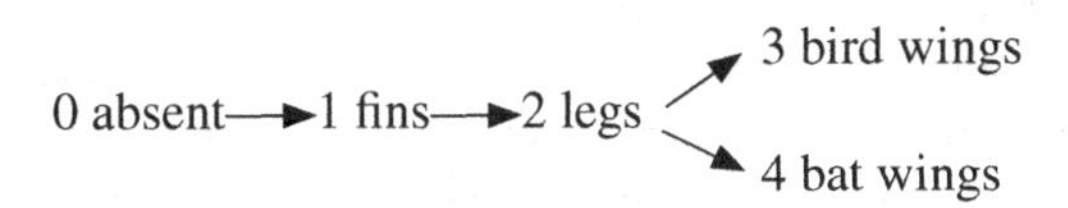

If the states are coded as 0–4, as indicated in the diagram, then the character state tree can be simply drawn with dashes, bars, and slashes with the `cstree` command:

```
cstree N =
    0-1-2-3
         \
          4
;
```

This set of transformation costs will be applied to character `N` (it can be a list of characters). The cost of transformation between any two states connected by a direct line is one step. Of course, the same scheme of costs can be represented in several different (but entirely equivalent) ways, e.g.:

```
cstree N =
      3
      |
    1-2
     \ \
      0 4
;
```

The costs of transforming between two states not directly connected is the length of the shortest path between the two states (e.g. three steps, between 0 and 4). As the `cstree` command needs to align the states (to establish what connects to what), make sure you edit the character state tree using a non-proportional font (e.g. `Courier`).

To establish multiple connections, you can use the symbol X (it can be lower case), which indicates a direct, single-step connection between its four tips:

```
cstree N =
   0 3-4
    X
   1 2
;
```

i.e. the cost of transformation between any two of 0–3 is 1, the cost between 3 and 4 is 1, and the cost between state 4 and any of 0–2 is 2; this set of transformation costs is not completely hierarchical, with some subsets of the costs (i.e. between states 0 and 3) corresponding to a nonadditive character, others (e.g. 2–3–4) to an additive one. The `cstree` command ends with a semicolon; a single character state tree can be specified per `cstree` command. When the scheme of transformations is fully hierarchical, it can be represented fully equivalently (when implied weighting is not in effect) with a set of binary variables (see Chapter 3, Section 3.4); by default, TNT will internally decompose fully hierarchical character state trees into binary variables, which can be processed much faster during searches and optimization; this is set with `cstree+`. This can produce differences under implied weights because every binary variable is (implicitly) weighted separately. That is, in the example prior, several transitions between fins and legs would not lower the (implied) weight of transitions between legs and wings; this may or may not be what you intend in your analysis. To have all character state trees always optimized without internal decomposition, as a single character (i.e. to have fin-leg transitions lowering the implied weight of leg-wing transitions, as well as all other transformations), use `cstree–`. Note that the option to not decompose characters that can be represented by binary variables applies only to character state trees, not to additive characters themselves (these are always decomposed). If you wish to perform implied weighting on additive characters without decomposing them into binary variables, you then have to specify that they have additive transformation costs with the `cstree` command (and apply the `cstree–` option).

All the options just discussed for setting transformation costs, in Windows versions, are also accessible from *Data > Character Settings*. Of course, it will usually be more practical to store the sets of costs in a file. If step-matrices have been defined with `smatrix` in the file, those will also be accessible (for application to specific sets of characters) via *Data > Character Settings*.

One of the restrictions enforced by TNT for transformation costs is that they have to obey the *triangle inequality*. This is, for any three states i, j, k, the cost $C_{i \to k}$ must be less than or equal to $C_{i \to j} + C_{j \to k}$. In words, the cost of transforming between two states via a third state cannot be *less* than the cost of the direct transformation. When the cost of the direct transformation is allowed to be lower, you can have cases where the optimal reconstruction for the ancestor of two species having an identical state has a different state at the common node. Consider as example the tree (0(0(0(0(0(22)), with all costs defined as unity, except $C_{0 \to 2} = 4$ (i.e. the direct transformation 0→2 costs more than 0→1→2). In that case, assigning state 2 to the common node of the two terminals with state 2 would have a cost of 4, while assigning state 1 would be cheaper (a change from 0→1 and two separate changes from 1→2, for a total cost of 3). Assigning a different state to a node leading to identical descendants is hardly a "parsimonious" conclusion, and in this case it results from the violation of the triangle inequality. When the triangle inequality is violated, TNT will issue a warning and fix the costs by changing some of the values. Different fixes may make a character obey the triangle inequality, and TNT will choose one arbitrarily—if you want to make sure the fix makes sense from your point of view, you will have to make the change yourself. In addition to "unparsimonious" character state reconstructions, violations of the triangle inequality also render the shortcuts used by TNT during searches and tree-length calculations inapplicable, and for this reason, running cases with violations of the triangle inequality is simply not allowed.

2.14.3 Deactivating Blocks of Data

In addition to its use to redefine data blocks (see Chapter 1), the `blocks` command can also be used to deactivate some data blocks with the = option:

```
tnt*> blocks = 1 5 8;
```

This will deactivate the taxa and characters not included in blocks 1, 5, and 8. Note that it will *not activate* taxa and characters that are included in the blocks; you have to activate them explicitly before executing the `blocks` command. A taxon will be considered to have been included in a block if it has a non-missing entry in one of the characters included in the block; if you have explicitly included the taxon in the block but as having only missing entries for the corresponding characters, the taxon will be considered as absent from the block anyway. Instead of using the block numbers, you can, of course, use the block names (if already defined with `cnames`, see Section 2.14.4). If the list of blocks is preceded by an ampersand (`&`), then the command will leave as active only those taxa that are included (i.e. shared) in all of the blocks listed; otherwise, only taxa absent from each of the specified blocks will be deactivated. The block deactivation can also be done, in Windows *GUI* versions, by selecting *Data > Character Groups/Blocks > Active Block(s)*. By using properly defined blocks, even if TNT does not allow reading multiple datasets, combining the individual datasets in a single multiblock dataset effectively allows handling them separately. Once some data blocks have been deactivated, the dataset behaves like a new dataset, so that you can easily explore different combinations of genes or character systems.

2.14.4 Character Names

In morphological matrices, it is highly advisable to name the characters and states. This greatly facilitates handling or editing the dataset and interpreting the results. The command used to name characters is `cnames`; it can only be executed after a dataset has been read. To define names, use the `{` symbol to define a new set of character names. If the data have been read as multiple blocks (see Chapter 1), square brackets instead of the braces are used to define names of blocks; block 0 cannot be named (it is always "all").

```
cnames
[1 morpho_block;
[2 DNA_block;
[3 Protein_block;
{1 charname1 state0 state1 state2 . . . stateN;
{2 charname2 state0 state1 state2 . . . stateN;
{3 charname3 state0 state1 state2 . . . stateN;
 . . .
;
```

The `{` or `[` symbol are followed by a character or block number, respectively. You can use `{+` to name the character subsequent to the last named character; after reading the dataset, the first instance of `{+` is interpreted to refer to the first character in the matrix, likewise for `[+`. Names must be single strings (i.e. no spaces; use underscores to emulate blanks within block, character, or state names). It is possible to leave some character(s), block(s) or state(s) unnamed or to name more state(s) than actually present in the matrix; name as much as you need. The list of names ends with a semicolon; execution of the `cnames` command ends with another semicolon (the `cnames` command thus differs from most other commands, which parse nothing beyond the first semicolon; `cnames` instead needs *two* semicolons for termination).

To change the name of previously named characters, which can be done only one character at a time, use `cnames+N` (where `N` is the name or number of the character to rename), followed by the new character and state name(s), terminated with a semicolon. Other options of `cnames` allow saving the names in a TNT-readable format (`cnames*;`) or saving only the active characters (so that the character numbers correspond to those of a matrix where inactive characters have been eliminated, with `cnames!;`). The options `cnames=` and `cnames–`control whether TNT outputs character and state names or numbers, respectively.

In addition to naming blocks with the `cnames` command, blocks can be named as they are read with the `xread` command. Remember (Chapter 1) that blocks of data in the `xread` command are separated by ampersands (`&`), possibly followed by additional options for the block within square brackets. One of the options that can be indicated within square brackets is the name of the block, preceded by a slash; thus, elaborating on the example dataset (`filelist.tnt`, see Chapter 1, Section 1.11.10) where the morphological and sequence blocks were kept in separate files:

```
macro=; ¶
xread ¶
'Reading data from different files'¶
14 9  ¶
& [ num /morphological ] @@ file_a.txt data ; ¶
& [ dna /DNA_COI ] @@ file_b.txt data ; ¶
; ¶
proc/; ¶
```

2.14.5 Taxon Settings and Taxonomic Information

Various taxon options are set with the commands `outgroup`, `taxcode`, `taxlabel`, and `taxonomy`. Taxon activation and deactivation are done with the `taxcode` command; all taxa are by default active. Any taxon but the currently designated outgroup taxon can be deactivated (at least four taxa must remain active). The outgroup taxon (designated with `outgroup N`, where `N` is a taxon name or number) is used to root the trees (by default, taxon 0; `outgroup;` simply reports currently designated outgroup). Inactive taxa are not included in tree searches. Trees stored in memory by TNT can have any taxon subset, but to be used as starting points for searches (e.g. branch-swapping), they need to have sets of taxa identical to the set of active taxa (non-matching trees are discarded when beginning the search). `Taxlabel`, if followed by a number, reports the corresponding taxon name and, if followed by a string, reports the numbers of the taxa matching the string.

The dataset can include taxonomic information in the taxon names (see e.g. Goloboff et al. 2009; Goloboff and Catalano 2012). This is very useful to check the results obtained (e.g. in analyses with different options) against an existing hypothesis of relationships. Multiple or individual taxonomic groups can be shown on a tree, with different colors indicating each group. Alternatively, TNT can report the degree to which the tree displays a group present in the reference tree by checking the taxon composition. For a group in the reference tree (say, a family) not recovered exactly in the tree *T* being checked, TNT then looks for the closest group, finding the group in the tree *T* for which the number of non-members of the family included in the group plus the number of members of the family *not* included in the group, divided by the number of members of the family included in the group, is minimum (it is possible to weight inclusion of non-members or exclusion of members differentially, but default weight is unity; see online help of `prunmajor` and `rfreqs` commands for details on how to change these weights). The rationale for checking group recovery in this way is that having to rearrange a single terminal taxon to perfectly retrieve a reference group with a total of three taxa is very different from having to rearrange a single terminal taxon to perfectly retrieve a group with hundreds of taxa—the group is recovered almost perfectly in the latter case but not in the former (Goloboff et al. 2009; this is the same idea subsequently expressed in Lemoine et al.'s 2018 modification of the bootstrap). Once the group closest to the reference has been identified, TNT can produce a quantification and a verbose report of the taxa that do not fit in the reference taxonomy. For example, in the analysis of Goloboff et al. (2009), the mammal family Soricidae and the subfamiliy Crocidurinae (shrews) are represented by 130 and 60 taxa respectively (from a total of 2,401 mammals). The best tree found in the analysis of Goloboff et al. (2009), tree 0 in the example following, does not display Soricidae or Crocidurinae as monophyletic groups, but it is close. The report is produced with the `&` option of the `taxonomy` command:

```
tnt*> taxonomy &0 /Soricidae /Crocidurinae;
```

Which produces as output:

```
Soricidae (130 taxa):
  Tree 0: 1 taxa added, 1 subtracted
```

```
    subtracted:
      Crocidura_paradoxura
    added:
      Uropsilus_sp.
Crocidurinae (60 taxa):
  Tree 0: 0 taxa added, 7 subtracted
    subtracted:
      Crocidura_foetida,
      Crocidura_mindorus,
      Crocidura_musseri,
      Crocidura_paradoxura,
      Myosorex_cafer,
      Myosorex_sclateri,
      Myosorex_varius
```

For a more verbose description, indicating in words the location of the misplaced taxa, use instead the ! option of the `taxonomy` command:

```
tnt*> taxonomy!0 /Soricidae /Crocidurinae;
```

which produces as output:

```
Soricidae (130 taxa):
 Removed from Soricidae (1):
  Crocidura_paradoxura goes outside
   * Goes out of Laurasiatheria, within Afrotheria,
     with Tenrecidae,
 Non-Soricidae added to group (1):
  Uropsilus_sp. gets inside (from Talpidae):
   * Goes with Soricinae,
Crocidurinae (60 taxa):
 Removed from Crocidurinae (7):
  Crocidura_foetida goes outside
   * Goes within Sorex, with Sorex_gaspensis,
    Sorex_ornatus, Sorex_thompsoni, Sorex_ugyunak,
 Paraphyletic group of Crocidurinae:
  various Crocidurinae
   * Paraphyletic in terms of Uropsilus_sp., Soricinae,
   * Goes with Talpidae,
 Crocidura_paradoxura goes outside
   * Goes out of Laurasiatheria, within Afrotheria,
   with Tenrecidae,
```

TNT also offers the possibility of coloring the branches of a tree, representing the taxonomic groups with different colors, and of labeling the branches of the tree automatically, indicating with a legend which is the group closest to a reference group in the taxonomy incorporated in the taxon names.

There are two ways to incorporate the taxonomic information in the taxon names. In both cases, prior to reading the data (with the `xread` command), it is necessary to enable the use of taxonomic information (with `taxonomy=;`) and longer taxon names so that they can contain the full taxonomic hierarchy of each terminal (with `taxname+N;` where `N` is the maximum number of characters in taxon names). For most hierarchies, names of 200 or 300 letters should suffice, but this depends, of course, on how much information you want to have in the names. The first way to incorporate taxonomic information is in the `xread` command itself; the list of taxonomic categories should be included for each terminal taxon, after an `@` symbol, with categories separated by an underscore. From the list of categories in each taxon name, TNT reconstructs the taxonomic hierarchy for the taxa in the matrix. As example, for the following list of taxon names:

```
Ambystoma_mexicanum_@Amphibia_Caudata_Ambystomatidae
Rana_boylii_@Amphibia_Anura_Ranidae
Gallus_gallus_@Archosauria_Theropoda_Aves_Galliformes_Phasianidae
Columba_livia_@Archosauria_Theropoda_Aves_Columbiformes_Columbidae
Mus_musculus_@Mammalia_Eutheria_Euarchontoglires_Rodentia_Muridae
Rattus_norvegicus_@Mammalia_Eutheria_Euarchontoglires_Rodentia_Muridae
Canis_lupus_@Mammalia_Eutheria_Laurasiatheria_Carnivora_Canidae
Homo_sapiens_@Mammalia_Eutheria_Euarchontoglires_Primates_Hominidae
Felis_catus_@Mammalia_Eutheria_Laurasiatheria_Carnivora_Felidae
```

TNT reconstructs the taxonomic hierarchy (displayed as an indented list with the `taxonomy/;` command, indicating in parentheses the number of terminal taxa within each group):

```
Mammalia,Eutheria (5)
   Laurasiatheria,Carnivora (2)
      Felidae (1)
      Canidae (1)
   Euarchontoglires (3)
      Primates,Hominidae (1)
      Rodentia,Muridae (2)
Archosauria,Theropoda,Aves (2)
   Columbiformes,Columbidae (1)
   Galliformes,Phasianidae (1)
Amphibia (2)
   Anura,Ranidae (1)
   Caudata,Ambystomatidae (1)
```

Then, any one of those taxonomic groups can be checked on a tree held in memory by TNT. Some categories are synonyms for the taxon selection in the example (e.g. Columbiformes = Columbidae). If you wish to contract the taxonomy, preserving only the highest-level category, use `taxonomy]`. Another convention used by TNT is that, when matching names for establishing taxon identity in multiple blocks, the part of the name that is beyond a quadruple underscore (i.e. ____) is ignored. Using this, it is possible to include the full taxonomy in the name of a taxon only in the first block where it occurs, using just the name (i.e. the part before the quadruple underscore) in subsequent blocks. Goloboff et al. (2009) also used this to include information on GenBank accession numbers in the names so that, for blocks representing different genes, it is possible to reconstruct the individual GenBank entry for each gene and taxon from the data files themselves. Further elaborating on the `filelist.tnt` example prior,

```
macro=; ¶
xread ¶
'Reading data from different files'¶
14 9  ¶
& [ num /morphological ] @@ file_a.txt data ; ¶
& [ dna /DNA_COI ] @@ file_b.txt data ; ¶
& [ dna /DNA_CytB ] @@ file_c.txt data ; ¶
; ¶
proc/; ¶
```

You can name a taxon with its taxonomic information only in the first block (`file_a.txt`):

```
Ornithorhynchus____@Mammalia_Monotremata_Ornithorynchidae
```

If the COI sequences have been taken from (say) GenBank accession number XXXX, then you can simply name the taxon in `file_b.txt` as:

```
Ornithorhynchus____XXXX
```

And in the third file, `file_c.txt`, you add the GenBank accession number of the CytB sequence (YYYY):

```
Ornithorhynchus____YYYY
```

In this way, the study can be fully self-documented, without the need to keep separate files with lists of the sources for each of the sequences.

For easily including information about taxonomy and data sources in this way, it is also possible to use the Gb-to-TNT program (Goloboff and Catalano 2012) for processing GenBank files. GenBank files include taxonomic information for each taxon, and Gb-to-TNT reads and processes this information, creating the TNT dataset automatically. Morphological information can be subsequently added to such a dataset, taking advantage of the fact that the categories are already named (for example, setting appropriate entries with the `xread=` option, as discussed in Chapter 1).

The second way to include taxonomic information in the dataset is by reading a parenthetical tree with labeled branches, then transferring the labels onto a taxonomy. The labels (=tags) are indicated in the parenthetical tree, after each group or terminal, preceded by an equal sign. Note that you still need to define `taxonomy=` and `taxname+N` before reading the dataset (i.e. prior to the `xread` command). For the example prior, the tree would be:

```
tread
((Ambystoma_mexicanum =Caudata Rana_boylii =Anura )=Amphibia
(Gallus_gallus =Galliformes Columba_livia =Columbiformes
```

```
)=Archosauria ((Homo_sapiens =Primates (Mus_musculus Rattus_
norvegicus )=Rodentia )=Euarchontoglires (Canis_lupus =Canidae
Felis_catus =Felidae )=Laurasiatheria )=Mammalia ) ;
```

In the Windows versions, the tree-tags can also be edited by first unlocking trees (so that editing is enabled, clicking on the padlock tool), then displaying the tree (with *Trees > Display/Save*) and double right-clicking on the node to be labeled. Once all the appropriate groups have been labeled, pressing T saves the tags for subsequent use. These tags can then be saved in parenthetical notation (as indicated in Chapter 1) for subsequent input into TNT. With the tags defined, typing `macro=;` (which enables use of some special commands, see Chapter 10) and then `copytree[;` will incorporate the hierarchy implicit in the tree labels as a reference taxonomy, changing the taxon names as well.

Once the taxon names have incorporated taxonomic information (or any other data, e.g. beyond a quadruple underscore), the taxon names can become exceedingly long, and displays (e.g. trees or taxon lists) with lists of taxa can be annoying. If you wish to truncate the taxon names (on output only), you can use the `taxname[` option. With `taxname[N`, taxon names are truncated to no more than `N` letters. With `taxname[0` (or with `taxname]`), taxon names are not truncated. To output only the taxon name itself (i.e. what precedes the `@` symbol which indicates the beginning of the taxonomic hierarchy or what precedes the quadruple underscore if present in the taxon name), use `taxname[!`.

2.14.6 Comparing and Merging Datasets

Many studies present modifications of previous datasets. These modifications may include new characters, changes in scorings, or (more rarely) elimination of previously used characters. The conclusions supported by the newer dataset usually differ from those supported by the earlier one, but the matrix differences causing the alternative results are very rarely discussed (Sereno 2009). Some of the changes in the matrix may be irrelevant to the differences between the resulting trees, while others may be crucial. This is less than ideal; deciding between the conclusions supported by the earlier and the newer dataset would be greatly facilitated by concentrating on the most relevant differences in characters and scorings, instead of having to consider every character and entry differing in the two matrices. TNT includes a command, `dcomp`, which allows making this type of comparison. The same functionality can be accessed, in Windows versions, from *File > Merge/Import Data > Pairwise Comparison/Combination*. The details of the comparisons allowed by `dcomp` are discussed by Goloboff and Sereno (2021); only the basics are discussed here.

The `dcomp` command requires at least three arguments: the name of the output file where to write the combined dataset merging both input datasets (by default, in case of differences between input datasets, the option in the newer dataset is chosen; the command can be set to ask for the action to take in the case of specific differences), the name of the newer dataset, and the name of the older dataset. The simplest function of the `dcomp` command is a simple comparison of two matrices, producing a report of the differences in taxon composition, differences in characters used, and (for shared taxa and characters) differences in individual matrix cells. TNT matches taxon, character, and state names with a string-matching algorithm (Needleman and Wunsch 1970), finding the most similar pairs for each category. For this, it is important to use taxon, character, and state names which are as consistent as possible (e.g derived from formal ontologies). Alternatively, specific taxon and character synonymies between the two datasets can be indicated manually to force a match even if the names are very different or to force a non-match between names with too similar a spelling. The differences between datasets are displayed in a color diagram (saved by TNT in *SVG* format to a file named `matrix_differences.svg`), where the differences between the two matrices are indicated using colors. Taxa, characters, or scorings exclusive to the newer dataset are indicated in red, those for the older in blue, and the shared options in black.

Another function of the `dcomp` command is evaluating the individual differences in the matrices that are responsible for the differences in results. For this, each of the individual data files must contain a tree (or set of trees) optimal for the dataset in question. These trees must be provided by the user, ideally resulting from a search for each of the individual datasets; TNT does not check the optimality of the trees contained in each of the files. Each of the two (sets of) trees must be included in its own data file (after the `xread`), using the taxon numbers (or names) corresponding to the dataset in question. If these trees are provided in each dataset, TNT then compares the fit for every character differing in the two matrices. This allows an approximate calculation of the individual differences responsible for the differences in results. TNT reports the characters that, for the groups found only in the newer result, constitute synapomorphies when scored as in the newer matrix but not when scored as in the earlier matrix. It also reports the characters that have fewer steps on the newer results when scored as in the newer matrix but the same or higher numbers of steps when scored as in the earlier matrix. The opposite calculations are done as well (i.e. inverting earlier and newer data and results). Those characters are (in part at least), responsible for the different groups displayed in the earlier and newer results. Finding the individual matrix cells responsible for differences in results, or *critical cells* (as Goloboff and Sereno 2021 called them), is more involved. For that, TNT converts the older matrix into the new one in a stepwise fashion (by adding or removing the unshared characters or taxa and converting differing cells for the shared characters or taxa). Changes are done one at a time and in different combinations of individual cells, checking two conditions. First, for each group present in the newer but not the earlier results, those (combinations of) changes from the newer back to the earlier scoring which make differing synapomorphies for each group to not differ any more are marked as critically in favor of the newer results in the newer dataset; earlier and newer dataset and results are then switched and checked again to find cells critically in favor of the earlier results in the earlier dataset.[7] Second, those (combinations of) changes which make a character increase their step difference in favor of the newer results are marked as critically in favor of the newer results in the newer dataset; earlier and newer dataset and results are then switched to find cells critically in favor of the earlier results in the earlier dataset. The combinations to try are limited based on a mapping of the character, making only those changes likely to be connected. Roughly speaking, these are terminal nodes connected through internal nodes mapped as having specific character states (see Goloboff and Sereno 2021 for details). The individual critical cells are also shown graphically, in the `matrix_differences.svg` file, surrounding them with red ellipses (when the critical scoring is in the newer results) or blue ellipses (when the critical scoring is in the older results).

The previous description applies to differences in optimal trees; the `dcomp` command can also (with the `&` option) calculate whether changes in scoring (for entire characters) affect the values of support for the different groups (using the combined Bremer support, Goloboff 2014b, estimated via TBR [tree-bisection reconnection]). The combined Bremer support depends on both the absolute and relative Bremer supports and in simple cases produces evaluations of support most similar to those of resampling (see Chapter 8, Section 8.2.1.2); it is reported as a percentage between 0% and 100%. TNT checks whether changing the scoring of a character, or adding or removing the character entirely (in the case of unshared characters), affects the values of support for each node, and it reports all differences above a specific threshold (5% is the default). The differences are displayed in a diagram of the tree for the newer dataset (or its consensus; it is possible to exclude some taxa from consideration when calculating consensuses; these are listed in a file, given as a fourth file name to the `dcomp` command, after the name of the older dataset). The diagram is saved to a file called `bremer_differences.svg`, where the taxa or characters exclusive to one of the datasets are indicated with colors (red for the newer, blue for the earlier). A difference in combined support preceded by a plus sign indicates that the support for the group increases when switching from the older to the newer dataset, and a difference preceded by a minus sign indicates the opposite. For every group, two differences are indicated for characters, corresponding to the case where the character is changed (or added or removed) from the earlier to the newer file first and the case where the character is changed last. It may well be that a character decreases the support of a group when changed first and increases it when changed last, or the opposite—this indicates that whether the character supports or contradicts the groups depends on its interactions with other characters in the matrix. The effect of taxa on the support is always checked by adding (or removing) the taxa in question first (i.e. with none of the other taxa differing in the two datasets added or removed). Alternatively, the differences in combined Bremer support can be displayed in the form of a table (using the `dcomp|` option).

NOTES

1 Borkent (2018: 110) emphasizes that he started learning about cladistics in the 70s, perhaps implying that once one has classic training, one can avoid the confusion inherent to newer methods. If that is what he implies, my response is that *some* people trained under traditional views can change and consider newer developments as better justified than traditional views. I also started working in taxonomy in the late 70s (publishing my first paper, Goloboff 1982, only four years after Borkent published his; Borkent 1978). Despite having begun with a classical view of taxonomy, I have evolved.

2 Obviously, I do not mean to imply that those who defend homology by special knowledge are dishonest. They have just not fully followed all the ramifications of their position.

3 That is, when Q. Wheeler still believed parsimony to be a useful phylogenetic criterion. He subsequently seems to have embraced three-taxon statements (e.g. see refs. in Section 2.4, "Homology by Special Knowledge?").
4 Who are unlikely to be reading this introductory book.
5 Of course, I am aware that milk secretions are not *entirely* independent from bones; e.g. bones are a reservoir for calcium and other minerals contained in milk. Yet the main reason for the perfect nestedness of the two characters is historical and phylogenetic.
6 Slowinski (1993) was occasionally cited (e.g. Prendini 2000: 8) as recommending a preference for nonadditive characters, which illustrates to what extent authors can be misread and cited in support of positions they actually never took. Slowinski (1993: 163) in fact stated that he preferred the additive treatment when this is "clearly superior . . . (e.g., as with the morphocline small–medium–large)", using nonadditive characters otherwise. He also stated that one could use the transformation series "that implies the least amount of change between states".
7 Conversely, some changes to the earlier matrix may make the newer matrix favor the earlier results. Although those could be calculated in the same way, TNT does not calculate them (mainly because the graphic display of all those alternatives would be overly complex).

3 Character Optimization
Evaluation of Trees and Inference of Ancestral States

After laying down basic principles, this chapter moves onto more technical questions. The first issue that needs to be tackled involves problems related to evaluating trees. Calculating the number of steps required by a tree for a given character requires that we select, from among all possible reconstructions for the character, the best ones—this could be simply done by brute force but would be enormously time consuming. In a seminal paper proposing algorithms for quickly evaluating trees, Farris (1970) referred to these as algorithms for "optimizing" ancestral reconstructions, and the expression "character optimization" was subsequently adopted in phylogenetics as a synonym of methods for tree evaluation and character state reconstruction (the term *optimization* actually has a much wider meaning in computer science referring to the maximization or minimization of any quantity). This chapter discusses methods and algorithms to find optimal reconstructions much faster than brute force, as well as several subtleties and implications of using most parsimonious reconstructions to evaluate trees and produce diagnoses of groups. All algorithms are implemented in existing computer programs, so there is seldom a need to apply them by hand. Discussion of these algorithms, however, demystifies phylogeny programs and helps explain why phylogenetic analysis can be so computationally intensive. Some knowledge of the technical details of character optimization also facilitates understanding many of the properties of most parsimonious trees (e.g. invariance under rooting, impossibility of a polytomy having a better fit than any of its resolutions, etc.). Together with example scripts on the TNT web site, this discussion can also help users interested in doing some programming of their own.

3.1 FINDING OPTIMAL ANCESTRAL RECONSTRUCTIONS

Before any further discussion, it must be noted that the optimal ancestral reconstructions for a given tree can be defined only by reference to a specific set of transformation costs between the possible states. These transformation costs should be defined, ideally, on the basis of specific observational criteria, such as the relative morphological closeness between states. Once the transformation costs between possible states are fixed, the optimal ancestral reconstructions follow, regardless of the method used for finding or calculating them. That is, a given set of ancestral assignments either is or is not the set that actually produces the minimum number

DOI: 10.1201/9781003220084-3

of ancestor-descendant differences along all the branches of the tree; this is wholly independent of how the ancestral assignments have been made. This cannot be emphasized enough because it is common to see textbooks speaking of the algorithms for character optimization as "rules", as if those algorithms involved some arbitrary decisions or conventionalism. The only decisions and conventionalism involved in finding optimal ancestral reconstructions are in the decision to isolate the observed variation in a series of discrete characters and/or states (a reductionist but unavoidable step for any phylogenetic analysis) and in the criteria to decide costs of transformations between states (which can have an empirical basis, as discussed in Chapter 2).

The simplest method for finding optimal reconstructions consists of generating all possible reconstructions, counting the steps for each, and selecting the best. In computer implementations, the reconstructions can be easily generated recursively, as shown in Box 3.1 (see also example scripts ***worstoptalg.pic*** and ***badoptalg.pic***). This is very time consuming: it requires trying each of the s states for each of the $T - 1$ nodes of a tree with T taxa, for a total of s^{T-1} reconstructions; in Figure 3.1, for two states and a tree with 6 taxa, there are 32 possible reconstructions. The time needed increases quickly; if finding the optimal reconstruction for a tree with T taxa and 4 possible states takes 1 second, finding the optimal reconstruction for a tree with only an additional 5 taxa would take over 15 minutes.[1] Regardless of the time used, however, the exhaustive enumeration gives perfectly correct results.

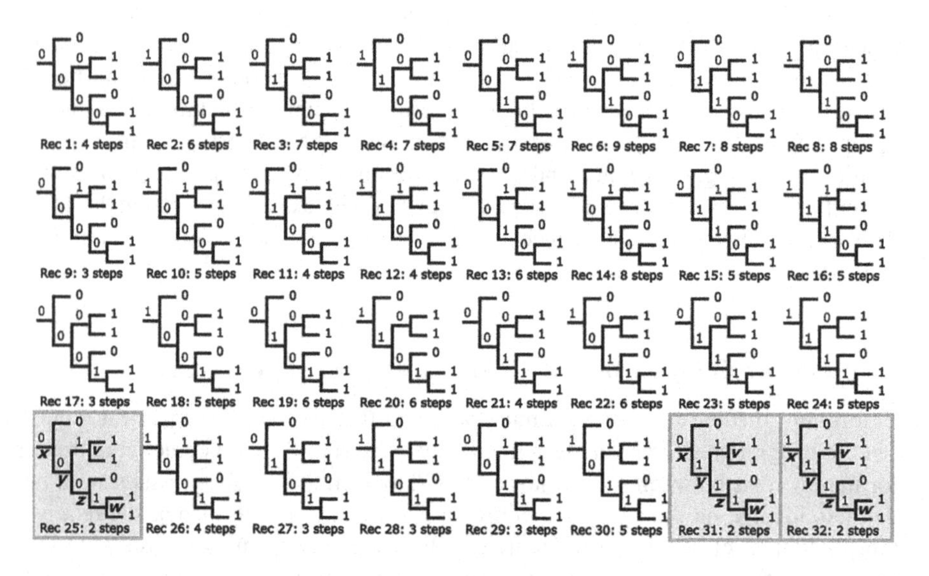

FIGURE 3.1 For a given tree and character state observation in terminal taxa, the most parsimonious reconstruction can always be obtained by exhaustively enumerating all possible ancestral state assignments and simply selecting those that minimize the score. Such a process, although extremely slow, would always produce the correct results.

3.2 GENERALIZED OPTIMIZATION: SIMPLE CASES

The previous discussion shows that the problem of finding optimal ancestral reconstructions must be solved with better (=faster) algorithms. In designing such algorithms, three important properties of most parsimonious reconstructions must be considered. The first is the most obvious one: which state is optimal at a given node depends on the states observed for the terminals descended from that node. Consider the simplest possible case, a tree with two taxa, A and B. The state that can be most parsimoniously assigned to the common ancestor x of A and B will be that which minimizes the difference along the branches leading to A and B. If the descendants have two different states (e.g. A = 0 and B = 1), then assigning x state 0 will imply a step along the branch leading to B and state 1 a step along the branch leading to A. Both states, 0 and 1, are equally parsimonious at x if all the available information is that A = 0 and B = 1; as more than a state can be parsimoniously assigned to a node, the states are indicated as sets, {01}. A step must occur in the subtree descended from node x; it could be on one branch or another, but there has to be a step *somewhere*.

The second property is less obvious: which state is optimal at an internal node x depends not only on the terminals included in the group that descends from x but also on the states observed for the terminals *outside* of group x. An example is in Figure 3.2a, where node x could equally well be assigned states 0 or 1 if considering only its descendants, but different states depending on what has been observed for terminal C, the sister taxon of x. If C has state 0, then it is clear that x should also be assigned state 0 and vice versa. It is as if taxon C, instead of being the sister group of x, were pulled down to be the ancestor of x (as shown in Figure 3.2a); in that case, the optimal state for x will be one that minimizes change along the three branches (i.e. the one coming from C and the two leading to A and B). Both properties imply that the optimal assignment for a node is a *global* instead of merely a local problem, simultaneously depending on all the taxa in the tree.

The third property is the one that allows designing efficient algorithms and is the fact that the determination of globally optimal states can be decomposed into a series of local subproblems.[2] This is shown in Figure 3.2b, where there is another taxon D, outside of group ABC, with common ancestor y between D and x. Locally optimal state assignments are those decided on the basis of the descendants of only the node. They can be determined by doing a pass down the tree in such a way that no node is visited unless its descendants are terminals (for which the states are data) or have been visited already (so that some locally optimal states have been assigned). In computer science, such a way to traverse the tree is called a *post-order*. The locally optimal state assignments decided for a node during a down pass are preliminary, in the sense that they may have to be modified by subsequent reference to terminals outside of the group. The down pass determines the minimum number of steps that must occur within the subtree descended from each node; when the root is reached, the number of steps is the total length of the tree. During the down pass, only the immediate descendants of a node must be considered, and if a descendant is an internal node, its preliminary states summarize the terminals included in that node. For example, to determine preliminary states for node y in Figure 3.2b (white arrows), the (observed) states of terminal C and the (preliminary) states of node x must be

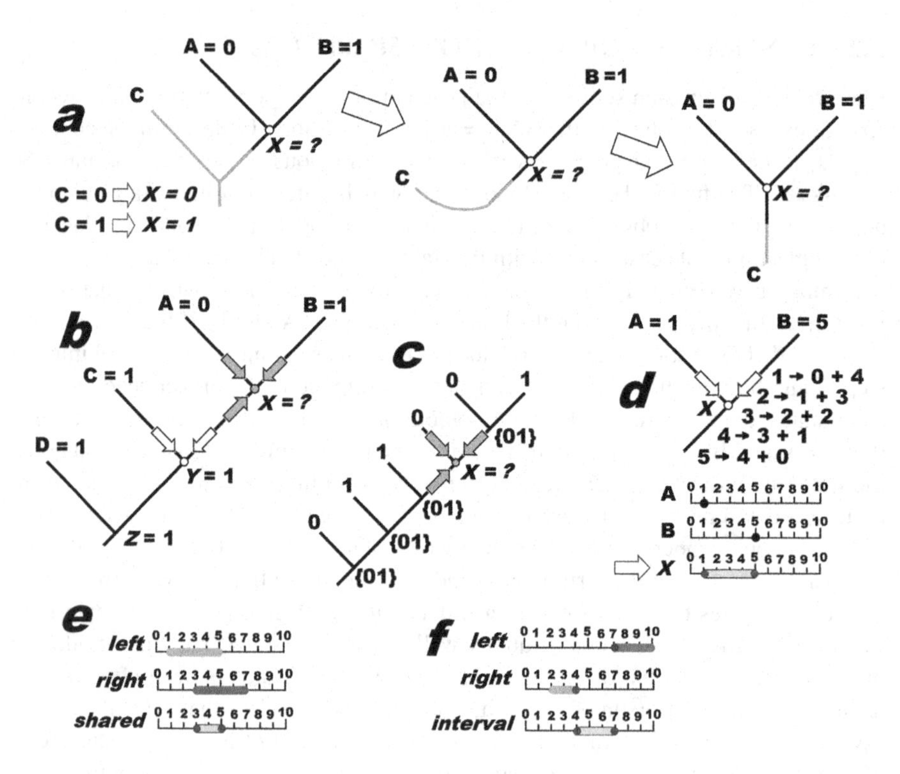

FIGURE 3.2 (**a**) The optimal ancestral states for a node depend not only on the states in taxa descended from the node (A, B) but also on the states in taxa outside of the group (C). (**b**) In a down pass, the locally optimal state(s) for each node can be determined from its immediate descendants, with internal nodes tractable in the same way as terminals. (**c**) In an up pass, the locally optimal states for a given node (e.g. x) can be determined from the ancestral and descendant states of the node. (**d**) In a lineal additive character, the cost of transforming between two states equals their numerical difference so that assigning any value in the interval between the two descendant states produces the minimum local cost. (**e**) When the two descendant intervals overlap, the overlapping range minimizes the local cost. (**f**) When the two descendant intervals do not overlap, the range between the largest value of lower interval and the smallest value of upper interval produces the minimum local cost.

considered, always choosing the states that minimize change along the two descendant branches. That the states of x have been synthesized from its descendants makes no difference at the time of determining optimal preliminary states for y; the set of states in the internal node x descended from y is treated in exactly the same way as if x was a terminal instead of an internal node.

The states determined during the down pass for the root of the tree are final: no further taxa can modify those assignments—the tree contains no further taxa to be considered. Once the down pass reaches the root of the tree, an up pass (called a *pre-order* traversal in computer science, one in which no node is visited unless its ancestor has been visited already) allows finding the final states for every internal node other than the root. In the up pass, the states of the immediate ancestor for each node are used just as we used taxon C when pulling it down in Figure 3.2a. For example, for node x, the globally optimal assignment is that which minimizes change along the three branches (i.e. the one coming from y and the two leading to A and B), marked with gray arrows in Figure 3.2b. The final states of ancestor y, when deciding final states for x, carry over information from the rest of the tree, thus allowing the assignment to x to be made on the basis of the global state distribution in the tree. The only proviso needed when doing the up pass is that each of the optimal states in the ancestor needs to be tried separately; thus, every state in x that minimizes change in the three branches relative to any one of the optimal states in the ancestor y must be added to the final state set x. This is shown in Figure 3.2c, where the optimal states for x must be determined from ancestor {01} and descendants {0} and {01}. One might be tempted to conclude that assigning state 0 to x implies no steps along any of the branches (given that $\{01\} \cap \{0\} \cap \{01\} = \{0\}$). But the states assigned to the ancestor of x are final, which means that there is at least one reconstruction where every one of those states occurs at that node. When the ancestor of x is assigned state 1, then state 0 at x implies one step (at the ancestral branch), and so does state 1 (at the left descendant branch). Both 0 and 1 are optimal at x when its ancestor is assigned 1, and then both states must be added to the set of most parsimonious states for x in Figure 3.2c.

To formalize this process more explicitly, let the node examined at every step (or its state set) be indicated with ***n*** (using upper case, ***N***, for the final states and lower case, ***n***, for the preliminary states), the ancestor (or its state set) ***Anc***, and the left and right descendants (or their state sets) ***L*** and ***R***, respectively. The two passes can then be defined as:

1) *Down pass*: Try every possible state at node ***n***, calculating the cost of transforming into each of the most parsimonious preliminary states of descendants. Those combinations with minimum cost are locally parsimonious. The steps increase with minimum cost at each node.
2) *Up pass*: Initialize ***N*** as an empty set, and for each of the states in ***Anc***, add to ***N*** every state ***s*** that minimizes the sum of costs from ***Anc*** to ***s*** and from ***s*** to each of the descendants, ***L*** and ***R***.

The two-pass process just described is illustrated, with a worked-out example, in Figure 3.3a. The transformation costs in that example are as in an additive character—where the cost of transforming between two states equals the numerical difference between the states. The down pass is shown above branches, with the lowest cost of transforming into any of the states in the first (upper) descendant as the first of the two terms of the summation, and the lowest of transforming into any of the states in the second (lower) descendant as the second term; the states with lowest sum (shown in darker color) are optimal during the down pass. For the up pass (shown below branches,) each one of the optimal states in the ancestor is tried separately (third term of the summation in this case is the cost of transforming from the state selected in the ancestor), and the best combinations (final states) are marked in black. The script ***generalopt.pic*** exemplifies generalized optimization. This method is many orders of

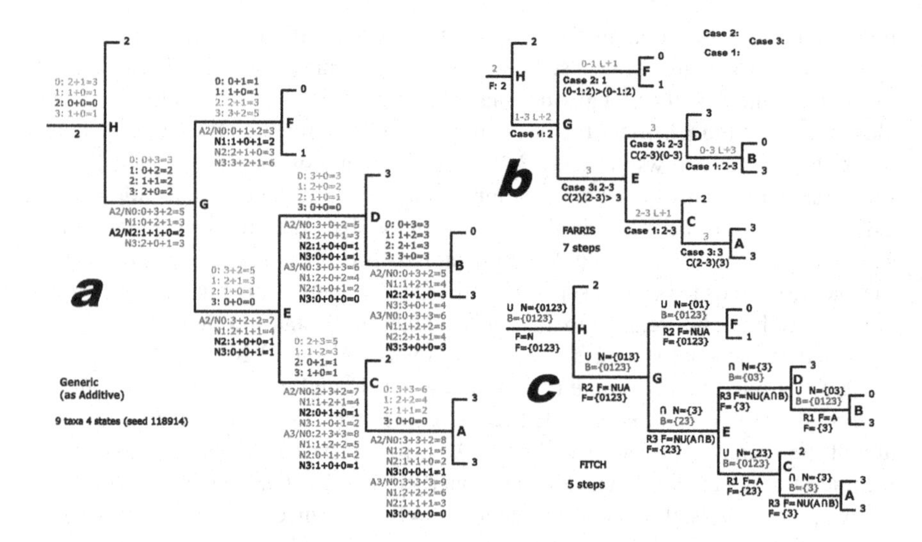

FIGURE 3.3 (**a**) A worked out example of generalized optimization, with additive costs. Values above branches indicate costs during down pass, and values below indicate costs during final pass. Darker values indicate optimal choices in the given context. (**b**) Farris optimization (i.e. using intervals) for the same character and tree as in (a). (**c**) The same character and tree but nonadditive transformation, with state sets resulting from Fitch optimization.

magnitude faster than a brute-force enumeration of all reconstructions and increases lineally with numbers of taxa. In a worst-case scenario (depending on implementation details), the two-pass method will require comparing each of the *s* states for the two descendants of each of the $T - 1$ internal nodes, for a time complexity of $O(Ts^2)$ per character. If adding only 5 taxa to a 100-taxon tree in the brute-force algorithm would make calculations 1,000+ times slower (as discussed prior), for the two-pass method, the increase in time would be only 5%.

The two-pass generalized optimization works for characters where the state transformation costs are simple and symmetric—additive (branched or lineal) and nonadditive characters. While much faster algorithms (described in subsequent sections) can be used for such schemes of transformation costs, the generalized optimization is useful for understanding the general principles of any two-pass optimization algorithm. When the transformation costs are different from additive or nonadditive (e.g. asymmetric), the process just described can produce incorrect results because it determines only the set of optimal states during the down pass. In that case, more elaborate algorithms are required (see "Step-Matrix Optimization," Section 3.5) to also take into account suboptimal state assignments.

3.3 OPTIMIZATION FOR NONADDITIVE CHARACTERS: FITCH'S (1971) METHOD

When a state can transform into any other one with the same cost (nonadditive characters), an algorithm by Fitch (1971) can be applied. For nonadditive characters, it is possible to express all the necessary operations in terms of set operations, which makes it easy to implement these algorithms in computer programs (see the following). For the down pass, Fitch (1971) noted that whenever there are states shared between the sets ***L*** and ***R*** (left and right descendant) of a node, using for ***n*** (the preliminary state set for the node) a set equal to the shared states will incur no steps along any of the branches, and when no states are shared, the states found in any of the descendants will require a step (along one of the branches). Every time ***n*** is formed by the union of ***L*** and ***R***, the number of steps is increased.

For the up pass, whenever the states in ***Anc*** are a subset of the states in ***n***, restricting ***N*** (the final state set) to those states (i.e. making ***N*** = ***Anc***) will ensure that minimum length occurs—whichever state is selected for the ***Anc***, the same state in ***N*** will imply no steps along the lower branch. Thus, when ***Anc*** ∩ ***n*** = ***Anc***, make ***N*** = ***Anc*** (this is case I of Fitch 1971). Otherwise, if ***n*** has been formed by a union, any state that is present in ***n*** and not in ***Anc*** will imply two steps (one along the branch leading to ***n***, the other to one of the descendants), which is the same number of steps implied by any state present in ***Anc*** but not in ***n*** (which will imply a step along each of the descendant branches, saving a step in the ancestral branch). Another way to view the situation when ***n*** is formed by a union is to consider e.g. the case of descendants with states 1 and 2, coming from an ancestor (and successive sister groups) with state 0; in that case, each of the possible transformations 0→1→2, 0→2→1 or independent transformations 0→1 and 0→2 are all equally parsimonious. Thus, when ***Anc*** ∩ ***n*** ≠ ***Anc*** and ***n*** = ***L*** ∪ ***R***, ***N*** = ***n*** ∪ ***Anc*** (this is what Fitch 1971 called "case IV"). Otherwise (i.e. when ***Anc*** ∩ ***n*** ≠ ***Anc*** and ***n*** = ***L*** ∩ ***R***), when choosing for ***Anc*** a state ***s*** absent from ***n*** (if any exists), no steps are required at the descendant branches for any of the states present in ***n***, but one is needed along the ancestral branch. In that case, if the state ***s*** chosen for ***Anc*** is shared with one of the descendants, using ***s*** for the node ***N*** will save a step along the ancestral branch (and place it in one of the descendant branches; note that the state of ***Anc*** cannot be present in both ***L*** and ***R*** because otherwise it would be present in ***n***). Thus, when ***Anc*** ∩ ***n*** ≠ ***Anc*** and ***n*** = ***L*** ∩ ***R***, ***N*** = ***n*** ∪ ***(Anc*** ∩ ***(L*** ∪ ***R))***. This is "case V" of Fitch (1971).

Hartigan (1973) provided mathematical proof of the correctness of Fitch's algorithms (the correctness of which should be obvious from the prior discussion) and proposed a simplification of the up pass, based on determining a backup state set, ***B***, during the down pass. To understand the use of ***B***, note that when ***Anc*** ∩ ***n*** ≠ ***Anc*** (i.e. when case I of Fitch does not apply), the states to be added to ***n*** to form ***N*** are those which are present in ***Anc*** but requiring only one additional step beyond the minimum if assigned to the node during the down pass; this is, in fact, the step that would be occurring between ***Anc*** and ***n***, which is saved in the ancestral branch and moved to one of the descendant branches. Thus, ***B*** must include all the states that require one step *beyond* the minimum during the down pass. During the down pass, if ***n*** = ***L*** ∩ ***R***, then ***B*** = ***L*** ∪ ***R*** (i.e. any of the states present in one of ***L*** or ***R*** but not both will require one step beyond minimum). If ***L*** ∩ ***R*** = ***∅*** and ***n*** = ***L*** ∪ ***R***, then **B** equals the set of all possible states (i.e. any state not present in either ***L*** or ***R*** will imply a step along both branches, instead of along one). During the up pass, ***N*** is obtained by adding to ***n*** any state present in the ancestor which locally requires a single additional step if assigned to the node—that is, ***N*** = ***n*** ∪ ***(Anc*** ∩ ***B)***. This saves time during the up pass (avoiding the need to check again the descendant sets ***L*** and ***R***). Figure 3.3c presents a worked-out example, for the same states observed in terminal taxa as in Figure 3.3a, but with transformation costs all unity. As is clear from comparing both figures, considering the character as additive (methods for which are discussed in the next section; examples in Figs. 3.3a and 3.3b, with 7 steps) or nonadditive (Figure 3.3c, with 5 steps) may imply different numbers of steps and state assignments.

As optimization of nonadditive characters can be performed with state set operations, it can be easily represented in computer programs with sets of bits. An example is in the TNT script ***graFitchopt.pic***. Computers internally treat numbers as binary (strings of 0s and 1s), decomposing every number into a series of powers of 2. Thus, state number 2 can be represented as $2^2 = 4$ and the state set {013} as $2^0 + 2^1 + 2^3 = 11$ (in 8-bit binary, 00001011). This has the advantage that sets of states can be stored as single numbers using very little memory. Also, as computer languages have operators that quickly calculate unions or intersections of bit sets (e.g. the operators | and & in the C language), the algorithms for Fitch/Hartigan optimization can be implemented with low-level instructions; as long as the number of states is within the numbers of bits of a word, the time needed to complete Fitch optimization in this manner does not depend on the number of states.[3] A snippet of code showing Fitch optimization is in Box 3.1.

BOX 3.1 EXAMPLE CODE FOR BRUTE FORCE AND FITCH OPTIMIZATION

Brute-force algorithm that enumerates all reconstructions and calculates number of steps for each. Note that partial reconstructions which exceed bound are abandoned to save time (line 12). After running, the array ***final_state*** contains a (single) state occurring in a most parsimonious reconstruction. Polymorphisms or missing entries not handled, only single-state terminals.

```
 1  int node_list [ Nnodes(0) ], state [ Nnodes(0) ],
        final_state [ Nnodes(0) ];
 2  int curlength = 0, best_length = 10e6,
        transfcosts [ 32 ] [ 32 ];
 3  void recurse_nodes ( int at ) {
 4      int node = node_list [ at ], L, R, n, j,
            initial_length = curlength;
 5      L = state [ firstdes ( 0, node ) ];
 6      R = state [ sister ( 0, firstdes ( 0, node ) ) ];
 7      for ( n = 0; n < num_states; ++ n ) {
 8          curlength = initial_length;
 9          if ( n != L ) curlength += transfcosts [ n ] [ L ];
10          if ( n != R ) curlength += transfcosts [ n ] [ R ];
11          state  [ node ] = n;
12          if ( curlength < best_length )
13              if ( node == root ) {
14                   best_length = curlength;
15                   for ( j = 0; j < Nnodes(0); ++ j )
16                       final_state [ j ] = state [ j ]; }
```

```
17          else recurse_nodes ( at + 1 ); }}
18  void brute_force_optimization ( void ) {
19      downlist ( node_list, 0 );
                 // write list of internal nodes to node_list
20      initialize_states ();  // function to initialize
21      recurse_nodes ( 0 );
22      printf ( "Best length: %i steps\n", best_length ); }
```

Fitch optimization using the ***travtree()*** function from the TNT library to traverse the tree. After execution, array ***state_set*** contains the full ***MP***-set.

```
 1 int state_set [ Nnodes(0) ], backup_set [ Nnodes(0) ],
      treelength = 0;
 2 void Fitch_down_pass ( int node ) {
 3     int L, R;
 4     L = state_set [ firstdes ( 0, node ) ];
 5     R = state_set [ sister ( 0, firstdes ( 0, node ) ) ];
 6     state_set [ node ] = L & R;
 7     backup_set [ node ] = L | R;
 8     if ( ( L & R ) == 0 ) {
 9        treelength ++;
10        state_set [ node ] = L | R;
11        backup_set [ node ] = missing; }}
12 void Fitch_up_pass ( int node ) {
13     int Anc = state_set [ anc ( 0, node ) ];
14     int n = state_set [ node ];
15     if ( ( Anc & n ) == Anc ) state_set [ node ] = Anc;
16     else state_set [ node ] |= Anc & backup_set [ node ];}
17 void Fitch_optimization ( void ) {
18     initialize_state_sets (); // function to initialize
19     travtree ( "down", "Fitch_down_pass", 0 );
20     travtree ( "up-", "Fitch_up_pass", 0 );
21     printf ( "Best length: %i steps\n", treelength ); }
```

3.4 OPTIMIZATION FOR ADDITIVE CHARACTERS: FARRIS'S (1970) METHOD

The algorithms described by Farris (1970) can be used for lineal additive characters. As the transformation cost between states equals their difference, the down-pass optimization can be effected by considering ranges at the internal nodes. The ranges can be visualized on rulers (Figs. 3.2d–f). Whenever the two descendants of a node ***L*** and ***R*** (Figure 3.2d) have a different state, any state that is between those two (inclusive) will minimize the cost, placing varying number of steps (which always sum up to the distance between the states ***L*** and ***R***) in each of the two branches. In the example of Figure 3.2d, the common ancestor of A and B can be assigned any state between 1 and 5, and the length is increased in 4 states (the difference between states) in every one of the cases. As the optimization moves further down the tree, the descendant ranges ***L*** and ***R*** may have some overlap, in which case assigning to ***n*** any of the overlapping states will imply no steps along any of the descendant branches (as in Figure 3.2e). If ***L*** and ***R*** do not overlap, the node must be assigned the smallest range that connects ***L*** and ***R*** (increasing length correspondingly), as in Figure 3.2f.

The up pass is considerably more complicated. Farris (1970) used a method which finds optimal states, but not necessarily all of them; many contexts in parsimony analysis require identifying all possible optimal states at every node. Swofford and Maddison (1987) proposed a method to find all possible states based on (implicitly) making each internal node the root of the tree, leading to three descendants, and finding the optimal "root" states from the preliminary states ***n*** of those three descendants. Goloboff (1993b) described alternative up pass algorithms that do not require rerooting; the method can be presented in three cases. Case 1 is the universal case (equivalent to Case I of Fitch 1971, applicable to any kind of transformations cost), when $\boldsymbol{n} \cap \boldsymbol{Anc} = \boldsymbol{Anc}$, and then the final state set for the node is $\boldsymbol{N} = \boldsymbol{Anc}$. Otherwise, Case 2 is when $\boldsymbol{Anc} \cap (\boldsymbol{L} \cup \boldsymbol{R} \cup \boldsymbol{n}) = \emptyset$ (i.e. the final range for the node's ancestor is beyond the values for any of its descendants), and in this case ***N*** is the range from the state in $\boldsymbol{L} \cup \boldsymbol{R} \cup \boldsymbol{n}$ closest to ***Anc*** to the state in ***n*** closest to ***Anc***, as in the diagrams:

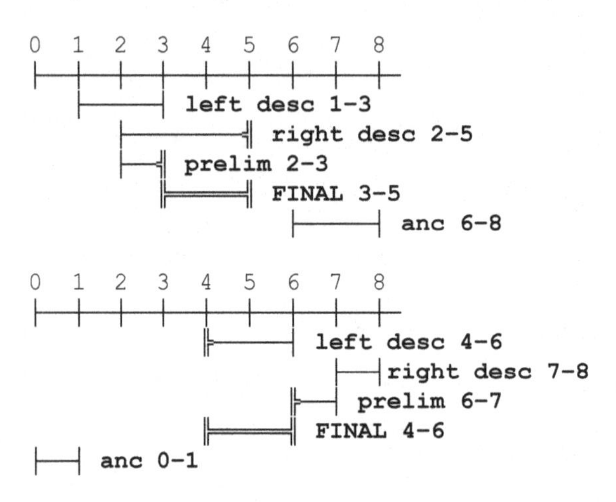

It is easy to confirm that, for each assignment in the final range (and for each possible assignment of ***Anc***), the cost along the three branches (i.e. ***Anc–N***, ***N–L***, and ***N–R***) sums up to a minimum. Case 3 is when ***Anc*** overlaps with $\boldsymbol{L} \cup \boldsymbol{R} \cup \boldsymbol{n}$, and then $\boldsymbol{N} = \boldsymbol{Anc} \cap (\boldsymbol{L} \cup \boldsymbol{R} \cup \boldsymbol{n})$, with the proviso (Case 3.b) that if $\boldsymbol{Anc} \cap \boldsymbol{n} = \emptyset$, ***N*** must be extended to the state in ***n*** closest to ***Anc***, as in the diagrams:

Case 3.a ($\boldsymbol{Anc} \cap \boldsymbol{n} \neq \emptyset$),

```
0   1   2   3   4   5   6   7   8
|---+---+---+---+---+---+---+---+---
|-----------|  left desc 0-3
    |-----------|  right desc 1-4
    |-------|  prelim 1-3
        ||======||  FINAL 2-4
        |-------|  anc 2-4
```

```
Case 3.b (Anc ∩ n = Ø),

0  1  2  3  4  5  6  7  8
|--+--+--+--+--+--+--+--+-
      |-----| left desc 2-4
         |---------|right desc 3-6
         |--| prelim 3-4
            ||=====|| FINAL 4-6
               |-----| anc 5-7
```

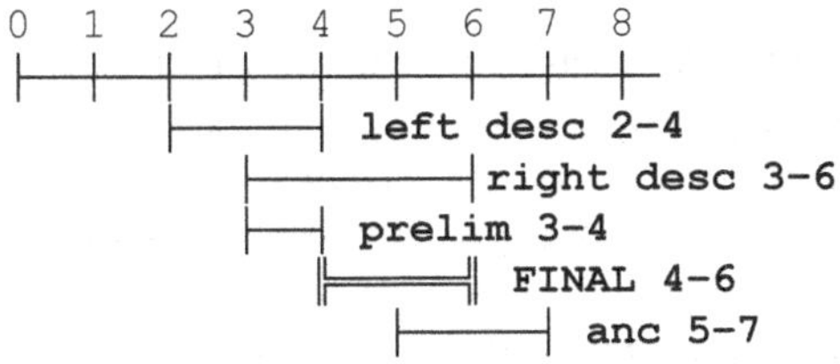

A fully worked example is in Figure 3.3b, with preliminary states indicated above branches and final states (with case) below. The final states are identical to those obtained with the generalized optimization because transformation costs are the same.

These rules can also be expressed by means of bitwise operations in computer programs. For example, the range 2–4 is $2^2 + 2^3 + 2^4 = 28$, or 00011100, the only difference with nonadditive characters being that the resulting bit strings have all the 1 bits contiguous. To calculate ranges, the C operators ^ (exclusive bit-oring, a bit is 1 if 1 in one of the numbers, not in both), ~ (bit complement, a bit is 1 if 0 and vice versa), and >> and << (bit shifting), together with the & and | operators explained prior, allow isolating bits and then extending above or below that bit. For example, the C expression $v = x \& (-x)$ produces a number v with only the rightmost bit of x, and $v\,| = v - 1$ adds all the bits below (taking advantage of the borrow operation from the lowest set bit in the subtraction). If the lowest bit itself is not needed in isolation, the same final result is produced by the more economic $v = x \wedge (x - 1)$. For an example (in 8-bit notation), when x = 01100000 (the range 5–6), both expressions produce v = 00111111. Bits can be extended to the left (higher order bits) of a value y, with the combination of $w = \sim (y \,|\, (y - 1))$, followed by $w\,| = w >> 1$ if the highest order bit of y itself is needed in the result. For an example, if y = 00000110 (the range 1–2), then w = 11111100. If x and y are non-overlapping state sets of left and right descendants of node n in the down pass (where, obviously, $x > y$; their values must be exchanged otherwise), then the preliminary states of n can be simply determined with v & w, which produces the correct range 2–5:

```
x = 01100000   →   v = 00111111
y = 00000110   →   w = 11111100
                   &  ----------
                   n = 00111100
```

The increase in number of steps can be calculated as the number of bits in n, minus 1 (e.g. with the TNT function ***numbits()***, which uses a lookup table). These operations are used in the example code for Farris optimization in Box 3.2. Another way to calculate step increases (from the individual lowest and highest bits of x and y respectively), using De Bruijn sequences, is shown in an example script at the TNT web repository, ***optimization.pic***. As in Fitch optimization, the time needed to optimize an additive character with the bitwise methods described prior does not depend on the number of states.

BOX 3.2 FARRIS OPTIMIZATION

The down- and up-pass functions should be called as in Fitch optimization (see Box 3.1), with the ***travtree()*** function (not shown), after initializing the array ***state_set*** from the terminals. The TNT library function ***numbits()*** (line 14) calculates the number of ON bits for its argument (internally using a lookup table). This code uses the operations of Goloboff (1993b), expressed as bitwise (advantageous because time does not depend on the number of states).

```
 1 int state_set [ Nnodes(0) ], treelength;
 2 void Farris_down_pass ( int node ) {
 3    int large, small, below, above, L, R;
 4    L = state_set [ firstdes( 0, node ) ];
 5    R = state_set [ sister ( 0, firstdes( 0, node ) ) ];
 6    if ( ( L & R ) != 0 ) state_set [ node ] = L & R;
 7    else {
 8      small = ( L < R ) ? L : R;
 9      large = ( L > R ) ? L : R;
10      above =  ~ ( small | ( small - 1 ) );
11      above |= above >> 1;            // extend bits above
12      below = large ^ ( large - 1 ); // and below
13      state_set [ node ] = above & below;
14      treelength += numbits ( state_set [ node ] ) - 1; }}
15 void Farris_up_pass ( int node ) {
16    int Anc = state_set[ anc ( 0, node ) ], large, above,
17        n = state_set[ node ], small, below, tmp, LRn;
18    LRn = n | state_set [ firstdes ( 0, node ) ]
19        | state_set [ sister ( 0, firstdes( 0, node ) ) ];
19    if ( ( Anc & state_set [ node ] ) == Anc ) // Case 1
20       state_set [ node ] = Anc;
21    else if ( ( Anc & LRn ) == 0 ) { // Case 2
22          if ( Anc < LRn ) {
23            below = n & ( - n );
24            below |= below - 1;
25            above = ~ ( LRn ^ ( LRn | ( LRn - 1 ) ) ); }
26          else {
27            below = LRn | ( LRn - 1 );
28            above = ~ ( n | ( n - 1 ) );
29            above |= above >> 1; }
30          state_set [ node ] = above & below; }
31    else { tmp = LRn & Anc;  // Case 3.a is default here
32           if ( ( n & Anc ) == 0 ) { // Case 3.b
33             small = ( n < Anc ) ? n : Anc;
34             large = ( n > Anc ) ? n : Anc;
35             above = ~ ( small | ( small - 1 ) );
36             above |= above >> 1;
37             below = large ^ ( large - 1 );
38             tmp |= above & below; }
39          state_set [ node ] = tmp; }}
```

Computer programs often treat additive characters internally by recoding them in several binary characters so that the operations prior are not always needed. In terms of numbers of steps, any additive character (even a branched one) can be represented by a series of binary variables. To represent an additive character with i states, $i - 1$ variables are needed; state s in the original character will be represented by variables 0 to $s - 1$ with state 1 and variables s to $i - 1$ with state 0, as in the table:

```
Original      Becomes
 State       0 1 2 3 4
   0         0 0 0 0 0
   1         1 0 0 0 0
   2         1 1 0 0 0
   3         1 1 1 0 0
   4         1 1 1 1 0
   5         1 1 1 1 1
```

This method can be used to recode *branching* additive characters as well; a binary variable is needed for each branch in the character state tree, with states in one side of the branch having 0 for the corresponding variable and states on the other side 1:

```
A   B
0\ /1
  C-D-F
2/ 3 4
E
```

The variables corresponding to each of the branches (numbered 0–4) are:

```
             0 1 2 3 4
state A      0 0 0 0 0
state B      1 1 0 0 0
state C      1 0 0 0 0
state D      1 0 0 1 0
state E      1 0 1 0 0
state F      1 0 0 1 1
```

Note that the cost of transforming between any two states in the original character equals the sum of costs of transforming between the variables, so the two representations are completely equivalent in terms of numbers of steps. The equivalence no longer holds when differential character weights are used—for example, under implied weights; the binary representation, however, seems to provide a more sensible approach, as it differentially weights the different transformations, so it is the default in TNT. The individual binary variables can be optimized with simpler and faster algorithms, so in the end, the time needed to optimize the multiple variables is the same or less as in the single-character algorithms used to optimize fewer characters. This representation by means of binary variables can also be used to reconstruct the full ancestral state sets (by tracing over equivalences) or to identify zero-length branches (see Chapter 6, Section 6.4).

3.5 STEP-MATRIX OPTIMIZATION

The algorithms by Sankoff and Rousseau (1975) and Sankoff and Cedergren (1983) can accommodate any set of transformation costs between states. In some computer programs, these transformation costs are specified in the form of a matrix, and then these characters are often called *step-matrix characters* (e.g. by Swofford and Maddison 1992, in their review of algorithms for character optimization). Given that these algorithms can accommodate any cost regime, they are useful as an aid for calculating (or approximating) optimal ancestral reconstructions in complex cases, such as position of points in a 2D or 3D grid (for landmark coordinates, see Chapter 9) or optimal frequency combinations in ancestors (e.g. Swofford and Berlocher 1987; Berlocher and Swofford 1997). They can also deal with cases where the costs are asymmetric—that is, when the transformation between two states does not have the same cost in both directions. Working code for step-matrix optimization is in the TNT script ***optimization.pic***.

The algorithms to optimize step-matrix characters are based on enumerating combinations of states in the node and the descendant, as in the generalized optimization used to introduce two-pass algorithms, except that instead of storing only the optimal values, they record the cost of assigning every state to every node, even for suboptimal states. This is needed, because storing only optimal states may lead to errors when transformation costs deviate from additive or nonadditive. Consider the example of Figure 3.4a. During the down pass, states 0 and 1 are optimal for node x (with a local cost of 3); state 2 is suboptimal (with cost 4). If the down pass records only the sets of optimal states (instead of their costs), then the best assignment for the root node is either 0 or 1 (each of which requires an additional two steps, to the state 2 in the terminal splitting first, for a total of five steps). It is easy to see, however, that setting x = y = 2 implies only four steps. The problem, in this case, is that state 2 is "intermediate" between states 0 and 1, and thus, even if it was suboptimal during the down pass, it may become optimal when the entire set of taxa is considered.

The down pass of the Sankoff algorithm, therefore, calculates the local cost C of assigning each state (with largest state m) to each node n by choosing the best transformation into any of the states in the descendant branches L and R, (considering the cost T_{ij} of transforming from i to j and the local cost of the two descendants):

$$C_{0,n} = min\ [T_{0,0} + C_{0,L}\ ,\ T_{0,1} + C_{1,L}\ ,\ T_{0,2} + C_{2,L}\ , \ldots ,\ T_{0,m} + C_{m,L}]$$
$$+\ min\ [T_{0,0} + C_{0,R}\ ,\ T_{0,1} + C_{1,R}\ ,\ T_{0,2} + C_{2,R}\ , \ldots ,\ T_{0,m} + C_{m,R}]$$
$$C_{1,n} = min\ [T_{1,0} + C_{0,L}\ ,\ T_{1,1} + C_{1,L}\ ,\ T_{1,2} + C_{2,L}\ , \ldots ,\ T_{1,m} + C_{m,L}]$$
$$+\ min\ [T_{1,0} + C_{0,R}\ ,\ T_{1,1} + C_{1,R}\ ,\ T_{1,2} + C_{2,R}\ , \ldots ,\ T_{1,m} + C_{m,R}]$$

. . .

$$C_{m,n} = min\ [T_{m,0} + C_{0,L}\ ,\ T_{m,1} + C_{1,L}\ ,\ T_{m,2} + C_{2,L}\ , \ldots ,\ T_{m,m} + C_{m,L}]$$
$$+\ min\ [T_{m,0} + C_{0,R}\ ,\ T_{m,1} + C_{1,R}\ ,\ T_{m,2} + C_{2,R}\ , \ldots ,\ T_{m,m} + C_{m,R}]$$

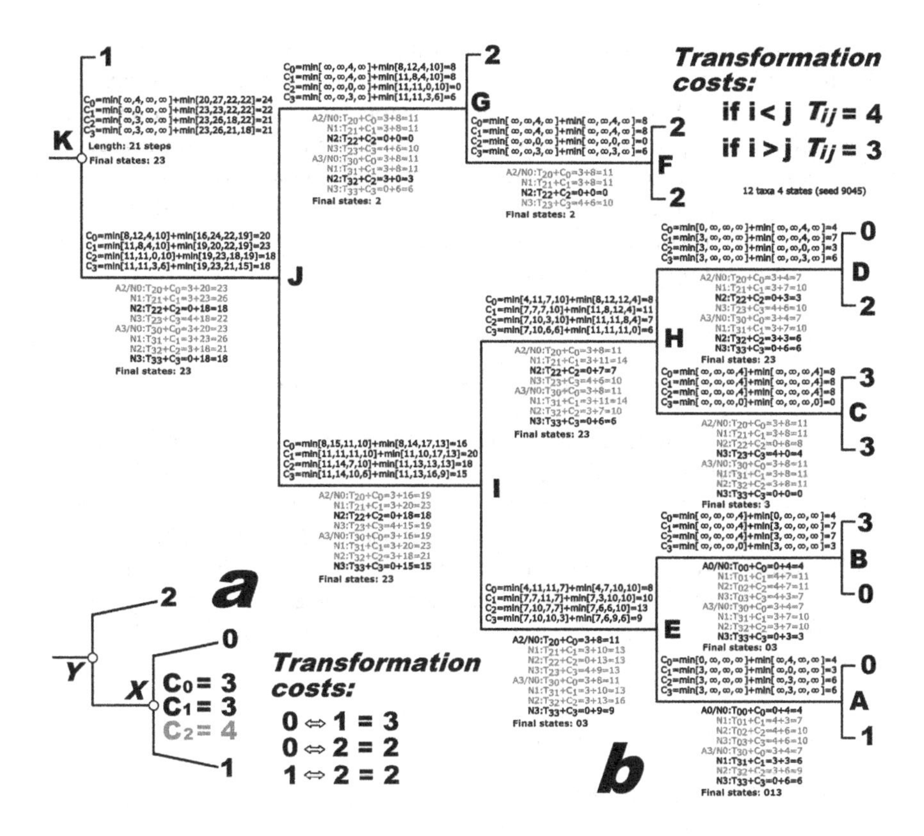

FIGURE 3.4 (**a**) When transformation costs do not follow the geometry of either additive or nonadditive characters, consideration of only optimal states during the down pass may lead to missing optimal assignments. See text for details. (**b**) A worked-out example of step-matrix optimization. Values above branches indicate costs during down pass, and values below indicate costs during final pass. For the final costs, darker values indicate optimal choices.

In terminal taxa, the cost of an observed state is initialized as 0 and the cost of unobserved states as ∞. Once the root is reached, the tree length is the lowest cost of all possible states, and all the states with such minimum value occur in the final state set for the root:

$$Length = min\ [C_{0,root}\ ,\ C_{1,root}\ ,\ C_{2,root}\ ,\ .\ .\ .\ ,\ C_{m,root}]$$

The final state assignments, in the up pass, are found using the C values stored during the down pass. For each of the states in the final state set of *Anc*, the up pass calculates the minimum cost M at the triplet of branches (the minimum sum of the cost of transforming from the selected state a in *Anc* into each possible state in the node n, plus the local cost of the state in node n):

$$M = min\ [T_{a,0} + C_{0,n}\ ,\ T_{a,1} + C_{1,n}\ ,\ T_{a,2} + C_{2,n}\ ,\ .\ .\ .\ T_{a,m} + C_{m,n}]$$

All the states with minimum value (for each optimal state in *Anc*) must be added to the final state set of the node, *N*. Figure 3.4b presents a worked-out example, with the down-pass costs shown above branches (as before, the first term in the summation corresponds to the upper descendant and the second term to the lower descendant). The up-pass costs are shown below branches; instead of considering summations of three terms (as in Figure 3.3a), they take advantage of the costs *C* stored in the down pass (the second term of the summation in Figure 3.4b). An example of working code to perform Sankoff optimization is in Box 3.3.

BOX 3.3 SANKOFF OPTIMIZATION

Prior to calling these functions, the variable ***num_states*** and the arrays ***state_cost*** and ***trancost*** must be initialized. The TNT library function ***minval()*** (lines 12, 14) calculates the minimum value contained in the array (first argument, it must be a ***double*** *) for a number of values specified with second argument. After reaching ***root***, the down pass (lines 15–16) sets cost of optimal states to 0, so as to easily skip suboptimal ancestral states (line 21) in the subsequent up pass (the same is done for nodes above root in the up pass, lines 29–31).

```
 1 int state_set [ Nnodes(0) ], num_states, treelength,
       trancost [ 32 ] [ 32 ];
 2 double state_cost [ Nnodes(0) ] [ 32 ],
          L_tmp [ 32 ], R_tmp [ 32 ];
 3 void Sankoff_down_pass ( int node ) {
 4     int i, j, k, min_L, min_R, this, rootmin;
 5     double * L_costs , * R_costs;
 6     L_costs = state_cost [ firstdes( 0, node ) ] ;
 7     R_costs = state_cost [ sister ( 0,
                         firstdes ( 0, node ) ) ];
 8     for ( i = 0; i < num_states; ++ i ) {
 9         for ( j = 0; j < num_states; ++ j ) {
10             L_tmp [ j ] = trancost [ i ] [ j ]
                   + L_costs [ j ];
11             R_tmp [ j ] = trancost [ i ] [ j ]
                   + R_costs [ j ]; }
12         state_cost [ node ] [ i ] =
               minval ( L_tmp, num_states ) +
               minval ( R_tmp, num_states ); }
13     if ( node == root ) {
14         treelength = minval ( state_cost [ root ],
                            num_states );
15         for ( i = 0; i < num_states; ++ i )
16             state_cost [ root ] [ i ] -= treelength; }}
17 void Sankoff_up_pass ( int node ) {
18     int i, j, best, tmpcosts[ 32 ], x = 0;
19     double * N_costs = state_cost [ node ],
           * Anc_costs = state_cost [ anc ( 0, node ) ];
20     for ( i = 0; i < num_states; ++ i ) {
21        if ( Anc_costs [ i ] ) continue;
22        best = 10e6;
23        for ( j = 0; j < num_states; ++ j ) {
```

```
24          tmpcosts[ j ] = N_costs [ j ] +
                                trancost [ i ] [ j ];
25          if ( best > tmpcosts [ j ] )
26              best = tmpcosts[ j ]; }
27      for ( j = 0; j < num_states; ++j )
28          if ( tmpcosts[ j ] == best ) x |= 1 << j; }
29    for ( j = 0; j < num_states; ++j )
30      if ( ! ( x & ( 1 << j ) ) ) N_costs [ j ] = 10e6;
31      else N_costs [ j ] = 0;
32      state_sets [ node ] = x; }
```

Step-matrix optimization is simple but very laborious and more time consuming than Fitch or Farris optimization. It is easy to see that the time needed increases with the square of the number of states, as it tries all combinations of states in the two ends of each branch. By not restricting the states chosen for *Anc* to the optimal ones (and considering their costs beyond the minimum), it is possible to calculate the extra cost beyond the minimum for every state at every node, which can be useful in certain contexts (e.g. to speed up searches with step-matrix characters; see Goloboff 1998b). In the implementation in TNT, whenever the program detects that the cost regime in a character could be handled by Fitch optimization, it saves time by internally treating the character in that way (regardless of whether the character has been set as step-matrix by the user). When the scheme of transformation costs deviates from nonadditive characters but remains ultrametric or additive, it is possible to use faster algorithms (Clemente et al. 2009) than the ones described here.

3.6 OTHER TYPES OF OPTIMIZATION

Special cases require yet other types of algorithms. For sequence data, the method of optimization alignment (Wheeler 1996) is an application of *dynamic homology* (Wheeler 2001a, 2001b), where possible ancestral reconstructions are judged on the basis of how they homologize states (De Laet 2005) over all the sequence (the concept of a single "character" not being applicable a priori). Some interesting proposals to use dynamic homology in cases where the homology between parts of morphological characters is uncertain (e.g. Ramírez 2007) have found little echo, mostly because of the involved computational complexity. Thus, dynamic homology (like the whole problem of sequence alignment) is of little relevance for morphologists. The methods for reconstructing ancestral landmark coordinates or continuous variables are discussed in Chapter 9. To allow transformation costs to be influenced by the tree topology (instead of being just defined in advance), special methods of optimization are required (Goloboff 1997), discussed in Chapter 7 from the point of view of character weighting.

3.7 AMBIGUITY, POLYMORPHISMS, MISSING ENTRIES

Any of the methods for finding ancestral state reconstructions (*MPR*s) can produce ambiguous results when more than a single reconstruction is most parsimonious under the given transformation costs. This ambiguity does not depend on the specific algorithm used to find the reconstructions, although is often discussed as if it did (e.g. Agnarsson and Miller 2008). Going back to Figure 3.1, where optimal reconstructions were found by enumeration instead of any specific algorithm, there are three alternative reconstructions that maximize parsimony. The most parsimonious state sets (or *MP*-sets) for the nodes that have alternative states is the union of all the states in the different reconstructions; for the ambiguous nodes x, y, or z, this is the set {01}. Those state sets are exactly the same ones that generalized, Fitch, Farris, or Sankoff optimization would have produced for that character at those nodes. The two-pass algorithms do not produce the individual reconstructions; they only identify the *state sets*, and ambiguity (i.e. the existence of more than a single *MPR*) is indicated by at least one node having a set composed of more than a single state.

Continuing with the example in Figure 3.1, reconstruction 25 implies that state 0 in the outgroup and the sister of w is homologous but the presence of state 1 in the descendants of v and the descendants of w is due to convergence. Alternative reconstruction 31 implies instead that state 1 in groups v and w is homologous but the state 0 in outgroup and sister of w is due to convergence. In earlier literature (e.g. Swofford and Maddison 1987), ambiguity in optimization was often resolved by reference to secondary criteria such as *Acctran* (accelerated-transformation) or *Deltran* (delayed-transformation). *Acctran* and *Deltran* are two extremes, often with alternative reconstructions in the middle that can be called neither "acc" nor "del". Reconstruction 25 "delays" transformation (*Deltran*) of state 0 into 1 as parallelisms further down the tree; reconstructions 31 or 32 consider the 1 state as homologous instead. There have been proposals (de Pinna 1991) to prefer *Acctran* over *Deltran* on the grounds that *Acctran* homologizes more instances of the apomorphic state. If the apomorphic state could indeed be defined in the abstract (see Section 3.9), that would be true, but it would also be no less true that the *Deltran* reconstruction homologizes more instances of the *plesiomorphic* state. The degree to which both reconstructions homologize similarities (weighted by the corresponding transformation costs) is exactly the same—that is how optimal reconstructions are chosen! If several *MPR*s exist, it is because they are equally good at homologizing similarities (with failures to homologize similarities weighted by their corresponding prior transformation costs). With a given set of transformation costs, the ambiguity (i.e. the existence of more than a single *MPR*) cannot be resolved by reference to secondary criteria (as proposed, e.g. by Agnarsson and Miller 2008). If there are grounds for preferring homologization of some similarities over others (e.g. gain or loss of a complex structure), this must be incorporated into the scheme of transformation costs and the tree optimized accordingly. Incorporating such preference into transformation costs, however, has no relationship to ambiguity—which similarities are preferably homologized must be considered when selecting from all possible reconstructions, not when multiple reconstructions are equally optimal.

Individual reconstructions are rarely needed for a phylogenetic analysis, as most of the operations can be carried out on *MP*-sets. If they are needed for any reason, *MPR*s can be enumerated with the appropriate combinations of states selected from the *MP*-sets. Not any combination of the states from the *MP*-sets will produce optimal reconstructions; for example, both x and y can be assigned states 0 or 1 for the tree of Figure 3.1, but if state 0 is selected for y, then x *must* be assigned 0 to obtain an *MPR*, and if state 1 is selected for x, then y must be assigned state 1. But when state 1 is selected for y, x can have either 0 or 1 in an *MPR*. Some choices of a state for a node, that is, make certain states mandatory at the surrounding nodes, and other choices do not. Enumeration of *MPR*s must recursively select each of the states at each ambiguous node with an *MP*-set with more than a single state. The individual reconstructions can be very numerous; consider the case of the tree with states (0(1(2(3(45))))) in a nonadditive character; we leave it to the reader to verify that 89 different *MPR*s exist. TNT includes options to enumerate *MPR*s (see Section 3.13); example code can be found in an example TNT script, ***reconstructions.pic***.

3.7.1 Polymorphisms

The reconstructions evaluated by parsimony all consist of assigning to the ancestral nodes a single state (the same is true of likelihood methods). Polymorphism (in the population genetics sense) is never considered. The reconstructions actually evaluated consider ancestors fixed at a single state. If optimization produces some *MP*-sets with multiple states, this should not be interpreted to mean that the ancestor was polymorphic. It means instead that there are reconstructions with one of the states assigned to the ancestor which are optimal and other reconstructions with the other state also being optimal. Maximum likelihood methods are similar in considering only single-state reconstructions for the evaluation of trees. There have been attempts to use the frequency of states (in polymorphic populations) as evidence for producing phylogenetic trees. Swofford and Berlocher (1987) proposed methods ("freqpars") where the difference in frequencies between ancestors and descendants is minimized, thus solving the logical problems (pointed out by Farris 1981) of analyzing such data as distances; Wiens (1995) and Berlocher and Swofford (1997) subsequently proposed faster approximations based on step-matrix characters. Clearly, the assumptions needed to interpret trees minimizing differences in frequency as phylogenetic trees are different from those required for more standard characters, and frequencies are rarely used as data for phylogenetic analyses.

The term *polymorphism*, in phylogenetics, usually indicates a taxon that is variable—e.g. a genus where different species have different states for the character. In that case, clearly, the frequency of occurrence of the different states is irrelevant. If the cladistic structure within the group were known, then the plesiomorphic state for the group should be determined and used to represent the group; the mere numbers of species with one or the other state give no indication of this. For instance, there are many more species with than without wings in Insecta, yet the obvious ancestral state for Insecta is absence of wings (silverfish and jumping bristletails are the first splitting groups).

In the absence of knowledge about internal cladistic structure of the polymorphic group, it is scored as having any of the observed states. The variable terminal taxon then requires no extra steps if placed within a group with some of the states it has but requires some extra steps if placed within a group where none of its possible states occur. The algorithms for two-pass optimization handle such cases naturally, without modification—a terminal with more than a state is treated just like an internal node in the down pass.

Doing this can create artificial tree scores when only some character state combinations exist in the group, not because of problems with algorithms for optimization but because the internal cladistic structure of the taxon scored as polymorphic is not being considered (Nixon and Davis 1991). Consider the case of a matrix including as terminal a genus with three species, for which the monophyly is well established (thus making it possible to consider the genus as just a terminal). For two binary characters, the states observed in the species of this genus are A = 00, B = 01, and C = 11. As each of the two characters is variable, the genus would be scored as having a missing entry (the same as a {01} polymorphism) for each. However, the relationships between A, B, C can be resolved in three possible ways; if A(BC) the ancestral character states for the genus are 0?, if B(AC) 01, and if C(AB) ?1. Note that we cannot know the ancestral state for any of the two characters, but we *do* know that if state 1 is one of the possibilities for the first character, then state 1 must be the ancestral state for the second character: no resolution of the species within the genus can have 10 as plesiomorphic states. With the genus coded as having missing entries for each of the characters, no steps would be added for these characters, regardless of where the genus is placed in the tree—but placing it as sister of taxa with states 10 should require more steps than placing it as sister of taxa with states 00! The only way to prevent this "hidden" cost in polymorphic characters is by including the three separate species (i.e. all the actually different character state combinations) as terminals in the matrix (Nixon and Davis 1991). Doing so, however, requires also that we add to the matrix characters for resolving the relationships between the species, and the analysis becomes more complex, so a tradeoff between precision and practicality is necessary.

3.7.2 Missing Entries

When a character has not been observed for a given terminal, the logical treatment is to have the same number of steps for the character regardless of where in the tree the taxon is placed. This is easily accomplished by internally treating the taxon with a missing entry as polymorphic, with all states observed for the character, {0123 . . . n}. Having the taxon with {0123 . . . n} as sister group of a taxon (or internal node) with any possible set of states will always produce a non-empty intersection, thus adding no steps for the character. In the case of step-matrix optimization, the cost of any of the states must be initialized to 0 in the case of a terminal taxon with a missing entry.

A special related case is that of a terminal having a unique state in a nonadditive character (i.e. a state not found in any of the other taxa in the matrix). Placing that taxon anywhere in the tree will always require one step, so the unique state effectively behaves as if it were a missing entry—it can, in fact, be replaced by a missing entry without affecting the results under parsimony.

3.8 MAPPING, SYNAPOMORPHIES, AND RECONSTRUCTED ANCESTORS

Optimization (i.e. most parsimonious reconstruction) on a tree is the only way to produce a map of changes for a character on a tree and a diagnosis of groups. Mapping character changes is useful for making hypotheses about character evolution. Note that Chapter 1 pointed out that preferring *MPT*s does not imply a claim that evolution proceeds parsimoniously because the parsimony score can be seen as providing only a lower bound for the homoplasy required by a tree. The situation is different when it comes to character mapping—the same justification cannot be claimed for preferring the *MPR* for a given character over less parsimonious reconstructions. The *MPR* is probably the most reasonable and logical explanation in most cases, but when used for hypothesizing the evolution of a single character, it does entail the assumption that the simplest explanation is probably the correct one (an assumption that was not required to justify the preference for most parsimonious *trees*). Inquiries as to the reliability of the ancestral states so inferred can be done only in the context of making specific assumptions about evolution, such as the Markov model used in maximum likelihood (e.g. Li et al. 2008; Herbst et al. 2019). Of course, a more empirical approach will consider the addition of new intermediate taxa (particularly fossils with transitional character state combinations) as a sort of "test" for the ancestral states inferred by optimization.

Lists of synapomorphies for the different groups can also be obtained only by means of character optimization. While some presentations of cladistics talk about synapomorphies as if they could be identified without reference to a tree, synapomorphies cannot be identified in the abstract. This is almost trivially evident: whether a character is a synapomorphy of a group not only depends on how the group is—it also depends on how its sister group is. The same group being present in two trees does not mean it can be characterized by the same synapomorphies; consider the case of taxa a–e with state 0 and F–I with state 1. A change 0→1 is a synapomorphy of group e–I if the tree for those taxa is (a(b(c(d(F(G(H(e I)))))))), but if the tree is instead (a(b(c(d(e(F(G(H I)))))))), it is not—even if group e–I is monophyletic on both trees. To translate this into a real-world example: we consider that Amphibians lacking wings is a plesiomorphy because we have a well-established hypothesis of vertebrate relationships, but if we were to accept a tree where Amphibians are a highly modified subgroup of Vultures, then their absence of wings would have to be interpreted as a synapomorphy.

A "synapomorphy" for a group then amounts to a state change in a character along the corresponding branch; some synapomorphies may be ambiguous when optimization is ambiguous and multiple *MPR*s exist. It is obvious that the number of synapomorphies mapped on the branches of a tree need not equal the length of the tree (as some transformations, e.g. 0→2 in an additive character, cost more than a single step). When only *unambiguous* synapomorphies are indicated on the branches of a tree, their number need not equal the absolute number of changes in an *MPR*. A typical example is the case of missing entries, which create ambiguity if placed in between two nodes optimized to have different states; in the tree (0(0(?(11)))), a change 0→1 might be a synapomorphy of the group formed by the two taxa for which state 1 was observed, but that same change 0→1 might as well be a synapomorphy of the group including also the taxon with an unobserved state. Although the character *must* have a change somewhere in the tree, no unambiguous synapomorphy can be identified. Since synapomorphies are defined to be those features that *distinguish* a group from its relatives, the same conclusion would be reached without a numerical analysis: the condition in the closest relative is unknown; depending on how the closest relative turns out to be (when finally observed), the character will be a synapomorphy of one or the other group. The conclusions reached by comparing numerically reconstructed ancestral states, or a mere a logical analysis of the possibilities, are thus completely equivalent. This applies also to discussions of character evolution with no quantitative analyses, often done by morphologists: having observed that some mammals and some birds have a character that is absent in some fish, the conclusion that the character is a synapomorphy of Tetrapoda is invalid—until the presence or absence of the character is confirmed in Amphibia, it cannot be ascertained whether the character is a synapomorphy of Tetrapoda or Amniota. The advantage of numerical methods is that they highlight such cases with clarity, but there is no conceptual difference in treatment.

Optimization therefore is not only a means to evaluate trees and choose the ones that best fit the data but also the means to propose optimal diagnoses of the resulting groups and to produce optimal mappings of character state evolution. The methods for finding *MPR*s described so far in this chapter work for a single character at a time, counting the number of steps. As characters are independent, they can be optimized independently, and the numbers of steps (or score, if using weights) can be summed across characters. Evaluating a tree thus requires that *all* characters are optimized, and thus the cost of evaluating a tree of T taxa with an optimization increases linearly with number of characters, as $O(TC)$, where C is the number of characters.

3.8.1 Reconstructed Ancestors

As all characters of a matrix are evaluated, the full descriptions of hypothetical ancestors are obtained. Reconstructed ancestors may well have character state combinations not found in any of the terminals; it is precisely the possibility of creating new character combinations that enables a cladistic analysis to better account for the data—i.e. smoother transitions between observed taxa, so to speak—by postulating the existence of intermediate ancestors. The situation is counterintuitive in the sense that only by adding more points (the *HTU*s) to the diagram (the tree) that interconnects the observed terminals is it possible to connect the increased number of points using a shorter path. Consider the matrix:

```
A  0000
B  0001
C  1010
D  1100
```

which produces (A(B(CD))), of four steps, as *MPT*; the list of (most parsimonious) character states for the common ancestor or *HTU* of (CD) is 1000. Before the advent of phylogenetic systematics, many discussions (particularly in paleontological studies) centered around whether or not a given fossil specimen could have represented an ancestor (the very notion of a higher taxon ancestral to another is problematic, but that is a different question; consider here only the case of species-level taxa, which *might* make sense depending on how the species taxon is defined). Note that if ancestors in this case were required to be identical to some terminal (instead of allowing new character combinations), then it would be impossible to have a tree of only 4 steps—ancestor (CD) has a character combination (1000) not found in any of the terminals. A terminal could be a candidate for an ancestor only in the case in which it is identical to the reconstructed *HTU* immediately following—for example, the root of the tree in the example prior is reconstructed as 0000, so taxon A (if a fossil) might conceivably be an ancestor. In the early days of cladistics, one of the most resisted ideas introduced by cladists was that individual ancestors (even if effectively collected) could not be identified as such (e.g. Engelmann and Wiley 1977; Platnick 1977). In the present context, taxon A, identical to its immediate ancestor, might have been the ancestor itself—*or not*. There is no (and there cannot be) any character-based evidence to prefer the idea that A is ancestral over the idea that it is not; there is no improvement in character fit by making A ancestral. Perhaps more importantly, whether or not a taxon is indeed ancestral—as opposed to merely "identical to the ancestor"—makes no difference in practical terms; all of the character states of the ancestor are recovered from its reconstructed *HTU* so that any conclusions as to the morphology of the ancestor are exactly the same regardless of the decision to consider A as ancestral or not. This last point seems to be lost in recent attempts to revive the idea that ancestors can be identified as such (e.g. Parins-Fukuchi et al. 2019).[4]

3.8.2 Branch Lengths

The sum of differences (taking into account their cost) between ancestral and descendant nodes of a branch is known as the "length" of a branch. The length is an estimate of the amount of evolution that occurred along the branch. This estimate is ambiguous whenever multiple *MPR*s exist. The most common course of action in the case of ambiguity is calculating minimum branch lengths, the minimum among all reconstructions. The sum of lengths along all branches of the tree is then equal to the length of the tree only when no character has ambiguous optimizations (rarely the case for empirical datasets).

3.9 THE MYTH OF POLARITY

The previous descriptions of how to evaluate ancestral reconstructions and how to measure the degree of parsimony of a tree are all that is needed to proceed with a cladistic analysis. The reader should have noticed that the "polarity" of characters was never discussed—at no point in the description of methods was there any discussion of whether state 0 or 1 is "primitive" or "derived". The direction of a given character state transformation can only be ascertained when the tree is rooted. The old notion that cladistic analysis requires a determination of polarity a priori is so ingrained that one can still find it in recent papers:

> Characters may be coded with respect to what is called polarity (de Queiroz 1985; Stevens 1991). In these cases, the phylogeny has informed the way in which the

> character is coded. The result of this is that one character state is designated pleisiomorphic (ancestral), and one is denoted apomorphic (derived) a priori. This is often seen in the form of the 0 state representing the state possessed by outgroup, or the purported ancestral state (Watrous and Wheeler 1981).
>
> **(Wright 2019: 5)**

Hopefully, the idea that polarity needs to (or can) be determined prior to the analysis will finally be laid to rest.

Note that changing the root of the tree, when transformation costs are symmetrical, has no effect on tree length and on states assigned to equivalent nodes. The only thing that can change when the tree is rooted is the direction of transformations; a state changing from 0 to 1 in some part of the tree may now change from 1 to 0, but either costs a step. The states assigned during down-pass optimization to equivalent nodes in different rootings of the tree will often differ (as they are based on only local information, i.e. the descendants of the node, which change with rooting); but once the up pass is effected, the final states will always be the same. This is illustrated in Figure 3.5.

This also means that, for symmetric transformation costs, it is not possible to choose different ways to root the tree on the basis of differences in character fit; rooting is "transparent". This is why it is necessary to include, as part of the data, at least one taxon known (on the basis of information external to the matrix itself) to be outside of the group under study—the so-called *outgroup* taxon. In the case of characters with asymmetric transformation costs, the assignment of states can change with the root of the tree. This does not mean that those characters are "polarized" prior to the analysis; the most primitive state can still only be identified once the tree is rooted and mapped.

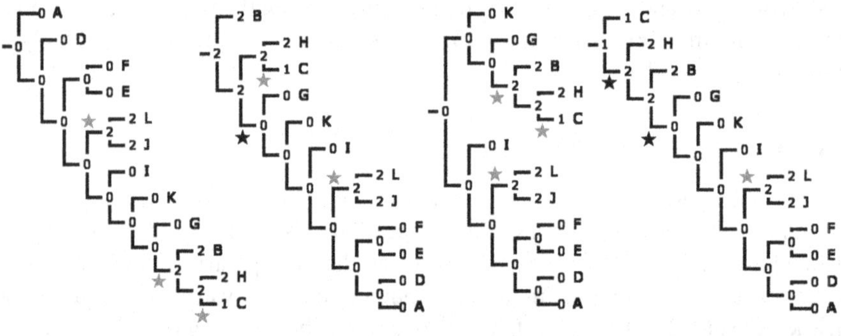

FIGURE 3.5 As long as all state transformation costs are symmetrical (e.g. as in additive or nonadditive characters), changing the root of the tree does not change the states that can be most parsimoniously assigned to nodes or the number of steps of the tree. A most parsimonious reconstruction continues being most parsimonious after changing the root, and the character state changes (stars) are located on the equivalent branches (at the most, they can change the direction of the transformation, as in black stars).

3.10 POLYTOMIES, MULTIPLE *MPTS*, AND CONSENSUS

The algorithms for generalized and step-matrix optimization were presented for binary trees, but they can be easily modified to handle polytomies (by adding more terms for each of the branches descended from a node). The Fitch (1971) and Farris (1970) algorithms work for binary trees only. For nonadditive characters, it is possible to use state-set operations for handling polytomies (which easily translate into bitwise operations and allow parallelizing operations into multicharacter algorithms; Goloboff 2002). In the case of trichotomies, the middle descendant node $\boldsymbol{M}$ must be considered in addition to $\boldsymbol{L}$ and $\boldsymbol{R}$ to determine $\boldsymbol{n}$ and $\boldsymbol{B}$ (the backup state set) in the down pass:

1) if $(L \cap M \cap R \neq \emptyset)$
  $n = L \cap M \cap R$
  $B = (L \cap M) \cup (L \cap R) \cup (R \cap M)$
2) else if $(L \cap M) \cup (L \cap R) \cup (R \cap M) \neq \emptyset$
  $n = (L \cap M) \cup (L \cap R) \cup (R \cap M)$
  increase length in 1 step
  $B = L \cup M \cup R$
3) else
  $n = L \cup M \cup R$
  increase length in 2 steps
  $B =$ **all possible states**

In the case of binary characters, case 3 is impossible, so the algorithm can be further simplified. For polytomies involving more than 3 nodes, the algorithms are considerably more complicated (see Goloboff 2002 for details). The advantage of the backup state set, $\boldsymbol{B}$, is that then the subsequent up pass can be performed by considering only the backup state set, without regard for whether the node is a bifurcation or multifurcation.

In the case of Farris optimization, no simple operations allow optimizing multifurcations. Either algorithms similar to those for generalized optimization must be used or (usually preferable) the characters must be recoded in binary form and optimized with simplified algorithms. This is the approach taken by default in TNT (see Section 2.14.2 for alternatives).

3.10.1 Length of Polytomies and Their Resolutions

The most salient feature of polytomies is that they always have a length greater than or equal than any of their possible resolutions. It is easy to see why resolving a polytomy may never lead to an increase in tree length. In a complete bush, some state will be assigned to the root (the only node) most parsimoniously; terminals with a different state connected to that node will then have independent parallel derivations of their state. As the tree is resolved, some of the terminals will now form groups; if grouping two taxa with the same state and different from the root, one fewer step will occur—what were independent derivations is now a single synapomorphy. If grouping two taxa with different states, their states will continue being derived independently. If grouping two or more taxa with the same state as the root, the length will either continue being the same or will be shorter if the grouping now makes a different state more parsimonious at the root. For the latter, consider the case of the polytomy (0001111), where state 1 is optimal at the root, implying 3 steps (3 independent derivations of state 0). Grouping two of the taxa with state 1 produces (00011(11)), with root now being either 0 or 1 (each option producing 3 steps). Grouping the two 1s that remain connected to the root produces (000(11)(11)). In the latter case, three taxa with state 0, and two groups with state 1, are connected to the root, obviously producing 0 as the most parsimonious assignment for the root (and two independent derivations of state 1, instead of three derivations of state 0 as in the original polytomy).

That the length of no resolution can ever be more than that of a polytomy also agrees with the idea that polytomies are only an expression of ignorance; positive evidence to prefer a polytomy over its resolutions cannot exist. Whenever a polytomy is of optimal fit (e.g. in the absence of data), any of its resolutions will also be of optimal fit and therefore cannot be ruled out as a possibility.

3.10.2 Polytomies as "Soft"

That polytomies are intended to include as potential possibilities their resolutions suggests that, in some contexts, it may be useful to calculate (Maddison 1989) the length that the characters (or the matrix, more generally) would have on the best possible resolution for the polytomy. This is what Maddison (1989) called the "soft" interpretation of polytomies (as opposed to the "hard" interpretation implicit in standard algorithms, which calculate the number of steps effectively required by multiple, simultaneous speciation). For nonadditive characters, state sets can be calculated in a down pass in the standard manner for bifurcations but with a modified procedure for multifurcations. For multifurcating nodes, the method chooses the combination of states with the minimum possible number of states such that no steps are implied in every one of the descendant branches; the number of steps is increased by the number of states minus 1. For additive characters, if no states are shared by all the descendants of the polytomous node, then an interval from the smallest state in the largest set to the largest state in the smallest set is assigned to the polytomous node, and the length increase equals the difference between those two limits.

Optimizing polytomies as soft has the peculiarity that (contra hard polytomies) the more polytomous the tree is, the shorter the resulting length—it potentially contains more resolutions. The complete bush optimized as soft provides the minimum possible number of steps for each character—no binary tree can have fewer than those steps. Keep in mind that optimization of soft polytomies calculates the minimum number of steps a character could have on the *best* resolution, but it does so *separately* for every character. The best resolution for character A may be very different from the best for character B, and then the lengths obtained by soft-polytomy algorithms may not be realizable on any actual tree. This is perhaps the most important caveat of soft-polytomy optimization, as noted by Maddison (1989). Another obvious problem is when the algorithms are applied to consensus trees—the length obtained for a given character may be obtainable by choosing resolution ***X***, but even if the consensus contains ***X*** as a possible resolution, the *MPTs* used to produce the consensus (i.e. the ones with actual optimal fit) were actually ***Y*** and ***Z***, not ***X***.

3.10.3 Informative Characters

The optimization of a complete bush with hard- and soft-polytomy algorithms produces the maximum and minimum possible number of steps for each individual character. No tree can have more steps than the maximum (hard interpretation), and no tree can have fewer steps than the minimum (soft interpretation). It is entirely possible that the minimum and maximum are the same. In such case, the character provides no grounds for preferring some trees over others, and it is then called *uninformative*. Whether a character is informative or not may depend on the transformation costs used. The most obvious case of an uninformative character is an invariant one, but it is not the only one.

In nonadditive characters and in the absence of polymorphic taxa, the maximum number of steps will equal the number of terminals that do not include the most frequent state. In that case, the minimum will be the number of states minus 1. If there are as many distinct states as taxa, the character is not invariant, but minimum = maximum, and thus the character is uninformative. Recall that Section 3.7, "Ambiguity, Polymorphisms, Missing Entries", discussed that single-taxon states of nonadditive characters can be replaced by a missing entry without affecting tree choice; when all but one of the states are unique, they all can be replaced by missing entries. If a single state remains unchanged to missing entries, the character (now composed of either a single state or missing entries) is also uninformative. In lineal additive characters (and no polymorphisms), the minimum possible number of steps on any tree equals the numerical difference between maximum and smallest state. The (sum of) minima calculated from the binary variables that result from recoding any additive character (see Section 3.4) is the same as for the original character, and thus this is the easiest method to determine minima in branched additive characters. When some terminal taxa include polymorphisms, determining whether minimum and maximum possible numbers of steps are the same is more difficult, requiring explicit application of hard- and soft-polytomy algorithms. The criterion to determine informativeness is, however, exactly the same as when there are no polymorphisms (i.e. the possibility of some trees being longer than others).

In the case of step-matrix characters, the maxima can be calculated by optimizing a bush, but determining the minimum is much more involved. The minimum possible number of steps is required for application of implied weighting, and TNT calculates the minimum for step-matrix characters either by (a) trying different mini-trees, including in the mini-tree-only distinct states (or state sets), and possibly making some states intermediate between others, or (b) performing a tree search for the character in isolation (including a single terminal for every distinct state or state combination).

3.10.4 Mapping Multiple Trees

In phylogenetic analysis, obtaining multiple *MPTs* is common, and an economical way to present the information common to those trees is by means of consensus trees, i.e. trees with polytomies. This poses the problem of how to obtain lists of synapomorphies, or summaries of character state transformation, for those consensus trees. It should be clear at this point that, if polytomies are optimized as hard, the consensus may be longer than any of the *MPTs* used—and thus should not be used to produce lists of synapomorphies—and if polytomies are optimized as soft, the resulting tree lengths can never be observed on any single tree (and for individual characters, possibly in none of the *MPTs*). While it is common to see papers where the consensus is simply used for producing diagnoses for groups, that course of action is far from ideal. A case where the synapomorphies concluded form the consensus are incorrect is shown in Figure 3.6a. Suppose that a parsimony analysis produces the two *MPTs*

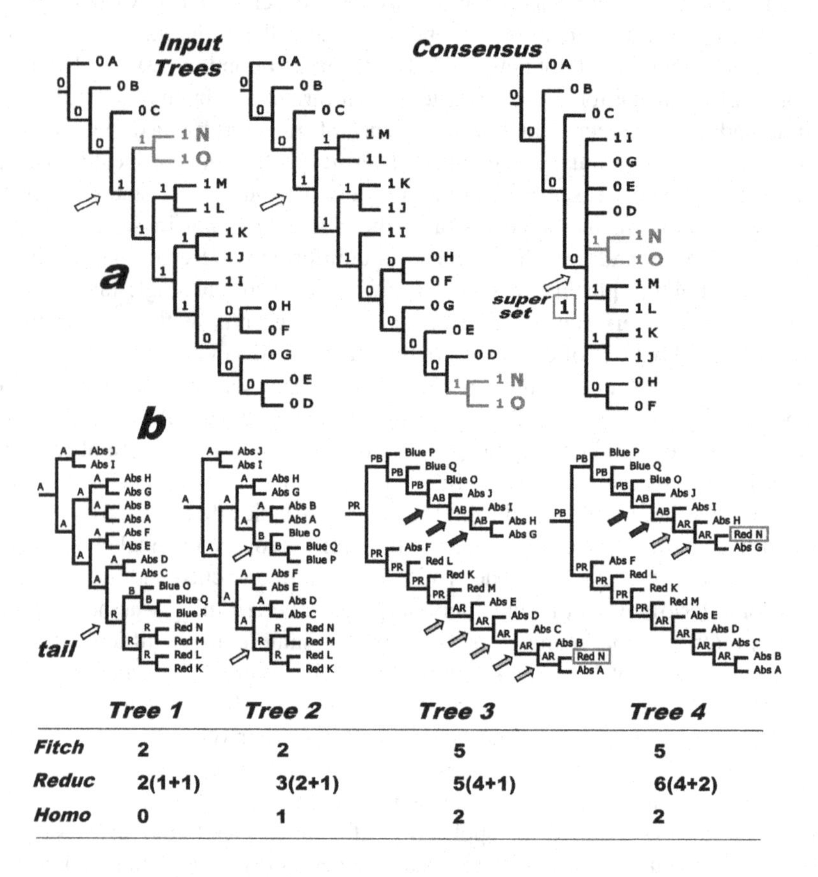

	Tree 1	Tree 2	Tree 3	Tree 4
Fitch	2	2	5	5
Reduc	2(1+1)	3(2+1)	5(4+1)	6(4+2)
Homo	0	1	2	2

FIGURE 3.6 (**a**) Proper diagnosis must be effected on the most parsimonious trees, not their consensus. Synapomorphies for a clade (arrow) present in two trees may not appear as synapomorphies when the character is mapped on the strict consensus of the input trees. The opposite case can occur as well (i.e. the character providing no synapomorphy for the shared clade is mapped as having a synapomorphy on the strict consensus). (**b**) A case where a tail can be present or absent and, if present, red or blue. Tail color is inapplicable when tail is absent. The results of different methods for scoring trees and selecting ancestral state reconstructions (Fitch, reductive coding, and counts of homoplasy) are shown below the trees. See text for additional discussion.

shown on the left. On each of those trees, the character shown is mapped as a synapomorphy for the group indicated with an arrow, with a change 0→1. But mapping the consensus as a hard polytomy has state 0 at the common node for the same group so that using the consensus for mapping would make us miss a change characterizing the group in each of the individual *MPT*s. It is easy to construct examples where the consensus has the opposite problem (i.e. to conclude that a group is characterized by changes occurring in none of the individual *MPT*s).

The best course of action, in such a situation, is to map the multiple, individual *MPT*s. The lists of synapomorphies can be recorded for each branch; synapomorphies occurring for each group in every one of the *MPT*s are then summarized on the consensus. The same can be done for character mappings, creating a *superset* of the *MP*-sets for each node occurring in all the trees and displaying it on the consensus, as shown in Figure 3.6a. Note that the supersets do not result from optimizing a tree; they are instead summaries of multiple individual optimizations. The process of optimizing multiple *MPT*s is fully automated in TNT.

3.11 INAPPLICABLES

Inapplicable characters are widely considered as one of the most problematic questions in phylogenetics. A vein in the wing can be observed only in those species of insect where wings are actually present; in insects with secondarily lost wings, neither the veins nor the wing size, folding, articulation, scaling, hamuli, color, or shape can be observed. All those wing characters are *inapplicable* to insects without wings. Inapplicable characters are often scored as a missing entry because the inapplicable character cannot be assigned any state. But clearly this is different from the usual implication of a missing entry. A missing entry is a state we have been unable to observe *so far*; those wing characters will *never* be observed in wingless insects.

Scoring inapplicable characters as missing entries would seem at first straightforward but may produce unexpected problems. Maddison (1993) first called attention to this and noted that no conventional treatment of the situation is satisfactory. Consider the case of Figure 3.6b, where a given structure (say, a tail) can be absent (A) or, if present, red (R) or blue (B). A possible treatment for this situation is considering a single nonadditive character (nonadditive because transitions A→B→R or A→R→B seem equally plausible). But this treatment is problematic in that states B and R imply a homology (the presence of a tail) which is not in A; the nonadditive costs imply that even on tree 1, reconstructing the character as two independent tail gains (A→B and A→R) requires only 2 steps (just like the A→R→B reconstruction shown together with tree 1). Thus, for a nonadditive coding, grouping taxa with tails separately costs the same as placing them side by side. Tree 2 separates the tailed taxa (thus requiring homoplasy in the tail origin, which tree 1 did not), and thus it should be judged as worse than tree 1. Because of that, a so-called *reductive* or *contingency* coding is often used, with two characters, the first indicating absence (A) or presence (P) of a tail and the second character (R, B) indicating color. For taxa with A, the second character is scored as a missing entry. As Maddison (1993) pointed out, this also creates problems of interpretation and inadequate comparisons between trees. Consider tree 3 first. The two characters are reconstructed independently; the nodes with tail absent or present are correctly labeled, but some nodes (indicated with arrows) are reconstructed as having a tail absent and red or absent and blue! Standard algorithms for optimization are based on attributing some state (the color which maximizes homology among terminal states) to each internal node; they cannot assign a non-state or missing entry to an *HTU*. Furthermore, because internal nodes must have *some* state, separating areas of the tree with different colors by intermediate taxa with no tail implies that some step in the color character must occur somewhere along the path between the two regions of the tree with color, even if the regions are interconnected by taxa with no tails, as in tree 4. This means that a gain of R costs a single step if happening deep within a group of tailless taxa surrounded by taxa with R (tree 3) but 2 steps if happening within a group of tailless taxa surrounded by taxa with B (tree 4). It is as if the second character had "memory"; the tailless taxa between the B and R states had no tail but somehow preserved the "predisposition" to have a B tail, not an R one. Tree 4 is then judged by the reductive coding as worse than tree 3, but both trees should be considered as providing equally good explanations.

A compromise solution offered by Maddison (1993) within the context of standard optimization methods is to define a single step-matrix character, where each combination of conditions is scored as a state. In that way, ancestral conditions cannot be artifacts (as in composite coding), for they are assigned only real states. If the cost of gaining a tail is more than both the cost of losing it and the cost of transforming between different conditions, then reconstructions implying homology of the tail will be preferable. However, Maddison (1993) could not satisfactorily solve the problem of what costs to apply to the different transformations because in his view there was no way to settle the question of how much more a tail gain should cost (2 or 100 times?).

Since Maddison (1993) discussed the problems with inapplicable characters, many papers have discussed the problem but few have contributed much clarity. Lee and Bryant (1999) did not go far beyond simply restating Maddison's discussion. Strong and Lipscomb (1999) argued that eliminating zero-length branches (see Chapter 6) with more stringent criteria solves the problem; their claim is incorrect: the problem has nothing to do with zero-length branches. Forey and Kitching (2000) examined numerous alternative scoring combinations, but without resorting to first principles their discussion is hard to follow and interpret.

The solution to this problem came with De Laet (2005, 2015), who used Farris's (1983) justification of parsimony to find a rationale for analyzing inapplicable characters. He proposed that the criterion used to select from among alternative reconstructions (and trees) should be that of maximizing homology. That is, to the extent that similar features occur in separate groups, the groupings of the tree are failing to explain that similarity by homology. Farris (1983) showed that, in the absence of inapplicables, total amount of change provides a direct measure of the number of similarities not explained by homology—that is, homoplasies. De Laet (2005) argued that, in the case of inapplicable characters, there is no such correspondence. He proposed to measure the degree of homoplasy implied by a given reconstruction with the sum of three quantities: the number of steps of the dependent character in the regions of applicability, the number of tail gains and losses, and the number of "subcharacters" or regions of applicability. The connection of the latter term with homology is, while correct, not immediately obvious. An alternative formalization was given by Goloboff et al. (2021b). Goloboff et al. (2021b) proposed to evaluate alternative reconstructions for the character complex, counting the number of similar features independently derived beyond the first (excess or *x-transformations*). Reconstructions implying fewer x-transformations are to be preferred. This allows choosing reconstructions on the basis of their requirements of homoplasy, without the need to consider that homologizing tail presences is preferable a priori to homologizing absences. This can be illustrated with an example where two tails, with three dependent features (e.g. color, texture, and shape), originate in far distant points of the tree, separated by tailless taxa. If the tails in both separate groups are similar (e.g. red, smooth, and sinuous), then the reconstruction which assigns a tail absent to all the ancestors intermediate between the two groups implies that we do not account for by homology the similarities in tail presence, color, texture, and shape—four instances of homoplasy. The alternative reconstruction that assigns a red, smooth, and sinuous tail to all intermediate ancestors accounts for by homology those four similarities, but it fails to account for by homology the similarity in tail absences in the intermediate, tailless taxa. Depending on how many tailless taxa (or groups) separate the two tailed taxa, one or the other reconstruction can be preferable; the choice between both can be done only on the basis of their implications on homology.

Consider now an alternative case where two tailed taxa are separated by several tailless taxa, and successive sister groups are also tailless, but where the tails *differ*—red, smooth, and sinuous in one taxon and blue, scaly, and straight in the other. Assigning a tail absent to intermediate ancestors will imply that the tail presence in both tailed taxa is not due to homology, but beyond that, there are no further similarities to explain between the two tailed taxa. Thus, just one tailless taxon separating the two tailed taxa is enough for the assignment of a tail to intermediate ancestors to allow the same amount of homology as the assignment of no tail. That would be just one similarity not accounted for by homology. If intermediate ancestors are reconstructed as tailless, the unaccounted similarity is in the independent presence of tails. If ancestors are reconstructed as tailed instead, the unaccounted similarity is in the independent absence of tails.

It must be emphasized that De Laet (2005) and Goloboff et al. (2021b) proposed the same criterion—only the way it is formulated differs. Finding the reconstructions that maximize homology given the dependence in character hierarchies is difficult. De Laet's (2018) program *anagallis* implements several experimental algorithms, but finding the optimal reconstructions can be a slow process in cases of several dependent characters and much homoplasy. Faster approximate algorithms (implemented in the program Morphy) were proposed by Brazeau et al. (2019), but those can fail to find optimal reconstructions (De Laet 2019; Goloboff et al. 2021b).

Understanding the problem of inapplicables as one about the degree of homology implied by different reconstructions allows reconsidering approximations by means of step-matrix recoding, which Maddison (1993) had discarded because he could not resolve the problem of how much higher gain costs should be. This can take advantage of well-defined algorithms for finding optimal reconstructions under complex transformation costs in a conventional character and shortcuts used to speed up calculation of tree lengths during searches (Goloboff 1998b). Goloboff et al. (2021b) proposed such an approximation and implemented it in TNT. TNT can automatically recode the character hierarchy, with the hierarchy embedded in the character and state names so that the recoded states represent logically possible combinations of features and assign appropriate transformation costs between recoded states. For single-level dependencies, where N nonadditive characters depend on a single one, transformations between two tails with every one of the N features having a different state must cost N steps—the number of similarities not accounted for by homology when two identical tails originate in distant parts of the tree. A change from no tail to any tail must cost $N + 1$ steps—one step for each of the dependent characters and one for the tail itself. The absence of a tail must cost a single step—two separate tail losses fail to homologize just the absence of a tail. This scheme might seem to overcount some transformations. For example, the regain (a) of a tail (previously lost several nodes down the tree) with each of the N dependent characters in a different state will cost $N + 1$ steps, just like the regain (b) of a tail with the N features in identical state. The relevant question, however, is the step *differences* with alternative trees in every case. In case (a), placing the taxon with the regained tail together with the taxa having a different tail will still require N steps (the transformations between each of the differing dependent features), so the *difference* in steps with the original tree is just 1 step—which correctly matches the homology cost of the tail being homologized or not. In case (b), placing the taxon with the regained tail together with the taxa having a different tail will require no steps to transform (tails are identical), so the difference in steps with the original tree is $N + 1$ steps—correctly matching the difference in homology between the two trees. Thus, the *ranking* of the trees produced by the step-matrix recoded character is exactly the same one as in *anagallis* (provided that the state at the root is forced to be a tail present, when the first splitting taxon in the tree is tailed; see Goloboff et al. 2021b for details and discussion). The *homoplasy* (i.e. number of independent originations of the same feature in the original character) is the difference in steps relative to the minimum possible number of steps in the recoded character. As the number of steps to assign to different transformations in the recoded character is unambiguously determined by considering the degrees of homology, this solves Maddison's (1993) dilemma of how to assign costs in the step-matrix. The goal of the step-matrix is not to make tail presences preferably homologized over absences but instead to properly count the numbers of features that are or are not homologized in a given reconstruction.

That approach also allows considering multiple levels of dependence, where a character is dependent on a character which in turn depends on another, thus defining primary, secondary, and tertiary characters. In that case, the costs of transformations in the recoded character representing gains in the primary character will differ, according to whether the secondary characters are absent or present (see Goloboff et al. 2021b). By properly considering the numbers of similarities being homologized, appropriate costs can be assigned in such a way that results identical to those of *anagallis* can be obtained, but with faster calculations. The drawback of such an approach is that the recoded character cannot have more than 32 states (because that is an implementation limit used in TNT), representing the possible combinations of features in the dependent characters.

3.12 REALIZABILITY OF ANCESTORS

An important aspect of evaluations under parsimony is that they are based on specific reconstructions of ancestors. The ancestral states and numbers of character changes reconstructed by parsimony may or may not be historically accurate, but they are certainly possible. This is also true of maximum likelihood methods (see next chapter).

The same is not true of some methods for inferring phylogenies, notably those (discussed in previous chapters) based on distances and three-taxon statements. Distances are often considered as a synonym of phenetics (i.e. of grouping by overall similarity, which amounts to ultrametric groupings). However, distances can be analyzed in such a way that ultrametricity is not assumed, using methods such as those pioneered by Fitch and Margoliash (1967) and the immensely popular neighbor joining method (Saitou and Nei 1987). Swofford et al. (1996) and Felsenstein (2004) provided detailed reviews of distances in phylogenetic inference. Chapter 1 briefly discussed these methods and showed that (under specific circumstances) they can approximate the results of cladistic analysis. That raises the question of how far the analogy between distance- and character-based methods can go. As shown in the following, the equivalence between distance fitting and cladistic analysis is not complete.

Distance methods are based on adjusting the lengths of the branches of a candidate tree in such a way that they are as close as possible to the observed distances. Distance data that can be better fit to a tree are generally considered to be more reliable. Some methods, such as those based on immunological distances (Sarich and Wilson 1967) or DNA hybridization (Sibley and Ahlquist 1984), very popular in the 70s and 80s, are based on data that directly come in the form of distances. Discrete character data can be converted into distances, of course, although it is usually understood that some information is lost by so doing (obvious when one considers that it is possible to go from discrete data to distances, but not the other way around, and that different discrete matrices may produce exactly the same sets of distances).

Farris (1981; see also 1985, 1986b) was the first to point out serious problem with distance analysis. He noted that, even in the most favorable cases, where metricity is guaranteed (e.g. as in the Manhattan distance), the values of branch lengths, and thus tree scores, obtained for some data inputs cannot correspond to any physical reality. He discussed the case of a single character, and three taxa X, Y, and Z, each of which is observed to have a different nucleotide, A, G, and C. The Manhattan distance between each pair of taxa is 1. Fitting these distances into a trichotomy (so that X, Y, and Z all connect directly to common ancestor w), where each branch has a length 0.5, has a perfect fit to the original distances. The problem comes in that there is no nucleotide that common ancestor w could have had such that it is 0.5 units of distance away from each of the terminals at the same time. In Farris's (1981: 21) own words, "If the branch lengths add up to the observed Manhattan distances, then they cannot themselves be Manhattan distances". The branch lengths and tree scores, in this case, cannot be interpreted as reflecting any physical reality. This is not to say that they are incorrect—they are *impossible*. A similar problem occurs when distances are used to infer phylogeny from data on allele frequencies (Farris 1981);

if three populations P1, P2, P3 are each fixed for a different allele (***A1***, ***A2***, ***A3***), the Rogers (1972) distance between each pair of taxa is 1. Again, fitting these distances to a trichotomy with branch length 0.5 produces perfect fit. If the distance between the common ancestor and ***A1***, ***A2*** is of 0.5 each, the ancestor must have had 50% of allele ***A1*** and 50% of allele ***A2***. But when the distances from common ancestor to ***A1***, ***A3*** are considered instead, the paradox arises: the ancestor must have had 50% of ***A1*** and 50% of ***A3***. Needless to say, 50% of ***A1***, 50% of ***A2***, and 50% of ***A3*** is impossible. And if the lengths of branches between the common ancestor and the terminals are interpreted to be something other than Rogers distances, what are they, and why are they to be minimized? In the words of Farris (1985: 68), the branch lengths inferred by distances are "not just an underestimate as that term is properly employed. The inferred value is outside the range of possibility". The freqpars method of Swofford and Berlocher (1987), mentioned prior, was designed precisely to avoid this problem by assigning specific (and realizable) frequencies to ancestors individually, instead of working with distances between terminals.

When molecular data are converted into distances, they are often corrected according to some assumed model of evolution. With such corrections (and if the data actually evolved under the model), then distances are likely to produce the true tree. But the problems of interpretation remain, and in the case of morphology, no model to produce the correction is generally accepted. In another example from Farris (1981), data which cannot be explained without homoplasy, such as,

```
N 1 1 1
O 1 0 0
P 0 1 0
Q 0 0 1
```

produce Manhattan distances of 2 among every pair of terminals so that fitting a bush with all four branches of length 1 has perfect fit. Perfect distance fit, therefore, does not mean that the underlying data are homoplasy free, for the data shown cannot have evolved with fewer than 2 steps of homoplasy. It is impossible for the data to have evolved with fewer than 5 steps, yet the length of the fitted distance tree is 4, thus showing that branch lengths fitted from distances cannot be interpreted as amounts of evolutionary change—or anything at all. And if the data had consisted of four characters, each with an autapomorphy in a different terminal, the pairwise distances (and the best distance tree) would have been the same. Once the data are converted into distances, distinguishing the two situations—data with and without homoplasy—becomes impossible. Another extreme example is provided by Huson and Steel (2004); with nonadditive characters such as,

```
A   0 0 1 0 0 1 1
B   0 1 0 1 1 0 0
C   1 0 2 2 2 0 0
D   1 2 0 0 0 2 2
```

the most parsimonious tree (AB)(CD), determined by the first (and only) informative character, can explain these data without homoplasy. The Manhattan distances between terminals are $D_{(AB)} = D_{(CD)} = D_{(AC)} = D_{(BD)} = 6$ and $D_{(AD)} = D_{(BC)} = 5$. Those distances can be fit without distortion to a tree (AD)(BC), with terminal branch lengths of 2.5 and middle branch of 1. As the distances fit the tree perfectly, the distances are "perfectly misleading".

The method of three-item analysis, although initially presented as a refinement of parsimony (Nelson and Platnick 1991), suffers from similar problems, as discussed in Chapter 1: the numbers of statements accommodated on a tree as "homology" correspond to ancestors that cannot have existed in a real three-dimensional world, ancestors that at the same time had and lacked the feature. These problems of interpretation arise because distances, or three-item analysis, do not work by actually assigning specific conditions to ancestors. The reconstructions used to evaluate a tree under parsimony may or may not be correct, but they are indeed possible.

3.13 IMPLEMENTATION IN TNT

The different commands for diagnosis and character mapping in TNT automatically take into account the type of character (including weights and transformation costs, as defined with the `ccode`, `cost`, `smatrix`, and `ancstates` commands, discussed in Chapter 2) and invoke the proper algorithms for reconstructions. A summary of the commands is given in Table 3.1.

3.13.1 Options for Scoring Trees

The number of steps for the trees in memory is calculated with *Optimize > Tree Length* (*GUI*) or the command `length`. Quick access to tree lengths in Windows is with the ruler tool. The individual character steps are calculated with *Optimize > Character Scores*; in the case of polytomies, they can be treated as hard (by default) or soft (by checking the corresponding box); the equivalent command is `cscores` (treating polytomies as soft with `cscores!`). As discussed already, you must keep in mind that the global lengths reported by soft optimization of polytomies may not be realizable on a single tree. The command `chomo` shows the homoplasy (the steps on the tree minus the minimum possible steps), either as hard (default) or soft (`!` option) polytomies.

These commands are applied to trees currently held in memory, regardless of where those trees came from (e.g. a tree search, randomly generated, or read from a file; recall that each of those options places the trees in tree groups with appropriate names, as per the `tgroup` command, so that tree lengths can be calculated for different sets of trees by referring to the number or name of the group enclosed in braces in the list of trees). The trees currently held in memory can be evaluated and sorted (by decreasing optimality) with *Trees > Tree Buffer > Sort Trees* or the command `sort`. The command `sort` also allows ordering trees by increasing numbers of nodes (i.e. resolution, with `sort=nodes`) or number of constraints violated (with `sort=mono`). The `best` command calculates the length of trees and discards suboptimal ones (with *Trees > Tree Buffer > Filter*).

For polymorphic terminals, the completion of ranges in additive and stepmatrix characters is automatic (e.g. a terminal with `[05]` in an additive character is

TABLE 3.1
Commands for Mapping Characters, Producing Lists of Synapomorphies, and Sorting or Handling of Trees

Command	Minimum Truncation	Action(s)
`ancstates`	`an`	For characters with asymmetric transformation costs, force outgroup states to root
`apo`	`ap`	Show synapomorphies, as tree plots or lists of text, for individual or multiple trees
`best`	`bes`	Calculate scores for all trees now in memory, discard suboptimal ones
`blength`	`bl`	Show branch lengths, in tree plots or tables
`change`	`ch`	Show min/max numbers of specific changes in possible most parsimonious reconstruction
`chomo`	`cho`	Show individual character homoplasy (i.e. steps beyond minimum)
`cscores`	`cs`	Show individual character scores
`fillsank`	`fill`	For polymorphic terminals in step-matrix characters, add intermediate states (default) or don't (– option)
`info`	`inf`	List informative characters
`length`	`le`	Report tree lengths, with polytomies optimized as hard (default) or soft (! option)
`map`	`map`	Plot most parsimonious state sets on trees (default) or state supersets (for multiple trees)
`minmax`	`mi`	Report min/max possible numbers of steps for each character
`nstates`	`ns`	Determine how states are read; convert data formats
`outgroup`	`o`	Designate the terminal (or group) used to root the tree (default: first taxon in the matrix)
`recons`	`rec`	Show all most parsimonious reconstructions
`reroot`	`rer`	Change tree rooting
`scores`	`sco`	Show tree scores (i.e. steps, by default, or weighted homoplasy, if using implied weights)
`smatrix`	`sm`	Read and apply step-matrices to characters; handle inapplicable characters
`sort`	`so`	Sort trees on the basis of scores (or size, resolution, or constraints)
`tequal`	`te`	Compare and report identical trees
`xcomp`	`xc`	Compare fit of trees on a character-by-character basis
`unique`	`u`	Compare trees in memory and discard duplicates
`usminmax`	`us`	For step-matrix characters, instead of minimum numbers of steps calculated by a heuristic method, use a value defined by the user

converted to `[012345]`); thus, placing the terminal within a group with either state 2 or 4 (inside the range) would cost no extra steps. The command `fillsank` sets whether polymorphisms in step-matrix characters are completed (`fillsank=`) or not (`fillsank-`). The command `cstree` allows defining state distances as in a lineal character (e.g. `0–1–2–3. . . n`) and sets whether character state transformations that can be expressed as sets of binary variables are so treated internally (`cstree=`, the default) or not (`cstree-`); this can be used to handle alternative interpretations of polymorphisms (i.e. counting steps to the closest state observed for the terminal, without filling ranges). For the `[05]` example, placing the terminal within a group with state 2 would then imply two steps (the distance 2-->0), while placing it in a group with state 4 implies one step (4→5).

The informativeness for the characters (which may depend on their additivity or step-matrix costs) is inspected with `info`. The option `info+` reports only the informative characters and `info-` the uninformative. A character is uninformative when the minimum number of steps on any possible binary tree equals the maximum; maxima and minima for each character are shown with `minmax` (or *Data > Show Min-Max Steps*).

The `ancstates` command, followed by a list of asymmetric step-matrix characters, serves to force the state of the outgroup to occur in the root of the tree; the same is done with *Data > Character Settings* and the choice *Make the selected characters . . . ancestral.* This is roughly equivalent to the `pset ancstates` option of PAUP*.

In step-matrix characters, sometimes a transformation via a state not observed in the matrix may produce a reconstruction as (or more) parsimonious, if the state is intermediate between the other (observed) states. Consider a tree (0 (0 (1 2))), with all transformation costs defined as 2 and transformation to (or from) the unobserved state 3 from (or to) any other state as 1. If the only candidates for assignment are states 0–2, the best reconstruction on that tree has 4 steps (e.g. with 0→1 and 1→2, each transformation of cost 2). If state 3 is a candidate as well, then a change 0→3, followed by 3→1 and 3→2, produces a total of 3 steps. By default, TNT will consider possible state assignments only between 0 and the largest observed state. This can be changed with the `usminmax` command. The need to do so will rarely arise in morphology, except when several inapplicable characters are combined into one by means of the step-matrix scheme explained in Section 3.11, "Inapplicables"—in which case reconstructions with unobserved combinations of features need to be considered (see Goloboff et al. 2021b for discussion).

3.13.2 Diagnosis and Mapping

Lists of synapomorphies are obtained with the `apo` command or with *Optimize > Synapomorphies* (*GUI* versions). With the *Optimize > Synapomorphies > Map GUI* options (or the default of `apo`), the numbers for the characters unambiguously changing along each branch are plotted on the trees (this does not show the actual state changes). A character that changes in every possible *MPR* is identified when the ancestral and descendant state set of a branch do not share any states. With the *Optimize > Synapomorphies > List* options (or `apo-`), the lists are shown as text; only in this case are the character state changes (and the state names) shown; the node numbers refer to the internal numbering of TNT (node numbers can be seen by plotting the trees under `naked-` or by clicking on the "naked" tool). To plot or list synapomorphies on multiple trees, select the *Common* options for *Optimize > Synapomorphies* or use the `apo[` option. This automatically calculates the (strict) consensus, optimizes all the trees, and plots the synapomorphies on the corresponding nodes; in the case of lists, node numbers refer to the consensus of the trees used to produce the synapomorphy lists. Note that the consensus corresponds to current settings (e.g. possibly considering temporary collapsing, see Chapter 6, Section 6.4.4, and, if so, according to character types and activities). Finally, an option that uses lists on tree diagrams is *Optimize > Synapomorphies > Text Lists on Consensus* or the `apo]` and `apo>` commands (the latter shows state changes, the former does not). This plots unambiguous changes on tree branches, one synapomorphy per line. Beware that for large datasets, the resulting trees may be tall and wide; choose concise character names!

Lists or plots of synapomorphies are useful for diagnosing groups, but they do not allow reconstructing character evolution (because ambiguous changes are not shown). Evolution of characters is best analyzed by mapping, with *Optimize > Characters* (in the *GUI* version) or the `map` command. As for synapomorphies, the *Optimize > Characters > List* options (with commands, `map-`) produce text lists of all the nodes at which there is a change in state sets, and the *Optimize > Characters > Character Mapping* options (default `map` options) show the *MP*-sets for each node; the *Optimize > Characters > Common Mapping* option (or `map[`) shows the supersets of multiple trees, plotted on the corresponding node of their consensus. By default, plots on trees use alphanumeric codes for the states, but with `cnames=` (or setting *Format > Use Character Names*), they can use the state names (if defined); in this case, to save space, the states are shown only at nodes for which the *MP*-set differs from that of its ancestor.

Branch lengths are a sort of "mapping"—the length of a branch indicates how many steps it costs to collapse the branch. Note that, in the case of ambiguous optimization, TNT calculates branch lengths as the minimum number of (weighted) character changes in alternative *MPR*s, so they can add up to less than the score of the tree if there are ambiguous optimizations. Branch lengths are obtained with *Optimize > Branch Lengths* or the `blength` command (the default gives a list, the `blength*` option plots branch lengths on a tree diagram). In the Windows version, it is possible to plot the branch lengths graphically by ticking the choice "*graphically*".

3.13.3 Diagrams for Publication

An introduction to saving graphic trees was given in Chapter 1. The character state mappings can also be displayed (in Windows versions) or saved (as *SVG* files, in any version) in color. To display the trees in color, tick on *Format > Map Characters in Color*; the option *Format > Preview Trees* must be ON for this. To save as an *SVG* file, you need to tell TNT first that tree-tags are to be stored (with *Trees > Multiple Tags > Store Tree Tags* or `ttag=`), then invoking the desired mapping option (via menus or commands; note that only a single character at a time can be mapped in color), then saving the tags (with the `color` option of the `ttag&` command or with *Trees > Multiple Tags > Export Tags as SVG > Color*). To save the next color diagram, you need to first clear and reconnect the tags (e.g. with `ttag-ttag=`), or they will be imposed over the previous ones. The general sequence is then (a) connect tags; (b) produce plot storing the tags; (c) save tags to *SVG* file; (d) clear tags and reconnect; and (e) go to next plot. This is summarized in Figure 3.7.

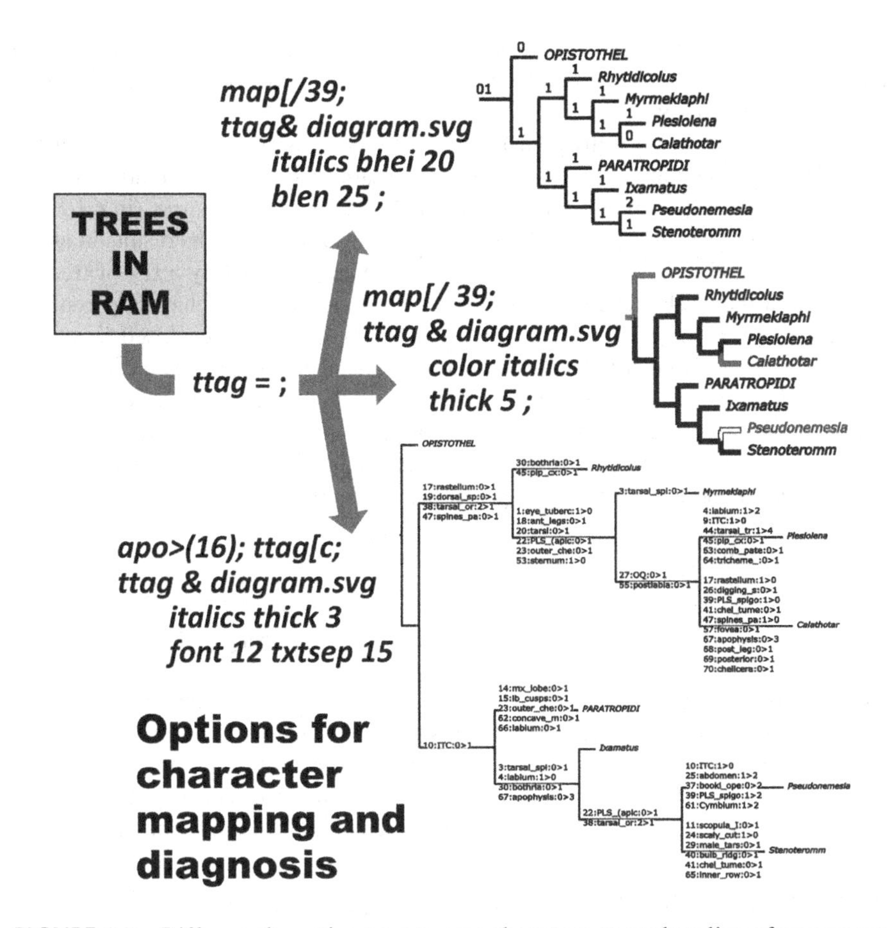

FIGURE 3.7 Different alternative ways to map characters or produce lists of synapomorphies, in graphic form. The option to store tree-tags must be enabled with `ttag=` and the diagram saved with the `ttag &` option. The `map` command saves character states to branch labels, which can be saved in numerical form (by default) or as a color diagram (with the `ttag & . . . color` option). The `apo>` option produces a list of synapomorphies for each branch of the tree (optimizing every tree and plotting on the branches only those changes that are synapomorphies on every tree).

The same process of storing the tags must be done prior to saving synapomorphies plotted on trees graphically; keep in mind that long tags may require you to change the maximum length of tree legends prior to reading the dataset with `taxname+N`. Details on how to handle branch lengths, thicknesses, and colors for the resulting diagram are given in Chapter 10 (Section 10.4.1).

3.13.4 Reconstructions and Specific Changes

The `map` command produces state sets, either in the form of lists or plotted on a tree diagram. In the case of ambiguity, the individual *MPR*s can be displayed with

Optimize > Characters > Reconstructions or the `recons` command. Individual reconstructions can also be processed (and manipulated) by means of the scripting command `iterrecs`.

While generating all individual reconstructions is rarely needed, counting the minimum or maximum number of steps of a certain kind of transformation may be more useful in studies of character evolution. The command `change` (in *GUI* versions, *Optimize > Count Specific Changes*) does this, presenting the results in tables of minimum and maximum values. The command is followed by a list of trees, a slash, and a list of characters, followed by a slash and the state change to count (a pair of states, `from to`). If either the `from` or `to` states are enclosed in square brackets, changes from or to any of the states indicated are counted; you can use `?` to indicate any state (never counting "changes" between identical states). For example, if a character `N` represents a structure that can be `absent`, or present shapes `oval`, `squared`, or `triangular` (with properly named states), the command `change` can be used to find out alternative types of transformations:

```
tnt*> change/N/absent oval                            absent to oval
tnt*> change/N/absent ?                               gains
tnt*> change/N/? absent                               losses
tnt*> change/N/[ov sq tri ] [ov sq tri ]              changes of shape
```

Rinsma et al. (1990) gave algorithms to count minimum and maximum numbers of changes, which work in three passes. TNT uses a different method, internally using a step-matrix which gives the transformations to minimize (or maximize) a slightly higher (or lower) cost. This has the advantage of allowing costs of ambiguous transformations (such as `absent` to any of the shapes), which Rinsma et al.'s (1990) method does not handle.

3.13.5. Selecting and Preparing the Trees to Be Optimized

The commands that optimize trees allow selecting the trees, but it is sometimes easier to just restrict the tree buffer to some trees. Several commands allow operating on the tree buffer, including changing or sorting the trees. The command `reroot` changes the root of subsequently specified trees (rerooting on current outgroup); `reroot =N` reroots on node numbered (or named) `N`. The trees are ordered from best to worst (according to current settings of character activities, additivities, etc.) with the command `sort`; the trees can also be sorted by resolution (`sort =nodes`), size (`=size`), or number of violated constraints (`=mono`, if constraints defined with the `force` command). The command `tequal` reports the trees that are identical; `unique` retains only the distinct ones (e.g. to eliminate duplicate trees from two different tree files). The command `best` retains only the best trees (according to current settings) and eliminates duplicates.

NOTES

1 The actual ratio of times, for s states and numbers of taxa T_1 and T_2, is s^{T2-T1} (i.e. for 4 states and 5 additional taxa, 1,024 times).

2 There are some exceptions to the decomposition into local problems. One is the optimization of 2D or 3D landmark data (discussed in Chapter 9). The decomposition is possible in principle but would require consideration of so many variables that it becomes impractical, requiring application of other methods. Another exception is the self-weighted optimization (described in Chapter 7, Section 7.8.3), where the (implied) transformation costs are determined from the reconstruction itself, thus precluding the use of a local decomposition and requiring evaluation of complete reconstructions. The algorithms discussed in this chapter (as well as those based on models, discussed in Chapter 4) deal with the simpler cases where the decomposition is indeed possible.
3 This needs some qualification when methods to store multiple characters in a single 32-bit or 64-bit word are used. It is correct in theory for single-character algorithms but not in practical implementation.
4 It is probably not a coincidence that Daniel Fisher (one of the developers of the "stratocladistic" method which gives fossil and stratigraphic data much importance; e.g. Fisher 2008) is one of the thesis advisors of Parins-Fukuchi and a coauthor of Parins-Fukuchi et al. (2019).

4 Models and Assumptions in Morphology

This chapter discusses the use of specific models of evolution for discrete morphological characters (*DMC*) which became common in the last decade. In the case of molecular sequences, models for phylogenetic analysis have been extensively used for a much longer time. General models of molecular evolution started with early proposals by Jukes and Cantor (1969), Neyman (1971), and Kimura (1980). Those models were extended into methods for evaluating alternative phylogenetic trees by Felsenstein (1981). The evolution of molecular sequences is much easier to model than that of morphology, and—details aside—the models used today for molecules seem to have a reasonable basis, both empirical and theoretical. Following Lewis (2001a), methods and models similar to those used for sequences began being applied much more commonly to morphology. For that reason, this chapter begins by introducing first the general ideas of models for molecular sequences, moving then to the problem of models in morphology.

The problem of phylogenetic reconstruction, from the point of view of authors defending the use of models (e.g. Felsenstein 2004; Swofford et al. 1996), is strictly a statistical one. From that perspective, a phylogenetic method can be evaluated *only* by the degree to which it is likely to deliver correct phylogenetic trees, and this necessarily requires reference to specific models of evolution. In many courses and introductory books, students and beginners are presented with a false dichotomy: would you prefer a method that is guaranteed to give you the right tree under the right circumstances, such as likelihood, or a method that can easily give you the wrong tree, such as parsimony? With the problem posed in that way, there is not much to think about, really. The catch is that the question itself is a serious oversimplification. While justifications based on the probability of finding truth are important, the peculiar situation of phylogenetic reconstruction involves some subtleties that make additional considerations about methods important. Science is not only about discovering truth; it is also about producing rational explanation and enhancing comprehension (see Emanuel 2020 for an interesting non-phylogenetic discussion of this topic). It is through theories that allow achieving rational explanation and comprehension that we believe a glimpse of reality can be obtained. A prime phylogenetic example where the goals of rationalizing and finding correct answers may depart is in the use of distances to obtain estimates of phylogenetic trees, a process which produces values that cannot correspond to any possible physical reality (see Chapter 3, Section 3.12). Yet distance fitting is a commonly used method for phylogenetic reconstruction, mostly because when large amounts of data are generated under specific models (and distances are corrected according to the model), distance data may produce reliable results—correct trees. Distance analysis can then be defended on the

DOI: 10.1201/9781003220084-4

grounds that it is likely (under certain conditions, at least) to produce reliable results, but there are problems with its internal logic under other (quite realistic) conditions. Distance analysis therefore cannot be said to enhance understanding under all circumstances—fitted branch lengths are incomprehensible in terms of any physical reality, regardless of whether the tree is correct or not. Standard maximum likelihood analysis is more sophisticated than distance analysis and does not suffer from such obvious lack of coherence, but the usual justification for the approach is also its ability to recover true trees. The point being made here is that this justification is short sighted, or incomplete at least, and that other considerations may be important.

The criterion of parsimony, under some possible conditions of evolution, can be justified in strict statistical terms; in addition, it can also be justified without reference to specific probabilities. For example, as discussed in Chapter 1, attributing as many observed similarities as possible to common ancestry with no other background assumption than descent with modification directly leads to parsimony (e.g. Farris's 1983 work), and so does maximizing the descriptive power of classifications (i.e. the ability to transmit and summarize information about taxa; Farris 1979b, 1982b). Given that evolution is, foremost, an explanatory device for understanding the pattern of observed similarities among taxa, the achievement of being able to discern that pattern in its purest form (i.e. without making specific assumptions as to the mechanisms by which evolution proceeds) is far from trivial. Such a justification cannot be easily summoned for model-based methods—they are typically premised *only* upon their ability to recover truth. Cladists prefer a broader justification of methods, but it is nonetheless true that many researchers are concerned solely with ability to recover truth and that such an aspect is not without importance.

Understanding the extent to which model-based methods are applicable to morphological datasets requires, first, considering the principles and properties of those methods (at least, the most ubiquitous) and, second, the properties of morphological data themselves. This chapter first provides a brief summary of the principles and methods used in model-based phylogenetics (much more detailed treatments can be found in Swofford et al. 1996; Felsenstein 2004; Bryant et al. 2005; Yang 2006) and then discusses the applicability of this approach to *DMC*s.

4.1 MAXIMUM LIKELIHOOD (ML)

Phylogenetic analyses using explicit models are based on the principle of maximum likelihood (*ML*; see Edwards 1992 for a classic treatise); this includes Bayesian analyses (as the posterior probability is a function of the likelihood, see Section 4.6). For many inferential problems, *ML* is generally capable of estimating parameters in a consistent manner. Statistical *consistency* is the property of an estimator to converge on the true value (the correct tree, in our case) when the data really evolve under the model and the amount of data tends to infinity. Consistency in the estimation occurs when the model is *identifiable*—that is, when (for any set of possible observations) there is only one set of parameters that maximizes the likelihood; some models (i.e. different ways in which characters could evolve on a tree) are identifiable, and others are not.

In the phylogenetic context, the *likelihood* L of a tree t is the probability of evolving the observations, D, given the tree:

$$L_{(t|D)} = P_{(D|t)}$$

The basis for *ML* estimation is that different hypotheses—trees, in our case—make the data more or less probable and that we should prefer those trees which make the data more probable. For a cartoonish phylogenetic example, imagine that there is a low probability of evolving wings, and we believe that pigeons and sparrows and owls have each evolved from different groups—from fish, toads, and mice, respectively. On that absurd theory of relationships, the parallel acquisition of wings in pigeons, sparrows, and owls would be an extremely unlikely event. Given a more reasonable tree placing pigeon, sparrow, and owl in the same group (e.g. next to hawks and grebes), the unique origination of wings is still an unlikely event, but one which would occur with a much higher probability than in the absurd tree. It is precisely that difference in the probability of actually generating the observations (i.e. wing presences) that makes one of the theories preferable to the other. Of course, $P_{(D|t)}$ can only be calculated by assuming a specific model of evolution; *ML* may lead to prefer one tree under some assumptions and a very different tree under others. Under some assumptions, parsimony is also an *ML* estimator of the phylogeny; under other assumptions, so is phenetic clustering. The model assumed is fixed during an analysis, but some of the assumptions needed for specifying a model can, to some extent at least, be tested with empirical data (e.g. the assumption of identical amounts of evolution from root to all terminals, needed to justify phenetic clustering, see Chapter 1, is quite strongly rejected by most datasets). Given that different assumptions lead to different methods, the model used for making inferences must be chosen carefully and should at least approximately reflect the conditions under which characters truly evolve. A very good aspect of the *ML* approach is that it forces one to be very explicit about what is being assumed about the evolutionary process. This is also its major weakness—for a process as long and complicated as phylogeny, choosing among alternative assumptions may be very difficult.

4.2 ASSUMPTIONS OF MODELS OF MOLECULAR EVOLUTION

The models used for morphology are applications, almost "verbatim", of models of molecular evolution. Many of the assumptions made by those models seem quite sensible when referring to molecular sequences. The molecular models used today were initially designed to describe the evolution occurring along the line connecting two sequences and were subsequently adapted (see the next section) for evaluating trees. Most models consider only the process of substitutions, i.e. the replacement of a residue (either a nucleotide or amino acid) by another. A few models consider insertions/deletions, inversions, translocations, or duplications, but none of those is used widely. Sites are assumed to be independent so that the calculations can be made for one site at a time. Multiple mutations can occur at a given position in a sequence, which implies that the same nucleotide at the two species may correspond either to the case of no mutation or to the case of a back mutation (i.e. a mutation back to the original residue). Some of the assumptions are required for tractability more than for realism—calculations would simply be impossible without them.

Both Chang (1996) and Rogers (1997) have proven that the basic (single-rate) methods now in use result in consistent estimation of tree topology when the data truly evolve under the model; Rogers (2001) extended the proofs to the use of a gamma-distributed rate variation. The proofs hinge on likelihood methods being able to properly estimate all the branch lengths—the amount of changes between all the terminals, taken simultaneously. These can coexist only on the correct tree, and this also means that dispensing with branch-length estimation (e.g. calculating the likelihood from summation of likelihoods with different branch-length combinations) might make the method inconsistent (as noted by Goloboff 2003). In addition to those proofs concerning substitutions only, Truszkowski and Goldman (2016) also proved that (under minimal assumptions made for the insertion/deletion process) multiple sequence alignments with gaps produce consistent results if gaps are treated as missing entries (that is, using only substitutions as information).

1) *Evolution is a Markov process.* A Markov process is one without "memory", i.e. a process in which the future state depends only on the present state, not on the past states. An example is Figure 4.1a, where several changes occur along the line from an ancestor to a descendant, starting with an adenine (A) and ending in a guanine (G). There is a certain probability of transformation between each of the nucleotides, but once a mutation occurs and the nucleotide gets first changed to a C, the fact that an A occurred before does not change the probability that the descendant will end up in the G (or any other) state. If many changes along a branch occur, the final states become randomized relative to the initial states and provide no genealogical information; the sequences are then said to be *saturated.*

The process of substitution assumed to take place along the branches of a tree is similar (e.g. as in Figure 4.1b); once the two branches diverge from their ancestor, the probability that the ancestral G will change to another nucleotide along one of the branches does not depend on whether or not the G changes to another nucleotide in the other branch. Ecological interaction between lineages can and does occur, of course (e.g. by competition), but no gene flow exists after cladogenetic events, and one branch does not "know" what is happening with the other—thus justifying the Markov assumption.

2) *Poisson distribution.* One of the consequences of the Markov assumption is that substitutions must follow a Poisson distribution. The Poisson distribution is the probability of a certain number of events along a fixed time (or space) when individual events (e.g. mutations) occur with a constant probability along time or space and independently of when (or where) the last event happened. In the phylogenetic context, this means that the probability of a certain number n of mutations (including, of course, back mutations) in an individual position of a sequence with an average proportion p of mutations per position is given by the familiar formula for Poisson distributions:

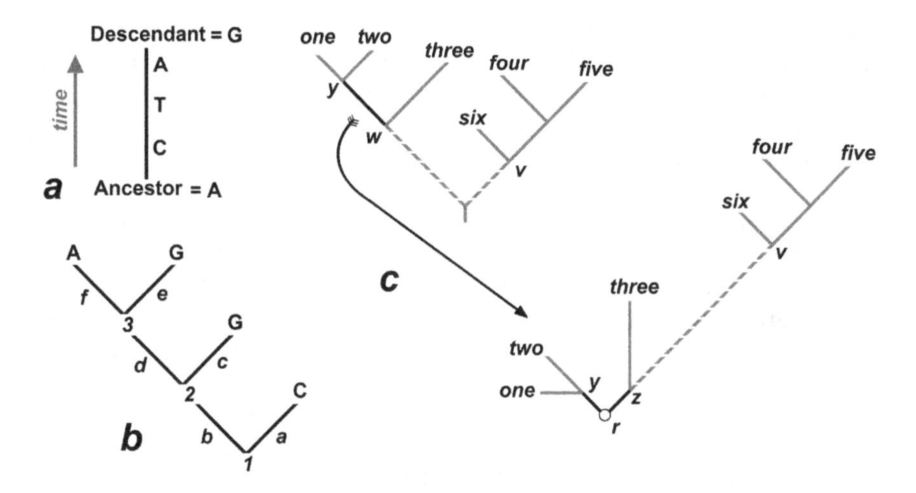

FIGURE 4.1 **(a)** Likelihood methods model evolution as a Markov process, which can possibly pass repeatedly through the same state. **(b)** The same process is assumed to occur in each of the branches of the tree independently of other branches of the tree. **(c)** The "pulley" principle (Felsenstein 1981) establishes that rerooting the tree along two nodes (e.g. y and w) produces the same likelihood as long as the sum of branch lengths between the two nodes (r–y and r–z) equals the original distance y–w). This allows finding, one branch at a time, the length that optimizes the likelihood fixing all the other branch lengths.

$$P(n) = \frac{\rho^n e^{-\rho}}{n!}$$

In the case of molecular sequences, the probability p depends on the rate of mutations (μ) and the elapsed time (t). Note (in the following) that both μ and t appear as exponent of e (Neper's number) in the formulae for probabilities of transformations between states. The possible rates of different transformations are often expressed in a matrix Q, and the probability of transformation for a time t is given by e^{tQ}, all of which is a consequence of the Poisson distribution. In simple cases, the expression e^{tQ} can be resolved analytically, as in the Jukes and Cantor (1969) model (JC69; generalized to n states by Neyman 1971), where the probability of transforming between states i and j as a function of μ and t is,

$0.25 + 0.75\ e^{-\mu t}$ *(if* $i = j$*, stasis)*
$0.25 - 0.25\ e^{-\mu t}$ *(if* $i \neq j$*, change)*

The product μt is generally referred to as the "length" of the branch, l. The longer the branch, the higher the probability of change and the lower the probability of stasis (reaching the lower limit of 0.25 when branch is longest: descendant will be in any one of the four states, randomly selected, regardless of initial state). Unless specific additional assumptions are invoked (e.g. a molecular clock), it is generally impossible to separate the two components, μ and t—a branch could be long because it has a high mutation rate or because it spans a long time (Felsenstein 1973). In complex cases (e.g. the GTR model), the probability matrix cannot be produced analytically from the Q matrix, so numerical solutions with eigenvalues and eigenvectors need to be calculated for every value of t (see Felsenstein 2004: 206).

3) *Homogeneity.* Another critical assumption of most molecular models, connected with the previous one, is that at a given branch of the tree, all of the sites evolve under the same rates (i.e. the same product μt). This means that, in a given branch of the tree, the probability of a change in a given position is exactly the same as in any other position. Put differently, branch lengths l are the same for all the positions in the sequence. This can be justified if most of the changes in the sequences are neutral or nearly neutral, under only weak selective pressure, with rates driven mostly by the basic mutation and fixation rates. This assumption of homogeneity (usually called a *common mechanism*, following Tuffley and Steel 1997) is a strong one, and it is the main cause of differences in results between parsimony and standard likelihood models. Within the standard paradigm, this assumption is sometimes relaxed by "unlinking" different genes or data partitions—making it possible for branches longer in one partition to be shorter in other partitions. In the case of sequences, partitions are typically considered to consist of the different genes. This relaxation still maintains the homogeneity within each partition.
4) *Evolution among sites is independent.* The probability of a mutation occurring at a given site is independent of mutations occurring in other sites and the states present in other sites. Although some models that do not make this assumption have been developed, they are rarely used, and their properties are not well known. The main implication of character independence is that the probability of evolving the observation at a given position in a sequence can be calculated without regard for the observations at other positions, and the possibility of evolving the full sequence is simply the product of the likelihoods of evolving each. Independence between sites is also an assumption of parsimony methods—almost all phylogenetic methods are alike in this regard.
5) *Fixed number of states.* The number of possible states that could occupy a given position (i.e. the size of the state space) is known and fixed throughout the tree. This is well justified in the case of DNA and proteins.
6) *Stationarity.* Given the instantaneous probabilities of change between different states (nucleotides or amino acids), running the system for a long time should result in specific frequencies for each state. This is said to be the assumption of stationarity—i.e. the frequencies have achieved their frequencies of equilibrium.
7) *Reversibility.* Most models assume reversibility—that is, in every branch of the tree, the probability P_{ij} of transforming from state i to state j is exactly the same as the probability of transforming in the opposite direction, P_{ji}. An important consequence of this is that the location of the root of the

tree has no effect on tree likelihoods, and this, in turn, enables the use of algorithms to facilitate branch-length optimization under likelihood (using Felsenstein's 1981 "pulley" principle, see Section 4.3.3). Some models that do not use this assumption are possible, but in that case, the branch-length optimization cannot use the pulley principle (see Boussau and Gouy 2006 and discussion of Bayesian methods for an alternative). The reversibility also implies that, just as in parsimony under symmetric transformation costs, the criterion of likelihood cannot by itself be used to root a tree—some external criterion must be invoked.

4.3 LIKELIHOOD CALCULATION

4.3.1 Basic Ideas

The models for the probabilities of change along a branch can be adapted to calculate the likelihood of trees. The probability of evolving the data can be computed one character at a time; the probability of evolving the full dataset is the product of the probability of evolving the observations in each of the characters (as they are assumed to be independent).

Keep in mind that the probability of evolving the data under the model can be quite small. Consider that, for any given tree, the probability of evolving every possible dataset must sum up to one. The number of possible datasets for modest numbers of taxa and characters quickly becomes astronomically large, so every one of the individual datasets has a very low chance of being the result of evolution on the given tree. Given that working with numbers so small becomes difficult, the *log* likelihood is usually presented instead of the likelihood. This also means that the product of the likelihoods for the individual positions (which results from site independence) can be handled as sums of log likelihoods.

For an individual character, the probability of the data given the tree is the sum of the probabilities of all possible ways the character could have evolved to produce the observations. This is in contrast with parsimony, which, although examining (implicitly) all possible reconstructions, uses only the best possible one to evaluate trees. Consider as example the tree in Figure 4.1b, with four taxa (two with a G nucleotide and one each with C or A). Tree nodes are numbered, and tree branches are indicated with lowercase letters. Only two sets of nucleotide assignments to the internal nodes are most parsimonious, GGG and CGG. The first implies that the ancestral node had a G, which transformed into a C at branch a; the second implies that the ancestral node had a C, which transformed into a G at branch b. Both reconstructions imply that the G transformed into an A at branch f, for a total of two steps.

Calculating the probability of evolving the data given the tree requires instead that all possible paths to the data on that tree are considered, along with their respective probabilities. We denote the probability of *change* between two nucleotides i and j for each branch b as $P_{b,ij}$. The probability of the same nucleotide occurring at both ends of a branch b, $P_{b,ii}$, is called the probability of *stasis*. Both the probabilities of change and stasis depend on the length of the corresponding branch. We return to the problem of how branch lengths can be estimated in Section 4.3.3; for the moment, assume they are known. The evolution of the character in Figure 4.1b might have been as in the first of the two most parsimonious reconstructions (GGG), and the probability of this happening is the product of the probabilities of change and stasis[1] along each of the branches of the tree:

$$P_{(GGG)} = P_{a,GC} \times P_{b,GG} \times P_{c,GG} \times P_{d,GG} \times P_{e,GG} \times P_{f,GA}$$

But it is also possible that the character started with a C, then changed into A at the next node (so that the shared Gs are a parallelism), denoted as CAA. This reconstruction has change along three branches (b, c, e) instead of the minimum two (in a most parsimonious reconstruction), so it implies stasis along three branches (a, d, and f):

$$P_{(CAA)} = P_{a,CC} \times P_{b,CA} \times P_{c,AG} \times P_{d,AA} \times P_{e,AG} \times P_{f,AA}$$

And it is also possible that the character followed an apparently erratic course, starting with A, changing into C, then into a T:

$$P_{(ACT)} = P_{a,AC} \times P_{b,AC} \times P_{c,CG} \times P_{d,CT} \times P_{e,TG} \times P_{f,TA}$$

This reconstruction implies change along each of the branches of the tree; this may seem absurd, but it is properly assigned a very low probability (especially when, along most of the branches of the tree, the probability of change is significantly lower than the probability of stasis). Thus, this reconstruction will contribute little to the total sum of probabilities, but considering it is necessary to calculate the exact probability of evolving the observations under the tree and model. Thus, all possible reconstructions must be considered, and the total probability of evolving the observed states for the character on the tree is given by:

$$\begin{aligned}
&P_{(AAA)} + P_{(AAG)} + P_{(AAC)} + P_{(AAT)} + P_{(AGA)} + P_{(AGG)} + P_{(AGC)} + P_{(AGT)} + \\
&P_{(ACA)} + P_{(ACG)} + P_{(ACC)} + P_{(ACT)} + P_{(ATA)} + P_{(ATG)} + P_{(ATC)} + P_{(ATT)} + \\
&P_{(GAA)} + P_{(GAG)} + P_{(GAC)} + P_{(GAT)} + P_{(GGA)} + P_{(GGG)} + P_{(GGC)} + P_{(GGT)} + \\
&P_{(GCA)} + P_{(GCG)} + P_{(GCC)} + P_{(GCT)} + P_{(GTA)} + P_{(GTG)} + P_{(GTC)} + P_{(GTT)} + \\
&P_{(CAA)} + P_{(CAG)} + P_{(CAC)} + P_{(CAT)} + P_{(CGA)} + P_{(CGG)} + P_{(CGC)} + P_{(CGT)} + \\
&P_{(CCA)} + P_{(CCG)} + P_{(CCC)} + P_{(CCT)} + P_{(CTA)} + P_{(CTG)} + P_{(CTC)} + P_{(CTT)} + \\
&P_{(TAA)} + P_{(TAG)} + P_{(TAC)} + P_{(TAT)} + P_{(TGA)} + P_{(TGG)} + P_{(TGC)} + P_{(TGT)} + \\
&P_{(TCA)} + P_{(TCG)} + P_{(TCC)} + P_{(TCT)} + P_{(TTA)} + P_{(TTG)} + P_{(TTC)} + P_{(TTT)}
\end{aligned}$$

In the case of the tree of Figure 4.1b (with taxa $T = 4$), every one of the terms in the summation is the product of six subterms; in larger trees, each of the terms of the summation would correspond to the product of a larger number of probabilities (for each one of the $T - 2$ branches of the tree). It is easy to see that evaluating the likelihood of a single tree (with known branch lengths) in this way requires simple but very numerous calculations. The number of possible ancestral reconstructions equals s^{T-2}, where T is the number of taxa and s the number of possible states, so that the time needed to compute the likelihood in this way quickly becomes unmanageable as the number of taxa increases (just as when parsimony reconstructions are evaluated by brute force).[2] When branch lengths have to be estimated, another layer of complexity is added to the problem. The solution to this apparent intractability was provided with the pruning algorithm described by Felsenstein in his now classic (1973, 1981) papers.

4.3.2 Pruning Algorithm

The pruning algorithm calculates the likelihood of a tree by considering local likelihoods, "pruning" away parts of the tree and concentrating only on the descendants of each internal node. This algorithm works in a down pass and has several similarities with the algorithms for calculating minimum steps under parsimony (which had been published earlier, by Farris 1970; Fitch 1971; see Chapter 3). The most important differences with parsimony are that the probability of transformation between two states (a sort of "cost") varies among branches of the tree and that stasis (i.e. a branch starting and ending in the same state) also has a "cost" (i.e. probability less than 1). In having to consider interconnections between states, they resemble more the algorithms for optimization of step-matrix characters (published some years later by Sankoff and Rousseau 1975).

The calculation proceeds by considering the *conditional* likelihoods of every state at a given internal node of the tree—that is, the likelihood $L_{n,i}$ of evolving the observations in the terminals descended from node n if the node had the state i. As the evolution along descendant branches is independent (the Markov assumption), the likelihood is the product of the results for each branch. For each descendant branch, d, the likelihood of state i is the probability of i transforming into any state k in the descendant times the likelihood of state k at the descendant; this is a summation of probabilities (i.e. state i can transform into any of the states in d).

$$L_{n,i} = \prod_{d=descendants} \left(\sum_{k=states} P_{d,ik} \times L_{d,k} \right) \qquad \text{[Formula 4.1]}$$

The likelihood of terminal taxa must be initialized to 1 for the observed state(s), 0 otherwise (every state must be assigned a likelihood of 1 for a taxon with a missing entry; e.g. Felsenstein 2004: 255). Each branch leading to a descendant d will have a given length, l_d (recall that, for the moment, we assume branch lengths as known and fixed), and that length is used to calculate $P_{d,ik}$ along the branch. This will depend on the model used. In the case of the Neyman-JC69 model (of most interest to morphologists, since this is the model adopted by Lewis 2001a for morphological data), the probabilities for s states are:

$$\begin{aligned} P_{d,ii} &= \frac{1}{s} + \frac{s-1}{s} e^{-l} \qquad (stasis) \\ P_{d,ij} &= \frac{1}{s} - \frac{1}{s} e^{-l} \qquad (change) \end{aligned} \qquad \text{[Formula 4.2]}$$

The likelihoods for each state at the root node r are final, and the likelihood L_c for character c is then the summation of the likelihoods for each state at the root times the probability that the initial condition was in that particular state. In the Neyman-JC69 model, all states are assumed to be equiprobable,[3] so this is simply $1/s$. Thus,

$$L_c = \sum_{i=states} \frac{L_{r,i}}{s}$$

The product of the character likelihoods is the likelihood for the full tree (using sums instead of products if converting to logs). It is easy to see that the time needed to calculate the likelihood for a fixed set of branch lengths is much smaller than the time needed for the brute-force approach of explicitly calculating the likelihood of every individual reconstruction, as described in the previous section. At every node, the possibility of every one of the s states transforming into every other state in the descendant must be considered, so this is a time $O(Ts^2)$ per character; thus, calculating the likelihood for twice the taxa takes only twice the time.

4.3.3 Pulley Principle

The pruning algorithm produces the correct likelihood for a (rooted) tree, with a fixed set of branch lengths. But the branch lengths are not known. The interest of phylogenetic inference is in determining the topology of the tree, with branch lengths being of secondary interest. However, branch lengths need to be estimated anyway because otherwise likelihood calculations are impossible (branch lengths are thus called *nuisance* parameters). Branch lengths are therefore chosen so as to maximize the likelihood for the tree in question; as discussed in the following, it is the branch-length estimation (via likelihood maximization) that makes *ML* consistent.

As in the case of all possible paths to the data, one could choose a huge number of combinations of branch lengths, retaining the best ones. Again, that would produce correct results, but it is computationally unfeasible. Felsenstein (1981) thus proposed to use the *pulley principle* as a way to optimize branch lengths, one at a time. He noted that changing the location of the root does not affect the likelihood when using time-reversible models. Every one of the branches of the tree can then be "pulled" down as a temporary root, without affecting the likelihood values. This is shown in Figure 4.1c, where a tree is to be rerooted by pulling down the branch marked in black, producing a new root r. As long as the corresponding branches, such as the black and dashed branches, have the same lengths, the results of applying the pruning algorithm are identical. For this, the individual lengths (in the new rooting) of the branches leading to nodes y and z are irrelevant; all that is required is that the summed length of r–y and r–z in the rerooted tree equals the length of the original branch, w–y. The root node r can be "slid" along the branch y–r–z (hence the "pulley" analogy). The conditional likelihoods for the nodes y and z can then be used to produce the conditional likelihoods at the new root r, varying only the (total) length of the segment y–r–z (recall that this is identical to the length of w–y), so as to find the length that will optimize the likelihood given the lengths of the *other* branches—without the need to do a down pass over the whole tree. In this way, each of the branches of the tree can be adjusted iteratively until no modification of the length of any branches improves the likelihood. This process does not guarantee finding the combination of branch lengths that truly maximize the likelihood (Steel 1994; section 6.3 of Tuffley and Steel 1997) but usually produces a good approximation.

Without the use of the pulley principle, finding optimal branch lengths would be extremely time consuming. The pulley principle can only be applied to time-reversible models, which is the reason the vast majority of programs for *ML* only consider these models (run times would be impossibly long otherwise). Boussau and Gouy (2006) have proposed extensions of the pulley principle that can be applied in the case of nonreversible models of evolution.

4.4 AMONG SITE RATE VARIATION

The formulation of likelihood just described considers that, at any given branch of the tree, all characters evolve at exactly the same speed. It is possible to relax that assumption to take into account that characters can evolve at different rates. A method often used is to assume a proportion of invariant sites (characters that cannot change at all); this proportion is set so as to maximize likelihood. Another common method, possibly used in combination with invariants, assumes that the rates have a certain distribution and that every character has a certain probability of belonging to the class with a specific rate value (Yang 1993, 1994). The gamma distribution is used to approximate this, as it can produce a wide variety of shapes by varying only two parameters. The exponential distribution (which the amounts of homoplasy in many morphological datasets approximate; Goloboff et al. 2017) is a special case of a gamma distribution, with many characters having relatively low amounts of homoplasy, fewer characters having medium amounts, and few characters having high amounts. To facilitate computation, the method of Yang (1994) discretizes the gamma distribution, dividing it into a number of rate categories, $r_1, r_2, \ldots, r_n$. For a character with s states and relative rate r, the probabilities of change and stasis are then obtained by multiplying the branch length l by the rate. For example, in the Neyman-JC69 model,

$$P_{d,ii} = \frac{1}{s} + \frac{s-1}{s} e^{-rl} \qquad (stasis)$$

$$P_{d,ij} = \frac{1}{s} - \frac{1}{s} e^{-rl} \qquad (change)$$

[Formula 4.3]

As the actual rate to which each of the characters belongs is not known, the insight of Yang (1994) was to use the probability that a character belongs to a given rate category to multiply the likelihood obtained under that category, summing likelihoods over categories. The use of gamma then adds only two parameters to the estimation, with the shape of the curve chosen so that overall likelihood is maximized. If all the characters evolve at the same rate, then a gamma distribution concentrated on a single value will be chosen (i.e. with a very high probability of each character belonging to that rate), and the likelihood will be similar to the likelihood obtained under a single-rate estimation. When characters have dissimilar rates, the use of gamma can increase the likelihood significantly.

Note that the use of among site rate variation does not change the fact that all characters coordinately increase or decrease their probability of change along different branches of the tree. The factor *r* is constant for each character throughout the tree. Thus, a branch twice as long as its sister branch under a single-rate method will also be considered to be about twice as long (and then, having a higher probability of change) for each of the rate categories under a variable-rates method. That is, using a gamma distribution does not make it possible that, at a certain branch of the tree, some characters increase their probability of change and others decrease it.

Another approach is that of Rosa et al. (2019), but that seems much harder to justify (particularly from the perspective of model-based phylogenetic inference). Rosa et al. (2019) proposed to do a parsimony analysis under implied weights (see Chapter 7), calculate the amounts of homoplasy for the characters on the trees resulting from that analysis, and set partitions so that each partition is composed of characters with similar amounts of homoplasy. Rosa et al.'s (2019) proposal has already been embraced by some (e.g. Matos-Maraví 2021). The worst problem with such an approach is that if the results of parsimony under implied weights are to be trusted to discriminate among character classes, then why cannot those results be trusted to establish phylogenetic conclusions? The subsequent use of a Bayesian approach (as done by Rosa et al. 2019) seems wholly contradictory. While implied weights may ameliorate problems with rates of different characters, it is still a parsimony method and hence does not solve the problem of long branch attraction (*LBA*)—the main reason why some researchers prefer model-based methods. If *LBA* is a serious problem in the dataset at hand, then implied weights may produce unreliable results and all downstream analyses are vitiated; if *LBA* is not a problem and implied weights produce accurate estimations of homoplasy, then why is a Bayesian analysis called for?

4.5 LINKED AND UNLINKED PARTITIONS

Likelihood analyses can also proceed without assuming that all characters belong in the same data partition. In the case of molecular sequences, partitions can be quite naturally determined from the different genes. In single-gene datasets, if coding, an alternative is to consider three partitions for all first, all second, and all third positions (again, these partitions can be recognized from the data themselves).[4]

When the data can be pre-partitioned (Nylander et al. 2004) on some reasonable basis, the rates can be estimated separately for each partition. In this case, each partition is assigned a rate on the basis of numerous characters (and then there is no need to use the probability that the partition belongs in a specific rate class). That partition rate can then be used to multiply the branch lengths that are common to all partitions; this is known as a *linked analysis*. As in the case of among site rate variation, the implication of linked partitions is that a branch longer for one partition must also be longer for all the others. This assumption of a common mechanism for all partitions can be dispensed by an *unlinked analysis*—letting different partitions have different, uncorrelated branch lengths. The branch lengths are therefore estimated separately for each partition. Note that, as the number of separate partitions grows, the results will necessarily approximate those of parsimony (this is a consequence of the no-common-mechanism model, see Section 4.10.4). While in that sense implying a weaker assumption, the problem remains that the quality of the estimation of branch lengths (and, thus, choice of a tree topology for the partition) decreases to the extent that fewer and fewer characters are included in the partition, so care is necessary to establish an appropriate tradeoff between number of partitions and likelihood. In the absence of external reasons to determine partitions (e.g. the existence of different genes), correctly partitioning the data becomes very difficult.

4.6 BAYESIAN INFERENCE

Likelihood methods, given certain conditions of evolution and adequate models, can in theory select tree topologies consistently. A related approach has become common in recent decades, one based on the work of Rev. Thomas Bayes (1702–1761). Widely used computer programs for Bayesian phylogenetic analysis are MrBayes (Ronquist et al. 2011, 2012), Beast (Drummond and Rambaut 2007; Drummond et al. 2012), and RevBayes (Höhna et al. 2016). Discussed here are only the basic aspects of Bayesian inference; detailed treatments can be found in Yang (2006) and Chen et al. (2014).

Bayes wondered how one could calculate the probability of a given cause having determined a certain effect (as opposed to calculating the probability with which a cause produces an effect). That is, the probability of a certain conclusion being true, given some observations; this is called the *posterior probability*, or *PP*. One of the components of the posterior probability is the probability of the hypothesis being true, regardless of any actual data. Note that such "probability" is a non-frequential probability—that is, it cannot be defined as the frequency of a certain event but expresses a degree of credibility instead. Bayesian approaches thus differ from *ML* and standard hypothesis testing in that the probabilities calculated cannot be defined solely on the basis of frequencies of events; they express something different. Bayes showed that $PP_{(H|D)}$, the posterior probability of hypothesis H given data D, is

$$PP_{(H|D)} = \frac{Prior_{(H)} \times P_{(D|H)}}{P_{(D)}}$$

$Prior_{(H)}$ is the prior probability of the hypothesis; some hypotheses may be more likely than others a priori. In the phylogenetic case, H is a tree, and the usual approach is considering that (in the absence of data) all trees are equally likely; this is called a *flat prior*, and it means in principle that sorting the trees according to their likelihood should order them exactly according to their likelihoods (the only term varying in the equation) so that the Bayesian approach would produce a selection of tree topologies identical to *ML*.[5] $P_{(D|H)}$ is the likelihood of the hypothesis, as explained in the preceding sections. $P_{(D)}$ is the probability of the data regardless of any hypotheses; in the case of phylogeny, the data must have originated from *some* tree, so $P_{(D)}$ is the sum of the likelihood of all possible trees, $\Sigma L_{(D,t)}$. Except for a handful of taxa, $P_{(D)}$ is incalculable, and then even the simplest approach to Bayesian phylogenetics would seem impossible.

The solution to the impossibility of calculating $P_{(D)}$ comes from using Monte Carlo methods to approximate the values of $PP_{(D|H)}$ without ever having to calculate $P_{(D)}$. The first researchers to propose using these methods in phylogenetics were Mau and Newton (1997), Yang and Rannala (1997), and Larget and Simon (1999), but their ideas were quickly adopted by other colleagues and became incorporated into computer programs. The idea is to generate trees by a process of random modifications of an initial tree, known as a Metropolis-Hastings Monte Carlo Markov chain, *MCMC*. At every step in the chain, a modification of the tree (in principle, topology, but see the following for branch lengths and other parameters) is proposed, which is (with a certain probability) either accepted or rejected. The probability of making different proposals is independent of the prior state of the chain (hence the "Markov" aspect of the chain). Every certain number of generations, the current state of the chain is stored, as a sample (this is used to avoid a strong autocorrelation of the sample, i.e. to make each tree sampled as independent as possible of the prior state of the chain).[6] If $PP_{(H|D)}$ was calculable for each tree, and the probability to accept or reject that modification depended only on the ratios of $PP_{(H|D)}$ before and after the change, the end result would be that trees with a higher $PP_{(H|D)}$ would be more frequent in the sample—in direct proportion to their *PP*. The value of $PP_{(H|D)}$ is incalculable, but the denominator in the expression, $P_{(D)}$, is a constant, which then has no effect on the ratios between posterior probabilities of two trees. This means that the chain can use just the ratios between the likelihood times the prior probability for the preceding and the new tree to calculate the acceptance probability. If the proposals of changes to the trees follow certain rules, then $PP_{(H|D)}$ can be estimated from the frequency with which a tree appears in the sample at the end of the *MCMC*. These rules include, e.g., the possibility that all trees are visited when the chain is run for a long enough time (simply fulfilled by using branch-rearrangement methods) and that if the probability of a tree i→j move is different from that of tree j→i, then the equations for the acceptance probability must include a correction term (called Hastings term; see Felsenstein 2004: 292).

Note that this process does not necessarily try to find the optimal tree; rather, it attempts to evaluate how posterior probabilities change across tree space. The chain may well have happened to find an optimal or near-optimal tree, then drift away from the tree. The initial tree for the chain may have been too far from optimal, so the trees used to make inferences are only those after the chain is expected to have *converged* (i.e. when the target distribution is expected to have been achieved); this is called the *burn-in*. In practice, many Bayesian programs use multiple chains running simultaneously (perhaps with chains at different "temperatures", i.e. with different probabilities of moves), periodically exchanging states between the chains as a way to improve convergence. After convergence, the expectation is that the frequency of trees (and other parameters) in the sample taken at regular intervals will exactly reflect their posterior probabilities.

There are numerous advantages to this approach, which easily explains why it has become so popular. It may allow evaluating not just the topology best supported by the data but also its degree of support, without having to actually perform heuristic tree searches (which, as explained in Chapter 5, can be extremely time consuming). It allows easy incorporation of many other parameters into the estimation, not just the tree topologies. As those parameters are changed in successive generations, the values that result in higher likelihoods will end up in the final sample more frequently so that the frequency of those parameter values in the sample also indicates their posterior probability. The posterior probability, $PP_{(t,b,\Theta|D)}$, is then proportional to $P_{(D|t,b,\Theta)} \times Prior_{(t,b,\Theta)}$, where t is the tree, b is the set of branch lengths, and Θ is the model and parameters (e.g. transition/transversion ratio, gamma, etc.). The decision on parameters for the model, or even the model itself, can then be determined in a single analysis. In the phylogenetic realm, and given standard likelihood models, the first advantage of this is that the branch lengths themselves, instead of being optimized to evaluate the tree in each generation, are simply varied; branch extension and shrinking proposals are then judged on the basis of their implied likelihood (and prior probability; different prior probabilities for the branch-length distribution can be used, e.g. uniform or exponential). This means that the branch lengths are *integrated* in the evaluation of tree topologies so that they are no longer a nuisance parameter (as they are in *ML*). Goloboff (2003) presented examples where summing up likelihoods for all branch-length combinations (under a uniform distribution of branch lengths) produces inconsistency for *ML*, but Steel (2013) showed that this integration creates no problems for Bayesian *MCMC*, as the probability of a given branch-length combination ending up in the final sample depends on the likelihood it confers on the trees. A huge advantage of not having to optimize branch lengths is that the pruning algorithm can be applied directly to the tree resulting from each generation, without having to iteratively optimize branch lengths by means of the pulley, and thus the computational cost of evaluating trees under models that are not time reversible is the same as for time reversible—allowing the use of probabilities of transformation between states that differ in both directions, which in the context of *ML* would require longer run times.

4.7 SOME DIFFICULTIES WITH BAYESIAN PHYLOGENETICS

Leaving aside the question of whether the models currently in use are appropriate for morphology (a problem discussed in subsequent sections), there are a number of caveats on the application of Bayesian methods in phylogenetics, even from the point of view of model-based approaches. Also left aside in the discussion that follows is the problem of the chains properly converging. Although several authors have raised doubts about the ability of current methods to converge (e.g. Mossel and Vigoda 2005, with rejoinder by Ronquist et al. 2006; Whidden and Matsen 2015), the problems pointed out exist even with correct convergence.

4.7.1 No Optimality Criterion

The first important difference between likelihood and Bayesian methods, at least in practice, is that the results of an *MCMC* cannot be judged on an independent criterion—there is no measure of optimality that can be applied to the results, other than whether or not they have produced the true phylogeny; the quantities used to accept or reject trees are themselves incalculable. As the estimation integrates over parameters that are simply optimized in the case of standard likelihood methods, the results of a Bayesian analysis may well be different from *ML* because of that integration, or they may be different because there is a problem with the priors, the proposal mechanisms in the Markov chain, the convergence of the chains, or the way results are summarized. Given results from two *ML* programs, the choice of results is simple—the results of the higher likelihood should be followed, and the approach is amenable to formal analysis (e.g. as in Chang 1996; Tuffley and Steel 1997). In the case of discordant results between two Bayesian programs, instead, there is no way to tell or to evaluate a specific result—you just take it or leave it. Formal analyses of *MCMC*-based Bayesian methods are thus very uncommon; a notable exception is Steel (2013), who showed that Bayesian analysis should, in theory (under certain conditions), produce consistent estimations—this, however, is not of much help when two Bayesian programs produce different results. The only way to show the superiority of a Bayesian program (and the one several authors have chosen, e.g. Wright and Hillis 2014; Puttick et al. 2017) is by means of simulations, but, of course, this critically depends on the conditions assumed for simulating the data.

4.7.2 Priors

The second problem with Bayesian methods is in the choice of priors. These, obviously, influence the final results, but there are usually no clear grounds for preferring some prior distributions over others. A review of priors in Bayesian phylogenetics was provided by Wang and Yang (2014). The use of priors is generally the cause of much concern for many model-based phylogeneticists; Felsenstein (2004: 301) stated that "a Bayesian is defined, not by using a prior, but by being willing to use a controversial prior". The use of flat (equal) priors is often used as a minimalist assumption, but the problem is not only in the distribution, it is also on the bounds to be placed; as put by (Felsenstein 2004: 303), "if a prior is to be flat from 0 to ∞, the density will be zero everywhere". For example, for priors on branch lengths, allowing branch lengths to vary between 0 and 1 may produce appropriate results, but allowing the branch lengths to vary between two different extremes may not even contain the *ML* tree in the confidence interval. In addition to that, a prior that is uniform for some aspects of the inference may necessarily mean a non-flat prior for other aspects of the inference, and this leads to the third problem: the problem of how the results are to be summarized.

4.7.3 Summarizing Results

Perhaps the most critical aspect of current Bayesian methods is the summary of results for tree topologies (Goloboff and Pol 2005; Wheeler and Pickett 2008;

Wheeler 2010). While summarizing single real-valued parameters may be simple, the situation is different with trees. Trees are complex structures, and there are many, many of them. Finding the optimal tree is a difficult task, and avoiding the need to do so was supposed to be one of the reasons why running a *MCMC* is beneficial (e.g. Lewis 2001b). But, when running a *MCMC*, unless the data are extraordinarily clean or there are only a handful of taxa, the actual tree of highest posterior probability (called *MAP* tree, from *maximum a posteriori*) may be found only once or very few times. This means that the posterior probability of even the most probable tree will be very low. These low values could be seen—depending on one's perspective—as properly reflecting the uncertainty plaguing phylogenetic analysis (Goloboff and Pol 2005: 159), but they seem unsatisfying to many model-oriented phylogeneticists who prefer the feeling of a strong inference. As a consequence (Yang 2006: 176; Felsenstein 2004: 299), the most common method to summarize the results is by calculating clade probabilities instead of tree probabilities (Larget and Simon 1999). For this, the *clade posterior probability* (*CPP*) is defined as the sum of the posterior probabilities of all the trees having the clade—which is the same as a majority rule consensus tree of the trees sampled in the *MCMC* with group frequencies indicated on its branches. O'Reilly and Donoghue (2017) defended the use of majority rule consensus trees based on simulations alone (and without citing any of the papers that discuss problems with the use of such method); Steel (2013) argued that the *CPP* tree will converge to the *MAP* (and hence, the underlying tree) for infinite numbers of characters but that the two may be different for finite datasets.

As in the case of support measures (Chapter 8), some undesirable consequences result from using majority rule consensus trees to summarize sets of trees. The first is that using a flat prior on *trees* automatically means that the prior on *groups* is non-uniform (Pickett and Randle 2005). Out of the total possible trees for T taxa rooted on the outgroup taxon, $NPT_{(T)}$ (see formula in Chapter 1), a group splitting S taxa will occur in $NPTr_{(S)}$ (the possible rooted topologies within the group) times $NPTr_{(T-S)}$ (the possible rooted topologies outside the group), as discussed by Carpenter et al. (1998). Therefore, the proportion of a group within all possible trees varies with its size; for $T = 10$, a group with size $S = 5$ will occur in 11,025 of the possible 2,027,025 trees, but a group with size $S = 2$ will instead occur in 135,135 trees. Thus, in the case of 10 taxa, having a uniform prior on *trees* means that the prior on *groups* of two taxa is 12.26 times higher than on groups of five taxa. Pickett and Randle (2005) showed empirically that these differences can bias the results of Bayesian analyses, and Steel and Pickett (2006) subsequently provided formal proof that it is in fact impossible to create a uniform prior on groups, so there is no solution for this problem.

Another problem resulting from using *CPP* was pointed out by Goloboff and Pol (2005), by analogy with the problems that arise from the use of the frequency within replicates (*FWR*) approach for summarizing the results of bootstrapping (see Chapter 8). The *ML* tree, or *MLT*, is known to be a consistent estimate of the phylogeny under the right model (Chang 1996; Rogers 1997), and Steel (2013) proved that (even when branch lengths are not optimized but instead integrated into the calculation as done in *MCMC* approaches) the *MAP*, given some conditions, will converge to the correct phylogeny and hence to the *MLT*. But things are different for *CPP* and sequences of finite length. Consider a case where the *MLT* displays a certain group, X, but many slightly suboptimal trees (i.e. with a likelihood only slightly inferior) display an alternative group, Y. In that case, the sum of *CPP*s for clade Y will be higher than for clade X, and the result will contain group Y—even when group Y is absent from the *MLT* and the *MAP*. By choosing suitable branch lengths of the model tree, the situation just described can be easily created. The example provided by Goloboff and Pol (2005) is reworked here (Figure 4.2a, b); in the model tree (Figure 4.2a), taxa T, U form a monophyletic group; all branch lengths are short, except the U branch, which is much longer. When datasets are generated (under Jukes and Cantor 1969) from this model tree, this provides a well-supported scaffold for most of the tree, except that the very long branch, U, can be placed anywhere in the tree with almost the same likelihood. The sum of posterior probabilities for the trees without the group T, U is then much higher than for the trees with, and the Bayesian result (Figure 4.2b) displays a very strong support for the wrong group, s, T (with U placed further down the tree). In this example, the incorrect group with a high *CPP* is a two-taxon group, and perhaps this fact is what led Yang (2006), Velasco (2008), and Wheeler (2016) to mistake the effect described by Goloboff and Pol (2005) with that described by Pickett and Randle (2005). The two effects are different, however, as pointed out by Goloboff and Pol (2005: 151–152)—the problems created by the shape of the likelihood landscape (i.e. the variation of likelihood scores across tree space) are different from those created by unequal clade priors. In the example of Goloboff and Pol (2005), an incorrect clade (s, T) has a spuriously high *CPP*, but the correct two-taxon group, T, U (with the exact same prior probability!), has a spuriously *low CPP*, so not even this example can be explained solely on the basis of unequal group priors. More elaborate examples show the difference between the two problems—unequal priors and frequency of clades in optimal and near-optimal trees—even more convincingly. Consider the case of the model tree shown in Figure 4.2c, with three partitions of 500 characters each; the branch lengths for each partition are shown on the tree. When the partitions are unlinked, taxon L can be placed anywhere inside the group with dashed branches with only a small difference in likelihood; the third partition makes taxon L significantly lower the likelihood if placed in the part of the tree with full branches; the dashed group is (the inclusion or exclusion of L aside) well supported relative to the rest of the tree (by virtue of the first and third partitions). While L is sister to K in the model tree, the Bayesian *CPP* tree places L within a group together with m–z because there are 26 possible locations of L within m–z and only one correct location, K. The *CPP* result (Figure 4.2d) is thus spuriously *low* for a clade with *high* prior probability (KL, with prior 0.0213) and spuriously *high* for a clade with *low* prior probability (L–z, shown with an arrow in Figure 4.2d, with prior 1.17×10^{-7}). This is exactly the opposite of what is expected by the unequal clade priors alone and results from the shape of the likelihood landscape, with many near-optimal trees displaying a group that is absent from the optimal tree.

While the most critical aspect is that of summarizing tree topologies, there may also be problems with other tree parameters, depending on how they are summarized. Yang (2006: 176–177) pointed out potential problems with the calculation of branch lengths done with the default settings of MrBayes. MrBayes calculates branch

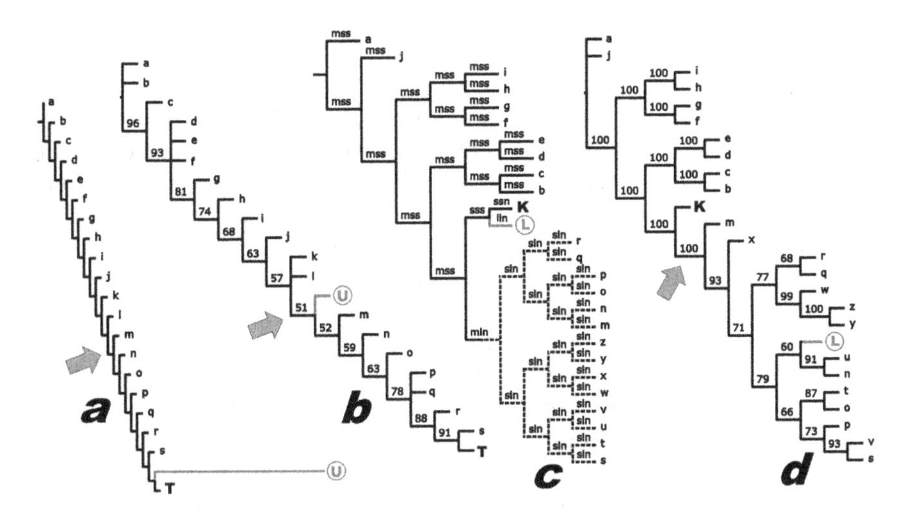

FIGURE 4.2 **(a)** A model tree where most branches are short and the branch leading to just one terminal (U) is very long; group m–U is monophyletic on the model tree. **(b)** Generating datasets with this model tree (example shown is for 1,000 four-state characters, with short branches of length 2×10^{-3} and long branch of 5) when using sums of clade frequencies to summarize the results of a Monte Carlo Markov chain (with MrBayes, 10^6 generations, no stop rule, sample frequency 100, and a burn-in fraction of 0.333; remaining parameters as default) produces high posterior probabilities for the wrong clades, placing U in the middle of the tree instead of the apex. **(c)** A model tree where three partitions evolve. For every partition, 500 characters were generated, with branch lengths indicated as negligible (n, 10^{-4}), short (s, 5×10^{-3}), medium (m, 10^{-2}), or long (l, 3). The dataset was analyzed with MrBayes (unlinked branch lengths, half a million generations, sampling frequency of 100, a stopping value of 0.02, and a burn-in fraction of 0.333). **(d)** The analysis with MrBayes indicates as *strongly* supported a group (arrow) which on the basis of tree topology alone has a *low* prior probability. This shows that unequal topological priors are not the only problematic aspect in Bayesian inference.

lengths by averaging the lengths of the branch splitting off the same clade in all the sampled trees, but (when the tree is allowed to vary) these are not all the "same" branch. For example, the arrows in Figs. 4.2a and 4.2b show equivalent clades, but the branch lengths implied by the two trees can be widely different—having the short branch m, or the long branch U, as the first split within the group is bound to affect the estimation of the branch length separating m and U. Averaging both lengths is then unjustified. Yang (2006) pointed out that, to produce appropriate estimates of branch lengths, the tree topology should remain fixed (e.g. at the *MAP*, once calculated) and a *MCMC* sampling of branch lengths on that topology should be run. MrBayes allows fixing the tree topology (with the `constraint` and `prset topologypr` commands) so that Yang's (2006) recommendation can be followed in a two-step analysis.

4.7.4 Sample Size and Frequency

A potential problem with *MCMC* is that the samples taken at regular intervals ought to be as independent as possible (to diminish autocorrelation). The topology of the tree changes during the run by means of branch-swapping, but the number of swaps needed to make two trees as different as possible changes with numbers of taxa. Two swaps suffice to explore the entire tree space for four taxa, but many more are needed for large numbers of taxa, suggesting that the sampling interval should probably be increased in that case. The same applies to branch lengths; the *MCMC* never optimizes them, only sampling combinations of branch lengths, but the representative number of combinations (needed for Steel's 2013 conditions of consistency for Bayesian analysis to apply) probably increases superexponentially with number of branches (i.e. taxa). This, in turn, also means that the total number of generations to run must be increased as we add taxa to an analysis. Thus, the initial hope that Bayesian methods would allow much faster phylogenetic analyses has vanished over the years, and well-designed search routines under *ML* (e.g. RAxML) turned out to be much more effective in saving time.

4.7.5 Proposals

The results of the *MCMC* strongly depend on the mechanism for making proposals. The probability and strength of different proposals (modification of branch lengths, branch-swapping, alteration of other parameters) must be adjusted carefully, but there is no general theory providing guidance as to how this must be done. Basically, the correct values are those which produce the correct results; these can be adjusted by carefully testing many cases where the expected output is known, but there is no guarantee that the resulting values are really general. Felsenstein (2004: 296) noted:

> A proposal distribution that jumps too far too often will result in most proposed new trees being rejected. A proposal distribution that moves too timidly may fail to get far enough to adequately explore tree space in the allotted time. At the moment the choice of a good proposal distribution involves the burning of incense, casting of chicken bones, use of magical incantations, and invoking the opinions of more prestigious colleagues.

While the many methods proposed since Felsenstein (2004) to test convergence of the chains (e.g. see documentation of MrBayes) are of some help here, the mechanisms for proposal continue to be much of a black box.

4.8 MODEL CHOICE

At this point, it should be clear that the model assumed for establishing phylogenetic inferences is expected to have some degree of approximation to reality, without necessarily being a perfect reflection of it. As described prior, standard likelihood methods assume that different branches of the tree can have different probabilities of change (i.e. lengths), and these become part of the estimation process, parameters. In addition to branch lengths, phylogenetic models postulate other parameters—some fraction of characters cannot vary, different characters have different propensities to change, the probability of transformation between different pairs of nucleotides differs, partitions where parameters are estimated separately—and this requires that each of those parameters becomes part of the estimation, too. Obviously, to the extent that more parameters are added to the inference, the inference becomes more realistic (in the sense of capturing evolutionary mechanism in more detail); as more parameters can vary, the likelihood will be higher and higher.[7] So, one could, in theory, start from a very simple model and then gradually add parameters. But the fact that the likelihood is higher does not always mean the estimation is better.

As a thought experiment, consider what would happen with a simulated dataset, evolved on a tree where all the branches have exactly the same length, and then trying to recover the original tree using two alternative likelihood models: one where the lengths of all the branches of the estimated tree are constrained to be the same and another where the lengths can vary (as described prior). The tree estimated with varying branch lengths will have slightly higher likelihood than the one where all branch lengths are the same—just because of random sampling, the numbers of characters changing along each branch will not be exactly the same, and then allowing small differences in branch lengths increases the fit of the estimated tree to the simulated data. These differences do not, however, reflect any reality—they simply represent noise. Estimating the length of each branch in this hypothetical case amounts to *overfitting* (also called *overparametrization*)—i.e. inferring from the data more than the data really allows inferring. Generating a new dataset from the same tree and reestimating branch lengths would produce slight differences in the lengths of all the branches; the estimation that forces all branch lengths to be the same will be more stable, as new data samples are taken from the same generating model. In principle, if all the branches could be estimated without error, they would be concluded to be all the same, and there would be no problem in using the more complex model. However, this never happens, and models that are too complex estimate more than can be sensibly estimated from the data, and they produce estimations that are unstable and unreliable.

Consider now a second thought experiment, one where the branch lengths of the tree on which the dataset evolves truly have important differences. The likelihood of the estimation where branch lengths are allowed to vary will be now much higher than the estimation where all branches are constrained to have the same length. The gain in likelihood by estimating the additional parameters is then significant; the error in the estimation of the individual branches incurred by averaging branch lengths is also substantial.

Those considerations form the basis for choosing models. As different models imply different sets of parameters, the number of parameters and the value of likelihood must be compared, establishing a tradeoff. To the extent that adding a parameter increases the likelihood enough, adding the parameter to the estimation is desirable. The problem, of course, is deciding how much is enough, and a number of methods are used to assist in this decision.

In the case of likelihood, one of the methods often used to choose between two models is the *likelihood ratio test*. This compares the value of the statistic, $-2\ (max\ L_{model_0} - max\ L_{model_1})$, with a chi-square distribution, rejecting the simpler model if the statistic exceeds the value for the upper (say) 5% of a chi-square distribution with degrees of freedom equal to the difference in number of parameters. Felsenstein (2004: 308–309) gave a clear derivation and intuitive explanation of this formula. The test is applicable to cases where the simpler model is fully nested within the more complex one, and the chi-square distribution of the statistic obtains only when the true model belongs in (i.e. is nested within) the more complex model.

The other method frequently used is the *Akaike information criterion* (*AIC*, Akaike 1973), which also compares the number of parameters (*NP*) and the likelihoods, but without requiring that the models are nested. This compares the value of $AIC = 2NP - 2\ ln\ L$ for both models; models with lower values of *AIC* are preferred. The *AIC* is related to the way in which models effectively capture information, and this (Burnham and Anderson 2002) is indicated by $e^{(AIC_x - AIC_y)/2}$, which measures the probability that the simpler model loses information relative to the more complex one.

Model selection in a Bayesian framework is based on similar principles, preferring models that explain variation well but at the same time avoid overfitting. In this case, the comparison between models is done by calculating Bayes *factors*, which are similar to likelihood ratios but use the *marginal likelihoods* instead of point estimates (marginal likelihoods being those that result from integrating across parameter space). The actual values of marginal likelihoods are difficult to calculate (recall that the advantage of an *MCMC* is precisely that it avoids calculating them), hence a number of methods are used to approximate them. Some of those are harmonic means (Newton and Raftery 1994), thermodynamic integration (Lartillot and Philippe 2006), and steppingstone sampling (Xie et al. 2011; Fan et al. 2011). Wang et al. (2018) recently reviewed these and proposed a new method, the partition weighted kernel. As in the case of likelihood ratio tests, it is important that one of the models tested must be "correct" for the selection to be adequate; Yang and Zhu (2018) pointed out that in comparisons between two incorrect models, the selection may be strongly polarized toward the least correct model.

4.9 ADAPTING MODELS FOR MOLECULAR EVOLUTION TO MORPHOLOGY

The general idea during the 80s and 90s was that morphology was too difficult to model for the application of *ML* methods. During that time, *ML* methods for the analysis of sequence data became widespread, particularly after implementation in PAUP (Swofford 1993) and PAUP* (Swofford 2001); the computational speed of RAxML (Stamatakis et al. 2005; Stamatakis 2014) was also instrumental in *ML* being used more widely. Even so, Steel and Penny (2000: 843), in reference to the model of "No Common Mechanism" (which produces results identical to parsimony, see Section 4.10), stated that "it might be regarded as the model one might start with if one knew virtually nothing about any common underlying mechanism linking the evolution of different characters on a tree (e.g. as with some morphological characters)".

The increased acceptance and interest in model-based methods for molecular sequences led Lewis (2001a) to propose using similar methods for *DMC*s. He used for this the Neyman-JC69 model, where a state can transform into any other equiprobably, but when applied to *DMC*s, he called the model Mk, with a variant to take into account acquisition bias toward variable characters, Mkv (see the following). Both the Mk and Mkv variant are implemented in PAUP* and RAxML; MrBayes implements this as well, allowing corrections for both invariant and parsimony-uninformative characters.

4.9.1 Mk Model

The general approach for setting transformation probability along a branch is exactly as in the case of sequences, determining probability from the length l of the branch and number s of possible states, with formula 3.2:

$$P_{d,ii} = \frac{1}{s} + \frac{s-1}{s}\, e^{-l} \qquad (stasis)$$

$$P_{d,ij} = \frac{1}{s} - \frac{1}{s}\, e^{-l} \qquad (change)$$

As example, the TNT script ***domklik.pic*** calculates the likelihood for two-state data (a single rate), optimizing branch lengths, illustrating the use of the pruning algorithm and the pulley principle. The model can also incorporate invariant characters and multiple rates (e.g. with a gamma distribution, as per formula 3.3; Harrison and Larsson 2014 suggested that a lognormal distribution may provide a better fit than gamma for morphological datasets).

4.9.2 Mkv Variant

One of the potential problems with the previous approach is that the characters for a morphological dataset are not "sampled" in the same way as sequences. A gene, a protein, or even a particular primer naturally delimit the characters that will and will not be sampled in the sequence, and all characters within those limits are considered, regardless of variability. In contrast, morphological datasets normally include only variable characters, but it is crucial for the model that all of the characters are used to estimate branch lengths. Even if one were to decide to include all possible characters in the dataset—those that vary within the group and those that do not—it is difficult to think of how to make such delimitation (Lewis 2001a: 917). The vast majority of the characters distinguishing subgroups of Arthropoda are invariant within Chordata; whether absent or inapplicable, characters such as wing venation or elytra are universally absent or inapplicable in Mammals. Every morphological dataset could potentially be expanded to include thousands and thousands of such invariable characters. This could be considered as an indication that morphology—when morphological variation known across all of life is considered—is extraordinarily conservative and that branch-length calculations should take into account that many more characters than have actually been sampled—all of them invariable—actually exist. Lewis's (2001a) approach is more pragmatic, borrowing from a molecular method earlier proposed by Felsenstein (1992) for restriction sites (restriction sites produce binary data, presence/absence, and some restriction sites in a sequence can typically be missed). This is a method to take into account *ascertainment* bias (or *acquisition bias*); it considers that, if the invariant characters occur according to the predictions of the model and are excluded, then the likelihoods and branch lengths could be calculated conditional on that exclusion. The advantage of Felsenstein's (1992) method is that it requires little additional computations, calculating likelihoods for every observed character *c* on tree *t* (with the pruning algorithm, formula 3.1) conditional on the character being variable, $L_{(c,V|t)}$. Thus, $L_{(c,V|t)}$ is proportional to $1/P_{(v)}$, where $P_{(v)}$ of is the probability of a character being variable. $P_{(v)}$ is simply $1-P_{(i)}$, where $P_{(i)}$ is the probability of the character being invariable. For a given set of branch lengths, $P_{(i)}$ can be calculated by considering an all-zero dummy character when applying the pruning algorithm for a given set of branch lengths, then dividing the final likelihood for the character by the complement of $P_{(i)}$. With this correction, optimal likelihoods are produced by shorter branch lengths. It is clear, however, that the correction assumes that the "true" frequency of invariant characters is exactly the one predicted by the model (and adjusted branch lengths), instead of approaching unity as earlier considerations would imply. As this method shortens all the branches of the tree, it changes the probability of identical derivations in sister branches, thus possibly leading to selecting different trees. Lewis (2001a) called this Mkv, an Mk method corrected to operate only on *variable* characters. Lewis (2001a) simulated four-taxon datasets (eliminating invariant characters) and found that the correction improves the estimation, but in empirical datasets for larger numbers of taxa, even the most severe bias (only informative characters collected) is associated with an insignificant correction of the likelihood values (Ronquist et al. 2011).

4.9.3 Assumptions of Mk/Mkv Models

The assumptions of the method are basically the same as when applied to molecular sequences. The strongest assumption of the method is in assuming a common mechanism (Lewis 2001a). With that, the probability of change for all characters is assumed to increase or decrease together by the same exponential factor—the length of the branch, common to all characters. Lewis (2001a) noted that this is a strong assumption, and as a consequence,

> The Mk model may strike many systematists as being highly unrealistic. One possible objection lies in the fact that the model predicts that the probability of observing a change along a branch in a phylogeny increases with the amount of time associated with the branch. This appears on the surface to be explicitly gradualistic.
>
> **(Lewis 2001a: 915–916)**

In the case of sequences, population size (common for all the genes in the population) will determine fixation rates; time is also common to all the genes. Assuming the same thing for morphology amounts to expecting that characters in different partitions evolve as shown in Figure 4.3a; the three partitions (skin, bone, muscle) may have different intrinsic rates of evolution, but all the branches of the tree (i.e. the product of time and mutation rate) change concomitantly. The Mk/Mkv model assumes this homogeneity not only at the level of (linked) partitions but also within all the characters of a given partition; the estimation of branch lengths from multiple characters is an essential component of the models. Demonstrations of consistency for standard *ML* methods require that branch lengths are homogeneous and estimated from infinitely long sequences of characters. In contrast, morphology has long been known (e.g. Farris 1983: 15) to be much more of a mosaic, with some character systems speeding up and others slowing down at a given region of the tree. This is more like the model shown in Figure 4.3b; for example, h is a longer branch than i for *skin* but similar for *bone* and shorter for *muscle*; in the group j,k,l, the longest branch for *skin* is j, but for *bone* it is l, and for *muscle* it is k.

Another significant assumption, which may not translate well from molecules to morphology, is that the number of possible states a character can take is fixed. Note that, in the case of sequences, this number does *not* depend on the number of observed states—i.e. it depends not on the number of states the character has actually taken but instead on the number it could *potentially* have taken. It is hard to envision

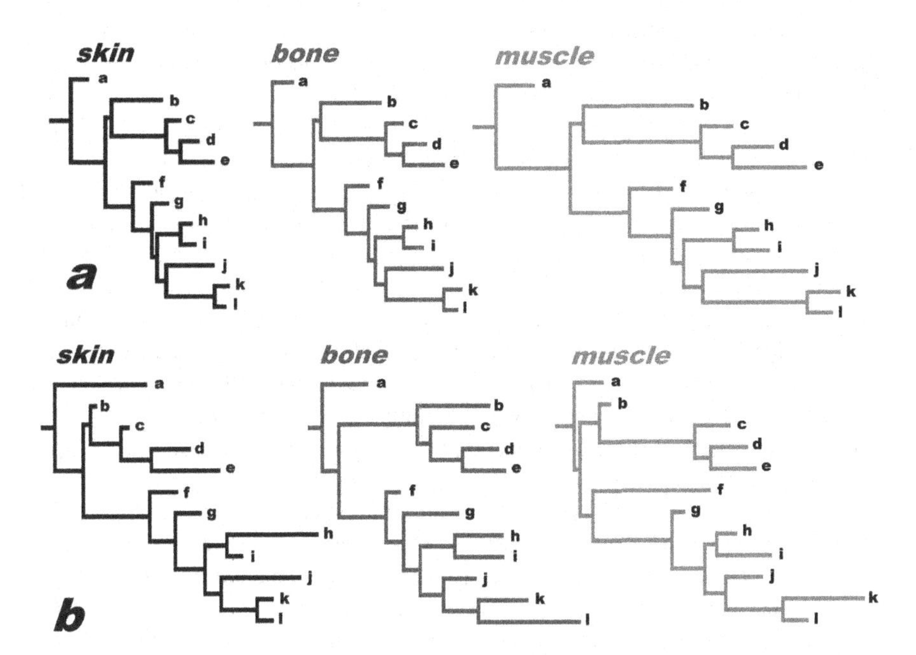

FIGURE 4.3 **(a)** The Mk/Mkv model of phylogenetic inference assumes that all characters (e.g. skin, bone, muscle) increase or decrease their rate of evolution together at the same branches of the tree. **(b)** Morphological data better seem to fit a mosaic model, where an increase in the probability of change for one character can be matched by a decrease in other characters.

any way to quantify the number of shapes a morphological character could have taken—this is potentially a very large number.

Some authors (e.g. Klopfstein et al. 2015; Wright et al. 2016) have expressed concerns about the equiprobability of transformations between all possible states being unrealistic, but this is no different from standard parsimony, and (just as in parsimony) it is possible to modify the Mk/Mkv model so that it considers different probabilities of transformation. For example, MrBayes allows taking into account the relationships between states in additive characters by setting to 0 the transition probability between non-adjacent states in the Q matrix, thus forcing transitions between intermediate states and effectively making transitions between non-adjacent states less likely. This, however, has the counterintuitive consequence that (for branch lengths larger than 0 and less than ∞) probabilities of stasis or change between neighbors of different states are different. In a five-state additive character, a branch is more likely to preserve states 0 or 4 (i.e. those at the tips) than state 2 (the middle state), and a change 0→1 is more probable than a 2→3.

4.10 PARSIMONY, MODELS, AND CONSISTENCY

Having examined the basic principles of standard models, I now turn to discuss their relationships with parsimony. While justification of parsimony by reference to explanatory power or descriptive ability assumes nothing in regards to evolutionary models, determining the probability of a method producing accurate phylogenies inevitably requires considering the mechanisms by which evolution proceeds. Neither claims that parsimony makes no assumptions at all nor claims that parsimony and every possible method must make some assumptions (e.g. Goldman 1990; Swofford et al. 1996) are entirely correct or incorrect in the abstract—it depends on what the goal of a parsimony analysis is taken to be. This section discusses the assumptions relevant for parsimony to be seen exclusively as a means of finding phylogenetic trees, with a high probability of being correct. Only the case with equiprobable transformations between all states is considered. A truly general discussion should consider, of course, the case of parsimony with characters optimized as additive, step-matrix, or auto-weighted (Goloboff 1997), where the probability of transformation between some states is higher—but I know of no papers carrying out such a comparison, and the equiprobable model already has enough complexities and subtleties.

There are two main aspects of the comparisons that are relevant here. One aspect is whether it is possible to assume for the inference some model where the parsimony tree is necessarily the same as the tree of maximum likelihood by modifying (e.g. constraining or generalizing) some aspects of a model similar to the standard model described in previous sections. Another aspect is to find the conditions under which the data must have evolved—the actual evolutionary mechanisms—for the parsimony tree to be a consistent phylogenetic estimator. There is a good deal of work that has been done on both aspects along the years, and much of it is relevant for the problem of inferring phylogenies from morphology. In what follows, the data-generating mechanism is indicated as M_{DG} and the model used for phylogenetic inference as M_{PI}. Ideally, the M_{DG} should be the same as the M_{PI}—only then can consistency be guaranteed, and in practice we never have certainty about the M_{DG}. To compound the problem, as discussed in the following, $M_{DG} = M_{PI}$ is a *necessary* but not a *sufficient* condition—there are models of evolution, M_{DG}, which, if true, make consistent estimation impossible.

The main alternatives, using Markov/Poisson models as a basis, are whether variation in the length of different branches of the tree can occur (and how much), whether branch lengths are common to all characters, and the number of states a character could take. A standard likelihood model (*SLM*) is one as described in the preceding sections, with the likelihood calculated from summing the likelihood from all reconstructions for each character, adjusting lengths of tree branches so as to maximize likelihood, and making these branch lengths common to all characters.

4.10.1 Low Rates of Change in the M_{DG} Make Parsimony Consistent

When rates of the M_{DG} are constant for all characters and vary among branches, but are all sufficiently low, $MLT = MPT$, and *MP* is consistent (Felsenstein 1973, 1981). Steel (2001) provided a more formal analysis, showing that parsimony is consistent when the mutation probabilities associated with different branches are relatively small and not too unequal (with parsimony being consistent when the mutation probability of any branch is greater than or equal to the square of the mutation probability over all the tree). In this situation, it is unlikely that multiple changes occur along any given branch of the tree; in this case, the bulk of the likelihood for the character will come from the most parsimonious reconstructions (Felsenstein 1973; Goloboff and Arias 2019); the remaining ancestral state reconstructions will make only a minor contribution. Both *SLM* and parsimony are consistent in this situation.

4.10.2 Inferring Trees by Fixing Branch Lengths and Using Best Individual Reconstruction Amounts to Parsimony

When the inference uses the likelihood of each character with the individual reconstruction of highest likelihood, and the length l for all branches of the tree is fixed to be the same, the likelihood is a direct function of the number of steps (Goldman 1990). The likelihood of a reconstruction with n changes over the branches of the tree and r branches without change is then $L = P_{(stasis)}{}^{r} \times P_{(change)}{}^{n}$. As long as $P_{(stasis)} > P_{(change)}$ (which, in the case of Poisson models, is true for any $l < \infty$), reconstructions with fewer steps imply a higher likelihood. While fixing l over all the branches of the tree is a special case of the model assumed in standard likelihood methods, Goldman's (1990) use of the value from the most likely reconstruction alone is an approach to estimate the likelihood of trees more than an actual "model" (i.e. it does not specify something about how evolution proceeds but only about how we evaluate its results). Goldman (1990) used this to explore relationships between likelihood and parsimony and to argue for the superiority of *SLM*s. Felsenstein (1978b) attributed the inconsistency of parsimony to the fact that parsimony uses the score for the single most parsimonious reconstruction to select trees and argued that doing so increases the number of parameters being estimated (i.e. a state for each character and node), causing parsimony to overfit; Goldman (1990) and other authors (e.g. Lewis 2001; Holder et al. 2010) were of the same opinion. Goldman (1990: 350) had admitted that the ancestral states "are not parameters of the evolutionary process, but random variables: particular realizations of parts of the process" but insisted nonetheless on the "convenience" of treating them as if they were parameters. Farris (1986a), Goloboff (2003), and Goloboff and Arias (2019) noted that the reconstructions are not properly parameters; the problem created by using the value of likelihood calculated from a most parsimonious reconstruction is that this does not produce the actual probability of evolving the observed character states on the tree. The correct value can be obtained only by summing up the likelihood of all reconstructions (as done in *SLMs*), and the differences in ranking by likelihood or parsimony increase when there are unequal branch lengths; as a consequence, a character evaluated as having a higher likelihood than another by means of a single reconstruction may actually be *less* probable under the corresponding M_{DG}. Thus, the inconsistency in Goldman's (1990) approach for calculating likelihood results from using the wrong values of likelihood, not from estimating many parameters.

4.10.3 For Data Generated with All Branches of the Same Length, Parsimony Produces Good Results

This is one of the most important situations in which data can be generated so that parsimony is an acceptable method. With Felsenstein's (1978b) demonstrations that under Poisson models, differences in branch lengths—i.e. change for all characters being much probable in some branches than in others—are the main cause of inconsistency for parsimony, it follows that making l the same for all branches should make parsimony less subject to inconsistency. With l constant for all branches, character changes for a given character are equiprobably located in any branch of the tree—instead of concentrated on some branches. That is why the vast majority of simulations (from Yang 1996 to Wright and Hillis 2014), done with the obvious intention of criticizing parsimony, have used model trees where branch lengths are unequal. Goloboff et al. (2017) used a model for generating data where character changes can be located equiprobably on any branch of the tree—that is, a distribution of character changes similar to the one generated by uniform branch lengths—and the results of their simulations were favorable to parsimony methods.

Parsimony analysis of data generated with l constant for all characters and branches may still produce inconsistent results under specific conditions (Steel 1989). This may seem surprising—there are no long branches that could attract in that case. The clearest example where this could happen was provided by Kim (1996), illustrated in Figure 4.4. For the parsimony results to be inconsistent under the Poisson model, it is necessary that long and short branches alternate, as in Figure 4.4a; to the extent that the branches connecting terminals M and N are long relative to the branches connecting A, B, and x–y, parsimony will produce inconsistent results. While that cannot happen directly with all branches of the same length, if M and N are actually internal nodes, each of which leads to a number of terminal taxa separated from N or M by several intermediate nodes (e.g. in symmetrical subtrees,

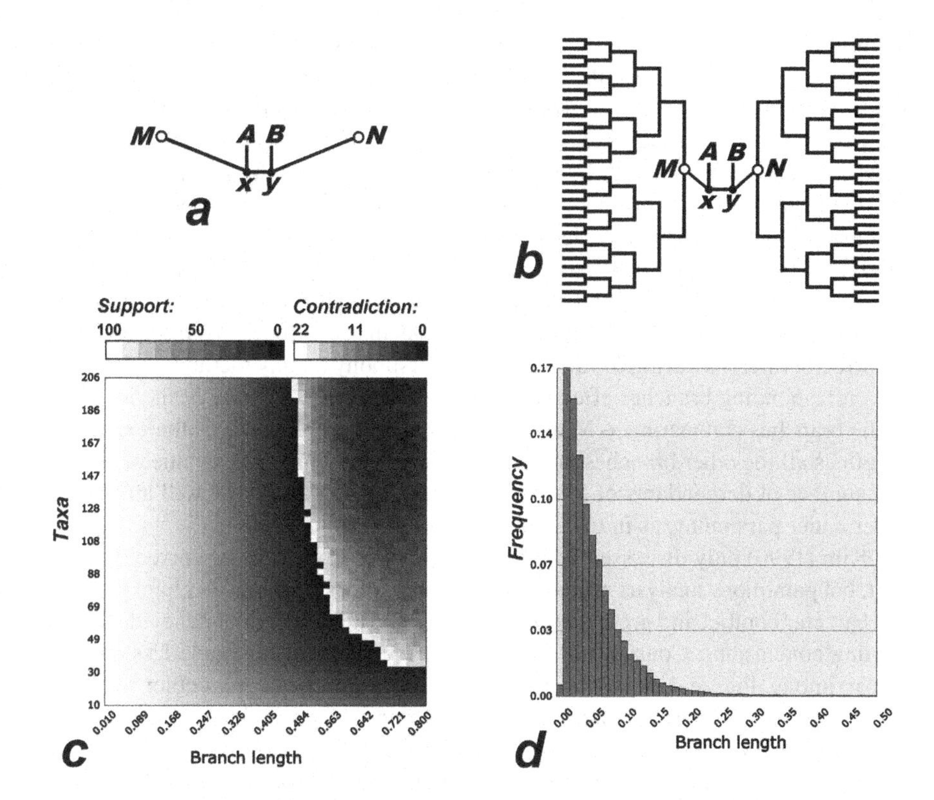

FIGURE 4.4 (**a**) The typical case of long-branch attraction, redrawn. (**b**) Even when all the branches of the tree have the same length, when a branch separates two groups, the first split of which contains a single sister taxon as sister of a symmetrical subtree, the cases where the common nodes of the symmetric subtrees are (by chance) assigned the incorrect state in a most parsimonious optimization produce the effect of "lengthening" the branches x–M and y–N. Therefore, this produces a "long-branch attraction" between M and N, possibly producing inconsistency even when all branches of the model tree have the same length. (**c**) Simulated datasets, on model trees as shown in (b), with 100,000 four-state characters, for increasing branch lengths and numbers of taxa (50 branch lengths and 50 numbers of taxa, with 10 datasets generated and analyzed for each combination of taxa and branch lengths, for a total of 25,000 matrices). Color indicates average value of relative Bremer support for the correct partition, AM|BN. When the strength of group support is taken into account (with relative Bremer supports), the parsimony inference is much less problematic when all branches become very long (as the support for the wrong group tends to zero). The support for the correct partition decreases as branch length increases, suddenly switching to the point of maximum contradiction (i.e. highest support of about 0.22 for AB|MN) with branch lengths of about 0.5 (a branch length for which expected number of changes under parsimony is about a fourth of the number of taxa; see Chapter 7, Section 7.10.4). As branch lengths continue increasing, the correct partition is contradicted more weakly until the two partitions (AM|BN and AB|MN) are almost equally supported. (**d**) Frequency of branch lengths when optimized as uniform throughout the tree, for 182 empirical datasets (a total of 37,240 characters examined). Branches long enough to produce inconsistency under parsimony never seem to occur in empirical datasets.

as in Figure 4.4b), then a strange phenomenon can happen. Consider how the parsimony score of the tree would be calculated for data generated under such model tree: a down-pass optimization would be used to infer ancestral states for nodes M, but in some of the characters (given that none of the terminals is directly connected to M), the ancestral state will be incorrectly inferred (see discussion in Herbst et al. 2019); the probability of this happening increases both with the number of taxa in the symmetric subtree (as then the terminals are farther from M) and the branch length (with extremely long branches, the states of every descendant are chosen almost at random, regardless of its immediate ancestral states, so that the reliability of inferred ancestral states is minimum). Likewise for N. As many of the states in M and N are incorrectly inferred during the down pass by parsimony (or any method, in fact), then the corresponding branches effectively behave as *if they were long branches*—even if the branches connecting x–M and y–N in the model tree were of the exact same length as all the other branches in the tree. Then, depending on the value of l, and as the number of descendants of M and N increases, nodes M and N will attract each other under parsimony, as in the classic *LBA* example.

Kim (1996) only discussed whether the parsimony tree will be correct or incorrect, but parsimony analysis needs to (and regularly does) take into account measures of character conflict and group support to determine the strength of the evidence supporting conclusions. Considering group supports makes this example less damning to parsimony. Figure 4.4c shows a plot of simulated matrices, with color indicating the strength of support or contradiction for the correct partition, AM|BN (as measured with the relative Bremer support, Goloboff and Farris 2001). As l increases, the support is weaker and weaker, suddenly switching to the point of maximum contradiction at branch lengths of about 0.5 to 0.65 (depending on the number of taxa; for fewer taxa, the branches need to be longer for the inconsistency to occur). The correct partition AM|BN is contradicted at the most by the conflict of about 10 characters contradicting the partition, in conflict with 8 characters favoring it (i.e. a relative Bremer support of –22%). One might think that, given that longer branches cause inconsistency, increasing l would continue worsening the results for parsimony, but that is not the case. As l is taken to extreme values, the resulting datasets approach a random collection of character states, and thus using support measures reveals that those resulting datasets can actually support no conclusion with any strength—either for the correct AM|BN or the incorrect AB|MN. When the M_{DG} has constant branch lengths, therefore, parsimony considering degree of group supports can produce misleading results only if the branches are long, but not too long.

Naturally, just as the M_{DG} can have all the branches of the same length, the M_{PI} can be constrained in this manner as well. The *SLMs* (where different branches of the tree have different lengths), known to produce consistent results when branches of the generating tree are different, are bound to produce consistent results when the M_{DG} has all branches of the same length (which is simply a special case of the model assumed in *SLMs*). However, the *SLMs* in that case estimate as many parameters as branch lengths—which are not truly needed. One can envision, then, an M_{PI} where the length is forced to remain constant for all branches, estimated from the data (i.e. chosen so as to maximize likelihood), and the sum of likelihoods for all reconstructions is used (instead of the value from the reconstruction of highest likelihood, used in Goldman's 1990 approach). There are no studies of the statistical properties of such an estimation model, but it seems logical to conclude that (if M_{DG} has constant l) the phylogeny can be inferred consistently—if *SLM* is a consistent M_{PI} in that case, so will be the M_{PI} with constant branch lengths, as branch length should be estimated even more precisely. Given that inferring trees under the assumption of constant branch lengths produces results closely approximating those of equal weights parsimony in most cases, Goloboff and Arias (2019) called such a method MP_{lik} and used it to further explore the relationships between parsimony and likelihood. Goloboff and Arias (2019) also showed that an extension of that method, adjusting l individually for each character (but still identical for all branches), takes into account that some characters evolve at faster rates than others and resembles the results of implied weights parsimony (Goloboff 1993a; see Chapter 7); they thus called the method *implik*.

4.10.4 If All Characters and Branches Can Have Different Lengths, MP Is Identical to ML

This is a consequence of the likelihood for a single character, under the *SLM*, being a direct function of the minimum number of steps required to map the character (i.e. the parsimony cost) and ancestral state reconstructions (i.e. the set of most likely states) being exactly the same as in a most parsimonious reconstruction (*MPR*). This equivalence (Tuffley and Steel 1997) results from the fact that, to maximize the likelihood, branches where changes occur in a *MPR* can be assigned $l = \infty$ (so that for s states, from formula 3.2, the probability of change along the branch tends to the maximum $1/s$), and branches where no changes occur assigned $l = 0$ (so that the probability of stasis is unity). The likelihood for a character with n steps is then $(1/s)^{n+1}$ (the exponent adds 1 to n to take into account the prior probability of the state at the root, $1/s$). Note that under such a set of branch lengths, the likelihood of any other reconstruction will be 0 (as $P_{(change)} = 0$ for branches where $l = 0$). When multiple most parsimonious reconstructions exist for a character, every one of those will imply a different set of branch lengths, each of which has the same likelihood.

The equivalence between states reconstructed as most parsimonious or most likely under the *SLM* does not occur when branch lengths are optimized for several characters at a time because, in that case, l is set to maximize the likelihood for all characters collectively, not just the character in question. For example, in the typical four-taxon case of inconsistency for parsimony (see Figure 1.4b), a character with state 1 in taxa BC and state 0 in taxa AD, the reconstruction that implies the highest likelihood (for the set of branch lengths shown in Figure 1.4b) is the one with state 0 at the common node of BC. Given the probabilities of change induced by those branch lengths, the explanation of identity by common ancestry is less likely than the explanation of independent derivations: change is very likely along branches v–B and v–C and very *un*likely along branch w–v, so the scenario where B and C share state 1 by common ancestry is very improbable. That can happen only when branch lengths are chosen to maximize the likelihood for all characters simultaneously; if the branch lengths are chosen so that the likelihood for the individual character to be mapped is maximized, a parsimony mapping *is* a likelihood mapping.

Tuffley and Steel (1997) also showed that calculating the likelihood for each of the characters in this manner produces results that are entirely equivalent to equal weights parsimony. As this does not assume that l is common to all characters, Tuffley and Steel (1997) baptized the model as "No Common Mechanism" (*NCM*). This is a consequence of the intimate relationship between parsimony and likelihood scores for an individual character: the total likelihood under *NCM* is the product of the individual likelihoods for each of the C characters, calculated as indicated prior:

$$L = \prod_{i=1}^{C} (1/s)^{n_i+1}$$

Applying logarithms to both sides of the equation produces an expression,

$$\ln L = \ln(1/s)\left(c + \sum_{i=1}^{C} n_i\right)$$

which makes it obvious that the (log) likelihood of the tree only depends on its number of steps under parsimony. Note that this equivalence no longer holds when the size of the state space differs among characters (Wheeler 2016; Goloboff and Arias 2019).

The *NCM* model, when used for inferring trees (i.e. as M_{PI}), needs to estimate a very large number of parameters: l must be estimated for every character and branch. With so many parameters, the resulting values of likelihood are the highest possible for any Poisson model—no other model M_{PI} can possibly improve those likelihood values. But, as the amounts of data grow, then so does the number of parameters to be estimated, and consistency under such a situation cannot be guaranteed. Because of this, the estimation of l jumps between extremes (either 0 or infinity); each l is "estimated" from a very low amount of data (a single branch, ancestor, and descendant state). This has been noted by Tuffley and Steel (1997), Lewis (2001a, 2001b), and many other authors. It is remarkable that the computations required for evaluating or searching trees under this model are much faster than for the less parameterized *SLM*, as more complex models typically involve more complex computations (Steel, in Sanderson and Kim 2000: 826). Holder et al. (2010) showed that model-selection methods cannot possibly choose *NCM* over *SLM*—the magnitude of the improvement in likelihood cannot possibly match the large number of parameters added when switching from the simpler to the more complex model (see also Huelsenbeck et al. 2011).

As *NCM* produces the same results as parsimony, many authors have simply considered that the two methods are entirely equivalent—i.e. that parsimony cannot be justified without overparameterization (e.g. Lewis 2001a; O'Reilly et al. 2018). However, there are other models under which parsimony is a maximum likelihood estimator; *NCM* is only one such model, but simpler ones can be devised. Thus, although the equivalence in results for *NCM* and parsimony is useful in exploring the relationships between parsimony and likelihood methods, the analogy cannot be overstretched to consider that the *NCM* is a *synonym* of parsimony (Huelsenbeck et al. 2008; Holder et al. 2010; Goloboff and Arias 2019). Ironically, while fashion has now swung to considering parsimony as overparameterized, it used to be considered by some authors (e.g. Yang 1996) as too simplistic a model—any argument seems to go when used for belittling parsimony.

The fact that calculating likelihood for each character with its own branch lengths is identical to *MP* also suggests—although I know of no attempts to test this on real examples—that a dataset with numerous unlinked partitions should produce results more similar to parsimony than when analyzed with branch lengths common to all partitions. Given that it produces results identical to parsimony, it is clear that *NCM* can be inconsistent under the exact same situations. This shows that likelihood inference cannot be consistent under every possible model: if the *NCM* model of Tuffley and Steel (1997) is the model truly generating the data, M_{DG}, as well as the model used for inferring trees, M_{PI}, the results can still be inconsistent. That is, the consistency of phylogenetic estimation crucially depends on the M_{DG}, not just on the M_{PI}, and there is nothing systematists can do about that—the data have evolved the way they have. If the data have truly evolved under certain types of models—such as *NCM*—consistency cannot be guaranteed, no matter how carefully the M_{PI} is designed.

The *NCM* is also interesting from two other aspects. One is the fact that the character distributions it generates are identical to the character distributions generated by the ultra-conserved model where all the branches of the tree have the same lengths, identical for all characters (Huelsenbeck et al. 2008; Steel 2011). Under *NCM*, if the length of a branch in the model tree is assigned independently of the length of other branches, any branch could be long or short, and then a character change (Figure 4.5a, left tree) could equiprobably occur in any of the branches of the tree; for the model with uniform branch lengths (Figure 4.5a, right tree), the same is true, as no branch is longer than any other. Thus, ironically, the simplest and most complex models are alike in producing the same net result: in the middle tree of Figure 4.5a, the three character changes might have occurred with equal probability in any of the branches of the tree.

4.10.5 Invariant Characters and a Large Number of States

For any dataset, when adding a large enough number of invariant characters, the trees inferred by assuming the *SLM* are identical to parsimony trees (Tuffley and Steel 1997). This is a natural consequence of all the values of l decreasing, thus approaching the condition explained in the first subsection.

Parsimony also becomes consistent when the number of possible states in the M_{DG} is large (Steel and Penny 2000, 2004); in that case, the probability of the same state being independently acquired in two different branches of the tree becomes very low. Likewise, when a very large state space is assumed for the M_{PI}, the results of *SLM* become identical to those of parsimony, for any dataset. Of course, if the state space assumed is larger than the actual one (e.g. assuming that many states are possible

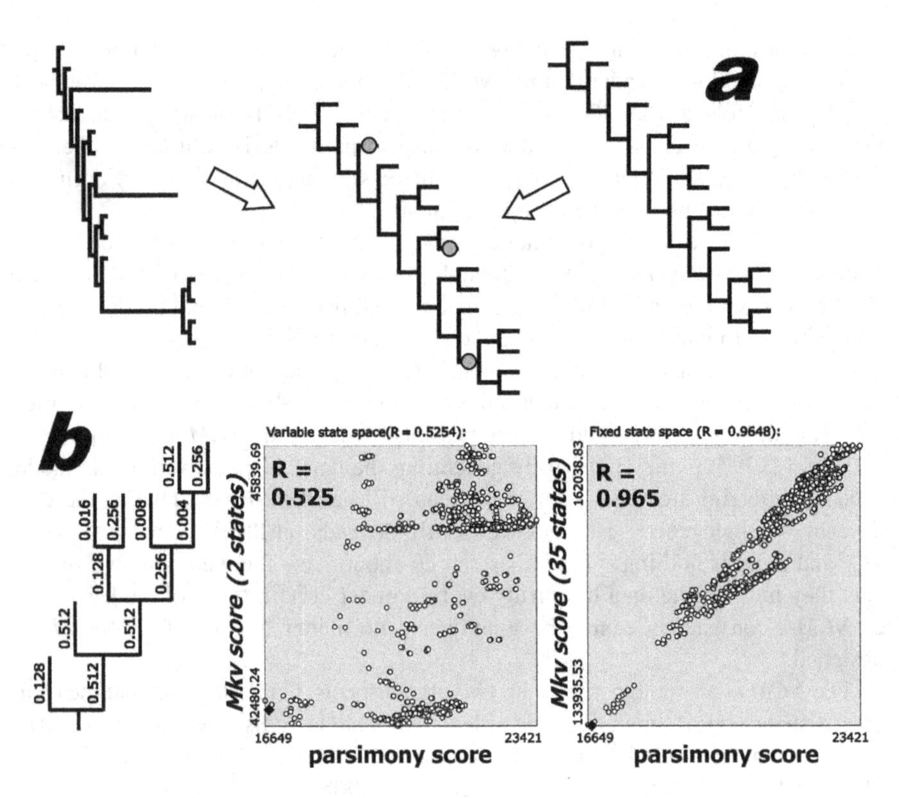

FIGURE 4.5 (**a**) The probabilities of different character state distributions under a model where every branch can have its own length and the model where all the branches have the exact same length are exactly the same. (**b**) A seven-taxon model tree with branch lengths, used to generate 9,999 characters which can take two states, plus a character with a state different for each taxon (N, O, P, Q, R, S, and T); all characters are variable (PAUP* thus defaults to the Mkv variant). The left plot shows parsimony (x-axis) and likelihood scores (y-axis) with an Mkv model where the state space for each character is the observed number of states (`lset mkstate=var` in PAUP*). The right plot shows the same, but assuming that the state space is common to all the characters and large (`lset mkstate=fix`). The plots show the model tree with a black diamond. For such a combination of branch lengths, the model tree is optimal under the Mkv model when inferring trees with a larger state space than used for generating the data. Pearson's correlation coefficient between *MP* and the Mk model with observed number of states is R = 0.525, and between *MP* and the Mk model assuming a large state space is R = 0.965.

instead of the four nucleotides ACGT), the results will become inconsistent (they are identical to parsimony, after all). On the other hand, when the true state space is unknown, it is entirely possible for a given sample of characters (for some combinations of branch lengths in the model tree), that assuming a larger state space produces trees that are closer to the model tree (Figure 4.5b). This also means that (in some computer implementations, such as PAUP*) the numerical codes chosen to represent states may influence the results: for example, in a binary matrix, representing the state 1 as 9 instead may produce different results under the Mk/Mkv model because the size of the state space is determined from the largest state in the matrix, not the actual number of distinct states.

4.10.6 Missing Data and Likelihood

The behavior of missing entries under *SML* is hard to predict, often producing counterintuitive results. Truszkowski and Goldman (2016) showed that certain patterns of missing entries will not affect the ability of *ML* to recover correct phylogenetic trees, but this is clearly not true of all possible patterns of missing entries. Goloboff and Arias (2019) discussed this situation for the MP_{lik} model. Simmons (2012) provided several cases where adding missing entries produces unexpected results under *SLM*. Goloboff and Wilkinson (2018) showed examples where a set of perfectly compatible characters (which can thus all be displayed without homoplasy on a single, most parsimonious tree) produces a tree which requires some homoplasy under *SLM* with JC69 (apparently as a result of the interaction between the branch lengths in the subtrees). Other situations with unexpected behavior for *SLM* can occur even for simulated data with blocks of missing entries added, as in Figure 4.6a. In that example, the model tree has two groups ("ONE" and "OTHER"), but once the matrix is generated, some of the taxa have their characters replaced by missing entries so that there is no overlap between the two groups (effectively making it impossible to establish any phylogenetic conclusion, something properly recognized by parsimony, which produces a bush as consensus tree). The analysis by means of *SLM*, however, produces the incorrect trees (mixing ONEs and OTHERs), with very high support (and regardless of whether the data partitions, generated on the same model tree and branch lengths, are analyzed as linked or unlinked). In general, it appears that missing entries do not imply full ambiguity under *ML* (unlike the case for parsimony) and that the "information" attributed to these missing entries under *ML* can be misleading.

4.11 STANDARD POISSON MODELS IN MORPHOLOGY

Finally, after having revised the basic premises and theoretical properties of likelihood methods and their relationships with parsimony, we can consider the problem of whether application of the Mk/Mkv model to morphology is preferable over the use of parsimony. This problem is partly theoretical and partly empirical—depending on whether the assumptions made by *SLMs* are appropriate for morphological data and on whether what is known about the evolution of morphological characters leads to a situation where *ML* and parsimony should produce similar results. Ironically, the Mkv model was proposed by Lewis (2001a) with no effort to test whether any of the assumptions of the model even remotely hold for morphological datasets; as many other likelihoodists, Lewis seems to have believed that the charge "guilty until proven innocent" (Felsenstein 1978b) is to be applied only to parsimony, not to *ML* methods.

The crucial assumption of the Mk/Mkv model is that the branch lengths l are shared by all characters (or all the characters in a partition), as indicated in Figure 4.3a.

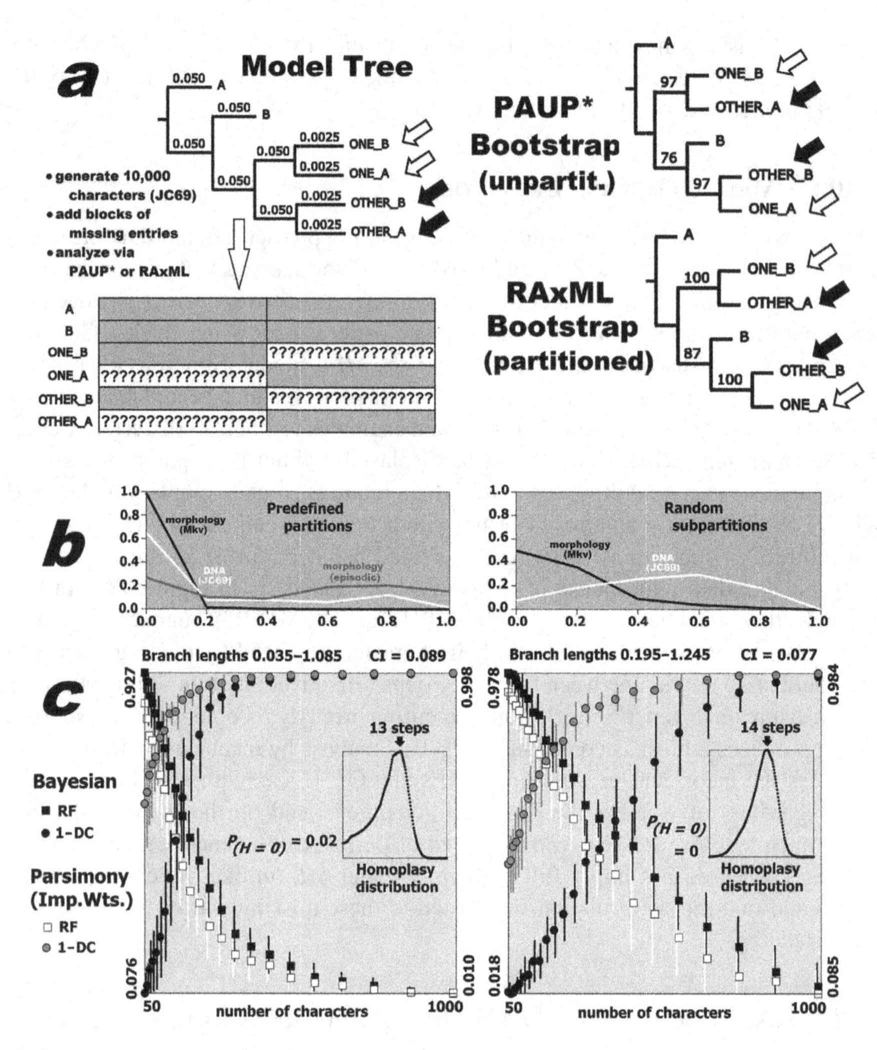

FIGURE 4.6 (**a**) Generating characters under the model tree shown (with JC69), and converting two non-overlapping blocks of characters into missing, should produce an unresolved tree as result—there is no information in common between taxa ONE and OTHER. However, maximum likelihood trees (found with either PAUP* or RAxML) display incorrect groups (ONE + OTHER) with high bootstrap support. (**b**) Frequency of datasets (on y-axis) showing probability of the observed degree of branch-length correlation under the Mkv model (x-axis), between predefined (left) or random subpartitions (right), in morphological (black line) and molecular datasets (white). Gray line shows probability of observed correlation under an episodic model (i.e. without a common mechanism) in morphological datasets. Data taken from Goloboff et al. (2018). (**c**) Datasets with 50 taxa, simulated with increasing number of characters and a Poisson model, with all branches of the tree with the same length, but differing among characters. For both character-length distributions, the resulting characters have a peak of homoplasy at 13–14 steps, but the homoplasy is not uniformly distributed across all characters. Both implied weights and Bayesian analysis converge to the model tree with increasing number of characters, but for realistic numbers of characters, implied weights is always closer, on average.

Morphology seems to evolve instead with a lot of mosaicism, as in Figure 4.3b. This was already noted by Farris (1983: 15) in his discussion of stochastic models:

> The general success of hierarchic summaries of characters of organisms seems to suggest that each feature undergoes evolution in only a restricted part of the evolutionary tree, and that different portions of the tree have different suites of varying features. That is, after all, what is implied by the observation that the synapomorphies of tetrapods involve different traits than do those of hymenopterans.

While some tree branches exhibit more morphological changes than others, it is often in particular groups of characters, and this does not seem to increase the probability that every other character changes along the branch (Goloboff et al. 2018). Goloboff et al. (2018) examined eight morphological datasets with predefined partitions (for character systems or body parts, allowing for 79 separate comparisons between partitions). Goloboff et al. (2018) showed that the branches longer for one partition are not the longer ones for the other partitions. The differences in branch lengths for the different partitions are much larger than expected under the Mk/Mkv model. In the tests of Goloboff et al. (2018), only 4% of the comparisons between morphological partitions produce a correlation between branch lengths for both partitions that is not below 99% of the correlation in partitions simulated with the same numbers of characters and branch lengths (which is the appropriate comparison; see Goloboff et al. 2018 for discussion). Goloboff et al. (2018) also used a subpartition test for 78 morphological datasets where partitions could not be defined[8]; they noted that, with datasets generated by means of multiple sets of branch lengths for different sets of characters, random subpartitionings increase the apparent homogeneity of the branch lengths between subpartitions (as this leaves both subpartitions with similar proportions of the characters in each of the original sets). This highlights difficulties with several methods proposed by likelihoodists to determine appropriate partitioning schemes (e.g. Duchêne et al. 2014; Lanfear et al. 2017)—it is easy to partition data generated with a homogeneous l into apparently heterogeneous partitions or to mix data generated with heterogeneous l in apparently homogeneous subsets. Thus, rejection of branch-length heterogeneity in pairs of randomly chosen subpartitions is especially meaningful (i.e. the test rejects homogeneity very conservatively); 86% of cases rejected homogeneity in the tests performed by Goloboff et al. (2018). This strongly contrasts with DNA sequences (tested under the JC69 model); contiguous partitions fail to reject homogeneity in DNA datasets for 42% of comparisons (over 10 times more frequently than predefined morphological partitions), and random gene subpartitions fail to reject homogeneity in 91% of comparisons. Figure 4.6b shows the frequency of different degrees of branch-length correlation (as measured by the probability of the observed correlation under a model with common branch lengths). The left graph corresponds to predefined partitions, with probabilities in the intervals around 0.2, 0.4, 0.6, and 0.8 each occurring in about 10% of cases of DNA sequences (white line), while basically none of the morphological datasets (black line) had a branch-length correlation even remotely approaching that frequency under the Mkv model. The right graph corresponds to random subpartitionings; for the majority of the morphological datasets, the probability of obtaining the observed correlation between branch lengths under a common mechanism is in the 0 interval, while for the majority of the DNA datasets, the probability of obtaining the observed correlation between branch lengths under the JC69 model is in the interval of 0.6.

The heterogeneity of branch lengths for different partitions or subpartitions of the morphological datasets examined by Goloboff et al. (2018), therefore, strongly rejects the Mkv model. Those datasets have a weak correlation between branch lengths, one that can easily occur if the data evolve without a common mechanism (and hence probably producing acceptable results under parsimony); Goloboff et al. (2018) tested a mechanism (which they called *episodic*) where character changes can occur equiprobably at any branch but only within certain regions of the tree, as seems to be the case of much morphological variation. The episodic "model" is reminiscent of the "covarion" model of Fitch and Markowitz 1970 but without a common mechanism (see discussion in Gruenheit et al. 2008). The episodic "model" creates a correlation between branch lengths which is just as observed in morphological datasets (with only 29% of the datasets rejecting the model and the rest showing correlations between branch lengths that are well within what is expected under such conditions of evolution).

Morphology, in other words, does not confidently allow identifying branches of the tree where morphological change is more likely to occur in all characters.[9] That being so, models such as the *NCM* model of Tuffley and Steel (1997) or a model with constant branch lengths seem indicated. Both models lead to changes being equiprobably located on any branch of the tree (at least within regions where variation is possible); this uniformity seems the best ground assumption for the evolution of morphology, where so few mechanisms can be hypothesized and so little extrapolation from one character to another is possible. Estimation models with constant branch lengths seem more desirable for inferring trees because they do not require estimation of enormous numbers of parameters of *NCM*. Goloboff and Arias (2019) explored *implik*, an M_{PI} which produces results similar (if not identical) to weighted parsimony, with all branches of a uniform length (optimized separately for each character). *Implik* estimates as many parameters as characters, instead of the $2T - 2$ for a tree of T taxa required for the *SLM*. Goloboff and Arias (2019) found that (when the number of taxa is large relative to the number of characters), model-selection methods (such as the *AIC*) tend to preferentially select *implik* over the Mkv model. As illustrated in Figs. 4.4b and 4.4c, uniform branch lengths in the model tree do not guarantee that parsimony is consistent for every value of l. The values of l that can produce inconsistency, however, are well beyond the branch lengths that are normally observed in morphological datasets; Figure 4.4d shows the distribution of l (optimized for each character at a time, as in *implik*, for 182 morphological datasets with 50 or more taxa; datasets taken from Goloboff and Arias 2019). When the generating tree has all branches of the same length, parsimony begins showing inconsistency (as shown in Figure 4.4c) for $l \approx 0.50$ (for 100 taxa) or $l \approx 0.45$ (for 200 taxa); this is the branch length when a mutation is expected for every other character in the dataset along every branch! In contrast with this, 95% of branch lengths in morphological datasets are estimated to be within 0.14; basically, no optimized branch length (out of the 37,240 morphological characters examined) is above $l = 0.3$. Equal weights parsimony can indeed be inconsistent if the model tree has uniform branch lengths, but only when all the branches are unrealistically long. In addition, with weighted parsimony (see Chapter 7), when the branches are longer for some characters, additional steps in those characters cost less, so those characters are expected to have less influence on the results anyway. Lest I be misinterpreted, I further clarify: parsimony is not guaranteed to be consistent by the mere fact that morphological datasets are well below the l values which make parsimony inconsistent when the data evolve with a Poisson model tree and constant branch lengths. Not at all. There are many other reasons that can produce inconsistency (i.e. many possible violations of any of the assumptions in Poisson models), and no morphological dataset has the infinite numbers of characters needed for consistency even in the ideal circumstances. But this is equally true of *ML* methods. Thus, in practice, no method can expect such guarantee of consistency, and no method is infallible for empirical datasets—different empirical analyses (for overlapping taxon sets) often show different results (for either parsimony or model-based methods; see e.g. Rydin and Källersjö 2002 for a case where MrBayes supports with 100% confidence mutually incompatible groups for different taxon subsets). The low values of l in morphological datasets only mean that parsimony is not guaranteed to be inconsistent, which seems to be the gist of the concerns of those who prefer likelihood methods.

The approach just discussed hinges on (a) expecting the probability of change in a character along a branch to be more or less independent of the probability of change in other characters and (b) having no grounds for expecting character changes to be concentrated on certain branches of the tree. Perhaps one can reject (a) and (b), if one believes that evolution is "explicitly gradualistic, excluding punctuated equilibria as a mode of morphological evolution" (Lewis 2001a: 916), and then "the probability of observing a change along a branch in a phylogeny increases with the amount of time associated with the branch" (Lewis 2001a: 915–916)—this would lead to an Mkv-like model producing acceptable results. But in that case, the properties of morphological characters are such that, if properly considered, they may lead to *ML* estimations producing—by necessity—almost the same results as parsimony. The most obvious of such properties is the extreme conservatism of much of morphology, resulting in many characters being invariant. As discussed in the previous section, explicitly adding invariant characters to a dataset leads the Mk model to produce most parsimonious trees. That Lewis (2001a) proposed a correction to determine what would be the expected proportion of invariant characters under the branch lengths inferred by variable characters only is beside the point because that takes the model for granted. The problem verges on the metaphysical: have those characters which vary in other taxa been conserved in the group under study because they *cannot* change or just because they *happen* to not have changed? If the characters cannot change, then estimating the proportion of invariant characters may correct the likelihood values of trees (and then change the ranking of trees under *ML* without affecting the proportionality of branch lengths). But determining the difference between those two alternatives seems impossible, and the notion that it is somehow physically impossible for the characters to change seems self-defeating—the characters do vary in other groups. Of course, by pooling all the characters known to vary somewhere in the Tree of Life in the estimation of branch lengths for a given morphological dataset, the low probabilities of change implied by such pooling are not compatible with the fact that some characters do show numerous instances of homoplasy—if the probabilities of change along all branches are so low, there would be no multiple changes in any character, ever. That this is observed empirically is as much a reason to not pool the characters as a reason to doubt the model. Note that these several conundrums are implied only by models with a common mechanism and tree branches varying in length—invariant characters have no influence on the analysis when branch lengths are constant throughout the tree and optimized on a per character basis (Goloboff and Arias 2019), just as in parsimony.

A similar situation arises when the size of the state space is considered. Poisson models require that the state space used reflects not just the number of states effectively observed in the dataset but the number of theoretically possible states—and for morphology this could be a vast number of possible states. A few studies have attempted to study the size of the state space (e.g. Wagner et al. 2006; Hoyal Cuthill 2015). The very fact that homoplasy occurs might seem a reason to suspect that the character space is finite, but Hoyal Cuthill (2015) showed that even very large state spaces, with a constrained model she called "inertial", may well produce homoplasy, and different empirical datasets seem to support one or another option. In the absence of positive evidence in this regard, a judicious course of action may well be assuming a large state space—which leads the Mkv model to converge with parsimony.

4.11.1 Simulations

In recent years, several papers (Wright and Hillis 2014; O'Reilly et al. 2016, 2018; Puttick et al. 2017, 2018) have defended the use of Bayesian methods for morphological datasets, based on simulations. Most follow a similar pattern, generating their data with a Poisson model with l unequal for different branches, then checking the distance between the trees inferred by alternative methods (parsimony with equal or differential weights and the Mkv model with MrBayes) to the model tree. Those are the precise circumstances known (since Felsenstein 1978b) to make parsimony perform poorly. This highlights the fact that simulations have only a relative utility for choosing methods of phylogenetic inference: depending on the model used to generate data, one or another method will perform better. For example, Goloboff et al. (2017) chose a model where parsimony outperforms Bayesian analysis, and parsimony—as if magically—performed better. Unless the conditions used to generate data in the simulations can themselves be validated empirically, the results of the simulations are of no help in the real world—and the homogeneity conditions used in all those simulations defending the use of Bayesian analysis are strongly rejected by empirical data (Goloboff et al. 2018). In addition, not even the task of measuring the distance to the model tree is simple (see Chapter 6, Section 6.7, for discussion), although it crucially affects the conclusions. In their initial comparisons, O'Reilly et al. (2016) and Puttick et al. (2017) did not collapse poorly supported groups in parsimony or likelihood, so the "inferred tree" was unnecessarily resolved for those methods. And, as noted by Smith (2019), O'Reilly et al. (2016) and Puttick et al.

(2017) used the standard Robinson-Foulds distance (*RF*, Robinson and Foulds 1981), which tends to favor unresolved results (i.e. two resolved trees may have a distance of 1, but the distance between a complete bush and any tree is at most 0.5) and therefore has the strange implication that datasets with no characters produce better inferences than datasets with only a few (see Figure 6.5d, in Chapter 6). When normalizing *RF* by the number of groups actually present in the trees (see Chapter 6, Section 6.7.1), Smith (2019) observed that even for the datasets generated under homogeneity conditions, the results of Bayesian analysis and implied weights produced results equally distant from the model tree.

A special mention must be made of the last paper in the series of simulations, Puttick et al. (2018). In that paper, they claimed to have generated datasets *without* a probabilistic model. But of course there are probabilities involved! They actually generated data from a model tree in such a way that there are a certain number of character changes located with equal *probability* on any of the branches of the tree.[10] This is exactly the same approach Goloboff et al. (2017) had used for their simulated datasets. Puttick et al. (2018) compared the performance of Bayesian and weighted parsimony methods for varying amounts of homoplasy and found that for low or medium amounts of homoplasy (i.e. low or medium number of changes randomly thrown on the tree branches), parsimony and Bayesian analysis perform more or less similarly (with parsimony even outperforming Bayesian analysis in some cases). Note that they still used the standard *RF* as their main statistic. They found that for extreme amounts of homoplasy, Bayesian analysis produces the best results—because it yields completely unresolved trees, and those have low *RF* distances to any tree (cf. Smith 2019). They stated that their "simulation procedure could not be accused of faithfully reflecting the process of morphological evolution" (Puttick et al. 2018: 16) but took this to be an advantage, proving that Bayesian methods are superior to all others (standard *ML* included) under any circumstances, regardless of how the data evolve. All the researchers (Felsenstein, Yang, Huelsenbeck, Ronquist, etc.) who painstakingly developed methods based on the most reasonable models they knew, and all those who provided formal analyses of specific models (Steel, Chang, Kim), were wasting their time, in the opinion of Puttick et al. (2018): *MCMC* as implemented in MrBayes reveals truth, regardless of models and regardless of how the data evolve—just by pure magic. Of course, not even Felsenstein (one of the most ardent defendants of *ML* methods) would maintain such a position because the success of methods to reconstruct phylogenies depends on the mechanism generating the data. Puttick et al.'s (2018) "finding" is even less justified when one considers that it hinges on the datasets with maximum homoplasy, those with consistency index 0.0 to 0.1. Their model trees have 32 taxa; for binary characters and that many taxa, that is almost exactly the consistency index one obtains when generating random two-state data—that is, datasets in which every taxon is assigned (for every character) either 0 or 1 completely at random, without any regard for the character state in its ancestor (see Goloboff 1991a, 1991b). Any of the methods discussed in this book to detect low group supports or character conflict will quickly detect that these datasets of consistency 0.1 or less do not lend themselves to confidently estimating phylogeny. Inferring a phylogenetic tree can never be a more futile exercise than for such data, yet it is the case Puttick et al. (2018) considered as universally validating Bayesian analysis.

When *RF* is corrected to not overfavor full polytomies, and the datasets are generated with variable numbers of characters and high amounts of homoplasy (of course, not reaching the absurdly high values used by Puttick et al. 2018), the picture becomes clearer. Figure 4.6c shows a simulation for 50 taxa with a Poisson model, branch lengths constant over all the tree, chosen for each character from a uniform distribution, and increasing numbers of characters. Puttick et al. (2018) only examined datasets with 100, 350, and 1,000 characters; the simulation of Figure 4.6c shows a much finer scale. The distance to model tree (measured by the *RF* normalized with number of observed groups and the complement of the distortion coefficient, *DC*, normalized to be symmetric, as in Goloboff et al. 2017; see Chapter 6 for details of these measures) is rather similar in Bayesian analysis and differentially weighted parsimony for very low or very large numbers of characters. Very few characters produce poor inferences (although only according to *RF*; *DC* suggests a clear superiority of parsimony), and many characters produce very good inferences. The difference lies in the middle: for medium numbers of characters, parsimony strongly tends to outperform Bayesian analysis. The partly overlapping error bars show that it is entirely possible that Bayesian analysis produces better results than parsimony in a particular case, but the pattern is one of clear superiority for parsimony.

4.12 CONCLUSIONS

Methodological developments by researchers focusing on the use of explicit models are indeed very interesting on their own but not necessarily relevant for practicing morphologists. Many properties of discrete morphological characters are such that estimations with parsimony will tend toward consistency, and estimation under *ML* will tend to converge to estimations under parsimony, for several reasons. First, the absence of known mechanisms driving evolution of most characters (leading to a difficulty in identifying branches of the tree where all characters could increase their probability of change) makes models where changes can be equiprobably located at any branch more reasonable. Second, the potentially high number of invariant characters in any dataset, and third, the potentially very large state space for morphological characters, lead *ML* to produce results approximating *MP*. The second and third factors affect Poisson models with either heterogeneous or uniform branch lengths. Thus, in the case of morphology, the conditions that make parsimony prone to inconsistency in the case of sequences are much less likely, and *ML* estimation considering the peculiarities of morphological characters tends to produce the same results as parsimony.

Why then not dispense with all the time-consuming and stringent assumptions needed for *ML* and just proceed with a parsimony analysis? It is true that by doing so, all the sophistication and accuracy attributed to *ML* methods (precise dating, calculation of evolutionary rates, etc.) cannot be invoked. Parsimony does not pretend to attribute specific probabilities to trees or events; it is just seeking the most plausible explanation—without probabilities attached. This seems to be the most that can be done with morphological data, and it seems wiser to accept this, rather than resorting to methods that use calculations based on unjustified assumptions.

4.13 IMPLEMENTATION

Most of the options discussed in this chapter can be implemented only in programs for model-based phylogenetics, such as PAUP*, MrBayes, and RAxML. TNT does not implement tree searches under likelihood, but it can calculate likelihood of given trees under several models. In most cases, it is possible to use a global state space (default; the maximum state in the matrix, specifically set with `lset gstatespace` before calculating the likelihood), a user-defined state space (`lset gstatespace N`), or the state space observed for each character (`lset nogstate`). To visualize the likelihood, as well as the likelihoods of the individual characters, the reports must be on (`report=`). To visualize optimized values of branch lengths instead of character likelihoods, use `lset display brlen` before calculating likelihood. The default value for `lset display` is `lik`.

To calculate the likelihood of tree number 0 (and save the value to user-variable 0; see Chapter 10 for options on naming variables):

```
macro=; report = ;
lset gstatespace ;
set 0 mklik  [ 0 ] ;
```

To use a different state space, substitute `gstatespace` for the appropriate expression. This calculates the likelihood under the Mk model. Keep in mind that all characters are treated as nonadditive (i.e. Neyman's 1971 model). The likelihood is calculated for active characters only (i.e. leaving a single character active, the *NCM* log likelihood can be calculated by summing up the character log likelihoods). Instead of Mk, several other models can be used by substituting `mklik` in the prior example:

`fixlik`. Calculate likelihood with branch lengths fixed, common to all characters. The branch lengths are taken from the tree-tags (`ttags`). Optionally, branch lengths can be given as an array (see Chapter 10 for declaration and assignment of values to arrays), the name (or number) of which is given as second argument to `fixlik` (i.e. before the closing square brackets).

`implik`. Calculate likelihood by optimizing branch lengths separately for each character, uniform across all branches of the tree. Likelihoods calculated by summing all possible reconstructions. The ranking of the trees when using this model, or implied weighting with mild concavities, is not identical but very similar (see Goloboff and Arias 2019).

`mplik`. Use the same length for all characters and branches, optimized to maximize likelihood. The ranking of trees when using this model, or equal weights parsimony, is not identical but very similar (see Goloboff and Arias 2019).

Adding an `s` to `implik` or `mplik` (i.e. `simplik, smplik`), likelihood is calculated from the single most likely reconstruction. The ranking of trees in that case is even closer to implied or equal weights parsimony, respectively (see Goloboff and Arias 2019). All those expressions, in conjunction with scripting options that allow primitive tree searches (e.g. `tbrit, sprit`; see Chapter 10), make it possible to do simple searches for most likely trees under the different models.

In addition to those options, TNT offers the possibility of mapping character states onto trees. For this, the proportional contribution to the total likelihood is calculated for each state. In trees with relatively short branches (i.e. most real datasets), the mapping is usually identical to the one obtained under parsimony. The command to map characters with likelihood is `mkmap`. The tree to map is given as first argument; this can be followed by a list of characters to map (note that when mapping only one or a few characters, the Mk model still requires that the branch lengths are optimized for all active characters together). If no list is given, the command maps all active, discrete characters. By default, the proportional contributions to the likelihood are displayed. Optionally, you can display the states with the highest contribution:

```
mkmap =0.80 0 ;
```

This will display on the branches each state with a proportional likelihood contribution representing 80% or more of the contribution of the state with the highest contribution. The `mkmap` command can be used to map characters under the `implik` model as well, with \ as argument before the tree number. The branch lengths can also be fixed by the user (if scripts are enabled) using `/array` (where `array` is a one-dimensional array containing values of branch lengths) after the list of characters. The values of branch lengths produced by optimization with the Mk model can also be displayed on the tree, if `*` precedes the tree number.

NOTES

1 For the time being, ignore the probability that a given state occurs as initial stage; this is considered in the next section.

2 That cost would be $O(Ts^T)$ for each character because each of the terms in the summation is composed of the product of a number of terms that increases linearly with numbers of taxa. Cost of evaluating a tree for a complete dataset also increases linearly with number of characters, $O(CTs^T)$.

3 In other models, different states can have different prior probabilities, either chosen to maximize the likelihood or empirically estimated from the data themselves.

4 At least in principle; separating in first, second, and third positions may be complicated by alignment, obviously.

5 In phylogenetics, this equivalence of results does not happen in practice—for a number of reasons explained in the following.

6 A typical sampling frequency is 500 iterations. This may depend on the number of taxa in the trees; for a tree with 6 taxa, 100 changes to the tree are likely to produce a very different tree, but for a tree with 1,000 taxa, 100 changes will only have moved 10% of the taxa somewhere else. I know of no studies to determine the best sampling frequency, but it should probably use larger periods for larger trees.

7 Keep in mind this refers to nested models—that is, when the more complex model differs from the simpler one in preserving all the parameters estimated and adding some new ones. For non-nested models, it is possible that the model with fewer parameters has a better likelihood.
8 No meaningful partitions could be defined in those datasets because the authors had not provided standardized character names in the data files.
9 This does not detract from the fact that synapomorphies which are specific adaptations are probably correlated with changes in other characters (e.g. anterior legs in raptorial insects are often larger and decorated with spines, and interdigital membranes often develop in terrestrial vertebrates that invade aquatic habitats); such correlations refer to specific circumstances and characters, not to the bulk of characters.
10 To be fair, they do not use the word *probability*: they say *equally likely*. Perhaps what they meant to imply was that they did not use a Poisson model, but even that is debatable.

5 Tree Searches
Finding Most Parsimonious Trees

Chapter 1 discussed the need for an optimality criterion; using parsimony as an optimality criterion leads to judging trees on the basis of how many steps they require. Chapter 3 discussed the problem of how to use character state optimization to evaluate any given tree. Even with efficient and well-known algorithms for evaluating any tree, finding the optimal tree is still a hard computational problem—the topic of the present chapter. The difficulty is just because there are so many possible trees and because there is no general way to derive the tree automatically from the data by means of any formula or simple deterministic algorithm. The only way to guarantee having found the optimal tree is to examine every possible tree, calculate the length for each by means of a character state optimization (i.e. finding optimal ancestral state assignments for each tree), and select the best trees. This is as true for parsimony as it is for any other criterion (e.g. likelihood), and most of the discussion of problems related to actually finding the optimal tree in this chapter applies equally well to likelihood, parsimony, or other optimality criteria in phylogenetics.

The process of finding the optimal tree then requires *searching* through the possible trees until we have examined them all—or a sample large enough to make it likely we have indeed come across the best. Note that a tree search must find trees of the best possible length, but finding just one tree of minimum length is still not enough to guarantee proper results. Even if a group AB is present on an optimal tree found, the conclusion of monophyly of AB is still not warranted. If any other tree can lack the group AB (say, by posing the existence of group AC instead) and also have optimal fit to the data, then the conclusion of monophyly of group AB is not inescapable. Only when a group is present in *every optimal tree* can we then say that the group is *supported* by the data—otherwise we are able to provide optimal explanation of the data (i.e. lowest homoplasy) without postulating the existence of the group. This also provides grounds for considering rejection of groups—we can only say that a group is *contradicted* by the data when postulating the existence of the group forces us to have a less than optimal explanation of the data, i.e. when the group is absent from every one of the optimal trees. When optimal trees can either have or lack the group (as in the AB or AC example), we have *indifference*—the data say nothing regarding monophyly of the group.

Given the vast numbers of possible trees, examining them all is impossible for medium or larger datasets. Optimality thus cannot be guaranteed, meaning that there is always the possibility that we have failed to find all (or any) of the optimal trees. Does this make phylogenetic analysis impossible? Certainly not. Hypotheses

DOI: 10.1201/9781003220084-5

(about phylogenetic relationships or about anything!) change as science progresses—the change may be due to finding new data or finding better hypotheses to explain existing data. Parsimony (or likelihood) provides a well-defined means to evaluate a phylogenetic hypothesis and compare it with others. Should a shorter (or more likely) tree for the present dataset be found in the future, then that tree should be embraced as a better explanation of the current data (of course, finding new data bearing on the subject can also change hypotheses).

While the optimality of a tree in regards to a dataset often cannot be guaranteed, methods for tree searching are available that can both come very close and give an idea of whether the optimal tree is likely to have been found. It is thus the responsibility of the user to make every effort to use appropriate algorithms for searching—a sloppy design in the search may well mean that some other researcher will quickly come along, reanalyze the dataset, and embarrass us by showing that a different tree provides a better explanation of our own data.

5.1 OPTIMIZATION

How to find an optimal answer to a difficult problem is an important branch of computer science known as *optimization*. Some problems can be solved with deterministic algorithms, which require polynomial time. An example is ancestral character state reconstruction itself, with complexity $O(TC)$ for Farris (1970) or Fitch (1971) optimization and $O(TCs^2)$ for Sankoff (Sankoff and Rousseau 1975) optimization. This is an example of polynomial time. Although evaluating a single tree is relatively fast, however, the number of trees that have to be potentially examined for guaranteeing optimality increases so rapidly with taxa that the size of the problem increases with the number of taxa in the exponent, approaching 10^{2T}. This makes parsimony a problem of a kind known as NP-complete: problems where verification of the score of a possible solution can be done quickly, but the number of possible solutions makes verification of all of them impossible. Foulds and Graham (1982) were the first to demonstrate that parsimony is an NP-complete problem, although this did not surprise anyone at the time (for it was widely known that finding *MPT*s is difficult).

The process of examining only a sample of trees in an attempt to find the *MPT* represents a *heuristic* approach. The main heuristics work either by creating de novo trees that may approximate *MPT*s, by successively swapping branches to improve them, by combining different solutions, or by combinations of those. There are several factors that make a heuristic search method *effective* (i.e. succeed in finding the *MPT*) and *efficient* (i.e. perform quickly), some of which can be controlled by the user, some of which cannot.

The first group of factors conforms to what can be collectively called the search *strategy*, or the higher-level design of a search. The strategy then determines how to sample trees to be evaluated as candidate *MPT*s. The sample is, of course, not random; a good search strategy judiciously chooses candidate trees so that they have higher chances of being (near) the *MPT*(s), thus reducing the number of trees that need to be evaluated. These are the most important factors for this chapter, especially because they are the ones the user can change. A good computer program for searching phylogenetic trees will offer a wide array of strategies and allow fine-tuning several options. Examples of such options are how many initial trees to create, how to rearrange them, and how many equally parsimonious trees to save. The strategy chosen may dramatically influence the effectiveness and efficiency of the search.

In connection to different strategies for resolving computational problems, it is important to consider a famous theorem in computer science, humorously called the "No Free Lunch" theorem (*NFL*; Wolpert and Macready 1997). The *NFL* states that no method for optimization can perform best across all datasets: for any specific dataset, it is possible to design a method that takes advantage of specific properties of the dataset to help find a good solution. This also means that tailoring a method to work better for some type of dataset must, by necessity, make it work worse for others. An easy-to-understand example is a matrix with no informative characters—the fastest method for finding an *MPT* is generating a random tree! Yet even if generating a random tree works wonderfully in that specific case, it works awfully in other cases. Phylogeny computer programs include search techniques that generally work well in real, empirical datasets. These are typically datasets with many congruent characters, yet exhibiting minor to medium degrees of overall conflict. When analyzing a given dataset, it is important to assess the type of dataset initially and then try to apply the search techniques that probably allow finding the *MPT*s in the shortest time. This can be done, in part, by running shorter superficial bouts of extreme techniques and observing the progress of results along time, subsequently setting search parameters to emphasize those algorithms or approaches empirically observed to decrease tree length faster or more effectively.

The second group of factors influences the speed of tree searches rather than the trees to try as candidates. These factors depend more on the internal design of the computer program used, allowing little influence on the part of the user. The most obvious one is the speed with which every rearrangement can be examined during a trial-and-error search. This also illustrates how effectiveness and efficiency are, in a sense, connected. Given a fixed amount of time available to search (e.g. results need to be submitted next week!), what we can do with a computer program that allows examining half a million trees per second is obviously very little compared to what we can do with a program that examines 20 million trees per second. Thus, a search allowed to proceed for (say) two hours with the latter program is much more likely to find an optimal tree than the first, at least if using comparable search strategies (i.e. if using similar methods to select the sample of trees to consider). Ways to speed up tree-length calculations during the search are discussed only briefly in this chapter; it is one of the aspects in which TNT excels, with algorithms finely tuned to work as fast as possible for real biological datasets. One of the methods TNT uses to speed up computations during tree searches is parallelizing character optimization (e.g. as shown in Box 5.1) by representing several characters in a single 32- or 64-bit word. Then, intersections and unions in character optimization can be done in tandem. These algorithms were first used by J. S. Farris in his "Jac" program during the early 90s (Farris et al. 1996) and were adopted by Goloboff (1999, 2002). White and Holland (2011) independently developed almost identical versions; Ronquist (1998) also published less efficient proposals for parallel algorithms.

BOX 5.1 ALGORITHMS FOR PARALLEL CHARACTER OPTIMIZATION

Pseudocode showing how TNT performs parallel multicharacter optimization. Every 32-bit word contains 8 sections of bits, each representing states 0–3 of a character. Variables `left[c]`, `right[c]`, and `prelim[c]` are down-pass states of node `node` for packed character `c;` `up[c]` and `below[c]` are up-pass states (pointing to `node` and ancestor, respectively). Variable `uni[c]` contains union states (used to speed up calculation of length increments if inserting a new taxon or clade at that position) and `back[c]` back-up states (to speed up the up pass). `LOW4` and `HIGH4` are magic numbers used to separate empty and occupied fields of bits. These algorithms, developed by J. S. Farris in the early 90s, are implemented in TNT; similar ones were independently discovered by White and Holland (2011).

```
#define LOW4    0x77777777
#define HIGH4   0x88888888

DOWN-PASS:
      x = left[c] & right[c] ;
      z = left[c] | right[c] ;
      y = HIGH4 & ~(x | ((x & LOW4 ) + LOW4 ) ) ;
      length += numbits [ 65535 & (y | (y >> 17 ) ) ] ;
      y |= y - (y >> 3 ) ;
      prelim[c] = x | (y & z ) ;
      back[c]  = z | y  ;

UP-PASS:
      x = below[c] & ~prelim[c] ;
      x = (x | ((x & LOW4 ) + LOW4 ) ) & HIGH4 ;
      x |= x - (x >> 3 ) ;
      if (node < root )    // if a terminal node
           uni[c] = below[c] | (prelim[c] & x ) ;
      else {  // if an internal node
           up[c] = (below[c] & ~x ) | (x & (prelim[c] |
          below[c]& back ) ) ;
           uni[c] = up[c] | below[c] ; }
```

Remember that fine-tuning for some type of dataset also means that a less than ideal performance for other types of datasets will be obtained (as predicted by the *NFL* theorem); for example, for randomly generated datasets, it is possible to design branch-swappers that proceed faster than the branch-swapper in TNT. Although the speed of rearrangement evaluation is mostly beyond the control of the user, some influence can still be exerted in specific cases. An obvious one is the use of continuous characters—these take much longer to be reconstructed on trees (see Chapter 9) than additive characters. If the values given to the continuous character are discrete integers in the range 0–32, this produces the same results as an additive character (at least if implied weighting is not in effect, because additive characters are recoded in binary form in TNT, weighting the different components of the character; using continuous characters is a way to force implied weighting to weight the entire character). The results are the same, but the efficiency is not—this slows down the program. Likewise, forcing character optimization by means of step-matrix algorithms for character state trees (e.g. with the `cstree` command; see Chapter 3), instead of the usual binary recoding, will also produce slower evaluations of trees during searches.

5.2 SMALL DATASETS: EXACT SOLUTIONS

Only for very small datasets is it possible to guarantee that all the optimal trees have been found. This process is known as implicit enumeration or branch-and-bound. This can be done by discarding many of the possible trees implicitly, without actually examining them, by deducing that their length would exceed that of the best tree(s) found so far (the *bound*). All possible trees for the dataset can be generated by first joining three taxa, then successively adding taxa at each of the available $2p - 3$ positions (where p = number of taxa already placed in the tree). The process then takes advantage of the fact that the length of the tree can never decrease as more taxa are added to the subtree. Algorithms for calculating tree length—e.g. with a full down-pass optimization—can be applied to subtrees as well as to complete trees. Thus, if a subtree is longer than the bound, this also means that every possible tree derived from adding more taxa to the subtree will also be longer than the bound. For example, if (A(B(CD))) is longer than the bound, that means that also (A(E(B(CD)))), (A(B(E(CD)))), (A(B((CE)D))), and (A(B(C(DE)))) will be longer than the bound, as well as any larger tree which contains (A(B(CD))) as a subtree, like (A(E((BH)(G(CF)D))))). Since there are only three unrooted trees for four taxa, knowing that (A(B(CD))) is longer than the bound allows discarding one-third of all possible trees by just looking at the length of one three-taxon subtree. Assume (A(C(BD))) also exceeds the bound; then only the trees derived from (A(D(BC))) remain to be examined in the same fashion. An example of exact solution is found in the didactic TNT script ***exactsol.pic***.

Of course, real datasets will rarely have so much congruence as to allow rejection of two-thirds of the trees from the addition of the fourth taxon; many more taxa will have to be added to the subtree(s) to exceed the bound, thus leading to examining many more subtrees and slowing down the process. The time to complete implicit enumeration therefore depends very strongly on the congruence between characters. In practice, exact solutions can be achieved for real datasets with 15–20 taxa, or perhaps 25 in rare cases. As the number of trees to potentially examine when adding a taxon to a tree with T taxa increases with $2T - 3$, just adding a few more taxa can turn an otherwise achievable exact solution into one completely infeasible; if the exact solution for a 20-taxon problem takes 1 hour, adding only 3 taxa may well take $(2 \times 20 - 3) \times (2 \times 21 - 3) \times (2 \times 22 - 3)$ times longer, or 6.75 years. Still within a lifetime, but one more taxon (for a total of 24 instead of 20) could take 290 years (instead of 1 hour). We could still be waiting for an analysis that Linnaeus had started to run back in 1758! Of course, these numbers could be over- or underestimates, because adding taxa to an analysis can make the data less or more decisive (depending on whether the new taxa contribute or decrease conflict), but they give an idea of the difficulty of calculating exact solutions.

The degree of conflict among characters is another factor strongly influencing the time needed for exact searches. As characters conflict more with each other, fewer subtrees will exceed the bound, and the paths leading to all possible trees will have to be followed almost all the way to the end (i.e. to complete or almost complete trees), examining many subtrees.

The number of characters in the dataset has a minor influence, with times increasing only lineally with characters. Another factor which has a relatively small influence on the times needed to complete exact solutions is the number of equally parsimonious trees—in most datasets, the number of optimal or near-optimal trees is a tiny fraction of the tree space that needs to be explored to complete an exact solution.

5.3 DATASETS OF MEDIUM DIFFICULTY: BASIC METHODS

When datasets are larger, only heuristic methods can be used. These give no guarantees on optimality but are, in practice, extremely reliable and work much faster than exact searches. The two types of method commonly used consist of creating trees de novo (with Wagner trees) and rearranging tree branches (with branch-swapping).

5.3.1 Wagner Trees

In the context of cladistics, the first computerized method to build trees was proposed by Farris (1970), who named them after Wagner (1961). Similar methods in computer science are called *greedy trees* (from the idea of accepting only the best option at every stage of the build). The idea of Wagner trees is creating trees by sequentially adding taxa to a growing subtree, adding every terminal at the best possible available position. This starts from a three-taxon subtree, on which three positions are available for reinsertion of the fourth taxon; for every taxon added, two new positions become available, so that the number of positions to try is $\Sigma 2i - 3$, thus approximating a complexity of $O(T^2)$. Remember that the number of possible trees is $\Pi 2i - 3$; $\Sigma 2i - 3$ is a tiny fraction of that, and Wagner trees can be built very quickly. A simple Wagner algorithm is used in the TNT script ***wagtree.pic***; TNT's own Wagner algorithms use many shortcuts for greater speed, but the general principle is the same as in ***wagtree.pic***.

Fast as they are, Wagner trees rarely produce optimal trees for datasets with some character conflict and more than a handful of taxa. At every stage of the build, the process makes the decision of where to insert the current taxon based on the taxa already added to the tree, but the taxa not yet added may have character combinations making preferable a different position for the taxon being added. The Wagner tree (at least as usually implemented) never backtracks to modify the position of previously added taxa. A simple example (from Goloboff 1998a) where the Wagner tree can fail to find the optimal tree is the dataset

```
        0 1 2 3 4
X       0 0 0 0 0
A       1 0 0 0 0
B       1 1 1 0 0
C       0 1 1 1 1
D       0 0 0 1 1
```

It is easy to see that the optimal tree is (X((AB)(CD)); two characters (1 and 2) contradict groups AB and CD, but three characters (0 and 3–4) win over two. However, adding the taxa to the growing tree in the same order in which they are in the matrix produces the wrong tree; at the moment of adding taxon C to the subtree (X(AB)), the data seem to be just:

```
        0 1 2 3 4
X       0 0 0 0 0
A       1 0 0 0 0
B       1 1 1 0 0
C       0 1 1 1 1
```

with characters 1 and 2 leading to place C as closer to B, producing subtree (X(A(BC))), rather than the "correct" subtree (X(C(AB))); characters 3 and 4 appear as uninformative at this stage, requiring a single step for every possible location of C in the subtree (X(AB)). The four taxa under consideration at this point are only *part* of the evidence, and the best location for taxon C should be determined from *all* the evidence. The Wagner tree saves time precisely by not considering all of the evidence, but just part of it, at every stage; what makes it fast is what makes it error prone.

Given that the sequence with which taxa are added to the growing tree may lead to different results, Farris (1970) examined several ways to determine the optimal sequence. In practice, Wagner trees are used mostly to provide a starting point for subsequent branch-swapping, and the most common approach is to use random addition sequences (*RAS*s), potentially generating a different Wagner tree with every sequence. In very specific cases, other addition sequences may produce results significantly better than *RAS* (e.g. Goloboff's 2014a "informative" addition sequences).

In highly ambiguous datasets, where there is a minimal difference in score among all possible trees, the order in which possible locations for a taxon being inserted are tried is important. Trying insertions in a specific sequence may produce (or increase; see Chapter 8) a bias, making it more likely that the Wagner tree (even if followed by *TBR*) finds some of the optimal trees more frequently than others. For example, trying locations on a bottom-up sequence (i.e. the sequence used by default in TNT) produces wholly pectinate trees in the case of uninformative data. Thus, in extreme cases, where the data are very poorly structured, it may be advisable to randomize the insertion sequence, instead of just the addition sequence (in TNT, with `rseed[`)".

5.3.2 Branch-Swapping

Given a preexisting tree, the branches can be rearranged, producing different trees. For every rearrangement, the tree score is calculated, either discarding or accepting the alternative tree. Three simple examples of branch-swapping (with different levels of detail) are included in the TNT educational scripts ***branswap.pic***, ***simple_swapping.pic***, and ***swapping_with_tntlib.pic*** (TNT's internal branch-swapping, of course, uses various methods to enhance speed, but the method for generating rearrangements is the same as in those scripts). Unlike the case of a Wagner tree, all the evidence (i.e. the full taxon set) is being considered when evaluating a tree produced by the rearrangement process. In computer science, such a process is known as *local search*: it explores the *neighborhood* of the best solution found so far, attempting to find better ones. One can think of the score of possible solutions as forming a landscape (in the vertical axis of Figs. 5.1a–c), with the space of possible solutions in the horizontal axes. A local search is the process of moving the candidate solution to the vicinity; in Figure 5.1a, it is like moving the little ball along horizontal axes and checking whether the score (vertical axis) is higher; if so, the candidate solution is accepted. In the phylogenetic context, neighbor solutions are obtained by rearranging branches, or *branch-swapping*. The three basic ways to rearrange branches are:

1) *Nearest neighbor interchange or NNI.* Cut every branch of the tree and move it to the two alternative distinct locations just one node away. This creates $2T - 6$ possible rearrangements (Semple and Steel 2003), for a neighborhood of size $O(T)$.
2) *Subtree-pruning regrafting or SPR.* Cut every branch of the tree and move it to each of the alternative remaining locations. This creates $2 \times (T - 3) \times (2T - 7)$ possible distinct rearrangements (Semple and Steel 2003), for a neighborhood of size $O(T^2)$. It tries every possible *NNI* move and many more.
3) *Tree-bisection reconnection or TBR.* Cut every branch of the tree and move it to each of the alternative remaining locations; if the branch cut delimits a subtree, produce every possible rerooting of the pendant subtree and move to each of the alternative locations. The exact number of rearrangements depends on tree shape, with maximum number of rearrangements for a given number of taxa occurring on pectinate (=imbalanced) trees and the minimum on symmetric (=balanced) trees. Humphries and Wu (2013) gave exact formulae for numbers of distinct rearrangements, $(2T^3/3) - 4T^2 + 16T/3 + 2$ for pectinate trees and $4T^2 log_2 T + O(T^2)$ for perfectly balanced trees. This produces *TBR* neighborhoods of size between $O(T^{2.5})$ and $O(T^3)$. *TBR* tries every possible *SPR* move and many more.

The programs Nona (Goloboff 1993d) and Pee-Wee (Goloboff 1993e) implemented multiple-cut branch-swapping, where the tree is cut into *n* pieces (instead of just two), recombining the pieces. For $n = 3$ and *TBR* reconnection, this produces all the *TBR* rearrangements for *each TBR* rearrangement of the tree. This generates vast numbers of rearrangements, $O(T^{3(n-1)})$. The implementation in those programs made reinsertions in an orderly fashion, checking tree scores at every step. If $n = 3$ and reinserting piece 2 into a location at piece 1 produces a tree 1 + 2 longer than the best one found so far, then there is no need to try any reinsertion of piece 3 into 1 + 2—all such reinsertions will exceed the bound. This is a hybrid between a branch-and-bound and *TBR*. It allowed implicitly rejecting large numbers of rearrangements, leading (implicitly) to examine very large numbers of rearrangement per unit time. Despite the large number of trees examined and the efficiency, this produced inferior results to the standard *RAS* + *TBR* or newer search techniques, and it is not implemented in TNT.

In parsimony, a big advantage of rearrangement methods based on swapping branches is that the score of the resulting trees can be easily derived from the lengths of the subtrees and the differences in state assignments in the nodes to reconnect, or *indirect calculation of tree lengths* (Goloboff 1993c, 1996, 1998b). In this way, the length of the resulting rearrangement is the sum of the length of the two subtrees (which does not vary for a given clip and rooting) plus the increase in length that would result from reinsertion. As the number of trees in the *TBR* neighborhood is up to $O(T^3)$ and cost of a Fitch or Farris down-pass character optimization is $O(TC)$, the complexity of *TBR* with a full down-pass character optimization would be $O(T^4C)$—evaluation of every rearrangement being slower, in addition to there being many more rearrangements, as there are more taxa. The use of indirect calculation of tree lengths allows (for a given number of characters) using a fixed time to evaluate a rearrangement, regardless of number of taxa. It is therefore at least T times faster than a complete down-pass character optimization. However, TNT can examine *TBR* rearrangements *faster* for trees with *more* taxa so that the time needed in practice to examine all the *TBR* neighbors for real datasets increases with about T^2 and somewhat less than lineally with C. This may seem counterintuitive, but it results from extensive use of indirect calculations and other shortcuts (e.g. "node clusters") that quickly reject many rearrangements in well-structured datasets (Goloboff 1996, 1999). This also means that *TBR* in TNT will run somewhat slower in random datasets (remember the *NFL* theorem!). Much of the work needed for indirect calculation of tree lengths is in deriving ancestral states assignments when dividing the tree; note that *TBR* can then make use of the resulting state assignments for generating a larger number of rearrangements than *SPR*.[1] Thus, *TBR* can evaluate rearrangements at a greater speed than *SPR* (Goloboff 2015).

Other types of rearrangements are possible. For example, both Lin et al. (2007) and Viana et al. (2007) proposed different methods, based on rearranging vectors describing the trees, which cannot be assimilated to exchanges between branches. A serious practical drawback of such methods, when compared to methods that exchange branches, is that indirect length calculations are not applicable. Therefore, the evaluation of the resulting rearrangements will necessarily be much slower than under an efficient implementation of standard branch-swapping, requiring a full down pass per rearrangement (Goloboff 2015: 214).

In the case of maximum likelihood, the score of a tree resulting from swapping branches cannot be derived exactly, but efficient programs (e.g. PhyML, RAxML) use approximations similar to those described in this section.

5.3.3 Multiple Trees

Branch-swapping can be used both to find trees of the best possible score and to multiply the number of equally parsimonious trees in the case of data ambiguity. When swapping on a single tree, only the neighborhood of that tree will be explored; trees of equal score in that neighborhood will be ignored. Obviously, those alternative trees of equal score have their own neighborhoods, some of which may contain better trees even when the neighborhood of the current tree contains no tree of better score. Thus, saving multiple trees during branch-swapping makes it more likely, in principle, to find better trees. However, saving the multiple trees requires much more time; the problem of the tradeoff between increased probability of finding better trees and increased search times is discussed in detail in Section 5.3.6.

5.3.4 Local Optima and Islands

The goal of a local search is to explore the vicinity of current solutions. If the landscape is smooth and has a single peak (e.g. as in Figure 5.1a), the local search will identify the *global optimum* (highest peak) in every case. The landscape, however, can be much more rugged (as in Figs. 5.1b–c), and in that case, moving in the neighborhood of current solution will not, for many points in the space of possible trees, lead to the global optimum—the search is then trapped in a *local optimum.* As with any local search method, this can happen with *SPR* or *TBR* swapping. In the context of phylogenetics, this is often called the problem of *islands of most parsimonious trees*, after Maddison (1991). Maddison's definition of islands, however, was somewhat more restrictive than the local optima, for he referred to the case when swapping on all the trees connected by a single *SPR* or *TBR* rearrangement fails to find other trees of better or equal score. Islands refer to branch-swapping saving multiple equally parsimonious trees, while a local optimum is less specifically defined, applying as well to the case where a single tree is being swapped. A related concept is whether the *TBR* neighborhood of a tree contains any better trees. When no better trees are within a single *TBR* move, the tree is said to be *TBR-optimal.* Likewise for *SPR* or any other rearrangement algorithm.

Trees in different local optima (or islands) generally have significantly different topologies; local optima are determined generally by character conflict and sometimes by ambiguity. Note also that the landscape itself is shaped after the type of moves used to travel along the search space. For example, *TBR* may succeed in finding a better tree where *SPR* fails, but *SPR* can never find a better tree where *TBR* failed (as every *SPR* move is also a *TBR* move). An algorithm capable of making radical changes to a tree would rarely get stuck in local optima, but it has the downside that the corresponding neighborhood is huge and exploring it in full would be time consuming. A real example of a case where *TBR* fails to find some of the equally parsimonious trees for a dataset is:

```
      0 1 2 3
X     0 0 0 0
A     1 0 0 1
```

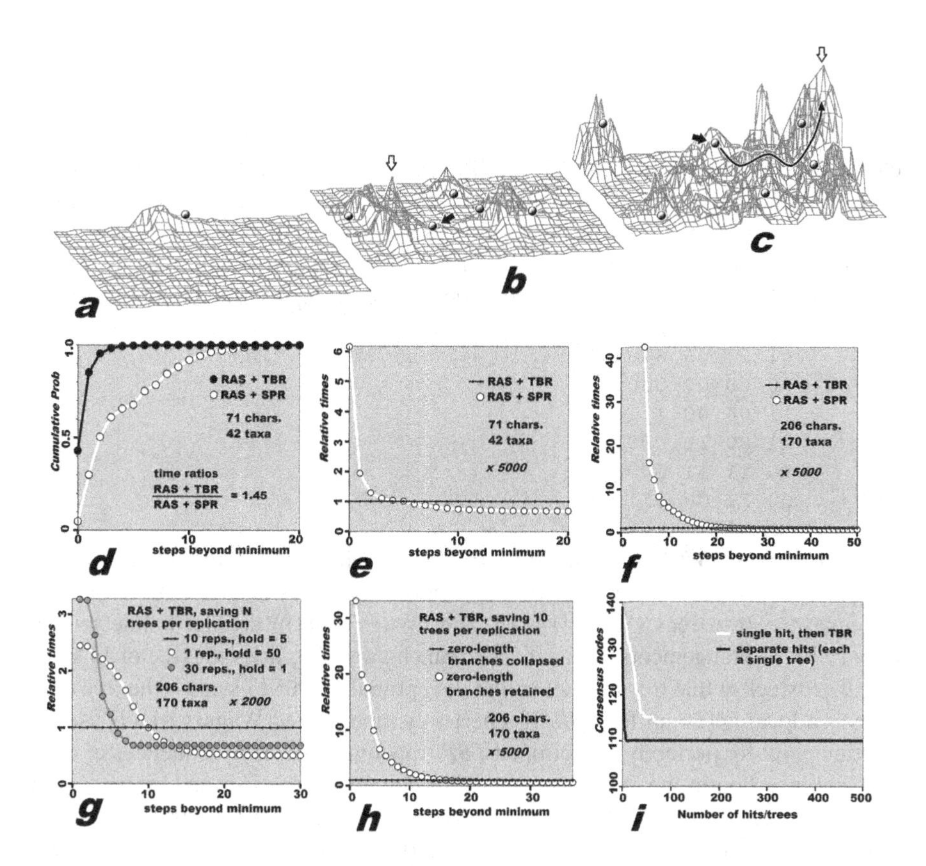

FIGURE 5.1 **(a, b, c)** Three different landscapes of trees. See text for discussion. **(d)** The probability (y-axis) of a search finding trees of a given length beyond minimum (x-axis) for *TBR* and *SPR*. *TBR* is more likely to find better trees, but whether it is more efficient in the long run (5,000 different addition sequences tried here) depends also on the speed ratios between *SPR* and *TBR*. **(e)** When the speed ratio is taken into account, the average time (y-axis, relative to the time used by *TBR*) to find trees of a given number of steps beyond minimum (x-axis) is longer for *SPR* (except for trees with five or more steps beyond minimum). Speed ratios between *SPR* and *TBR* increase as desired trees are shorter. **(f)** In a matrix with more taxa and much more homoplasy, the speed difference to find trees of a given length with *SPR* or *TBR* is much greater, with *TBR* finding trees close to minimum length at least 40 times faster than *SPR*. **(g)** Doing several replications saving few trees per replication (e.g. black line, *10 RAS* + *TBR* with 5 trees/replication) produces equivalent results faster than doing a single replication saving larger number of trees (e.g. white line, *1 RAS* + *TBR* with 50 trees/replication). However, saving too few trees (e.g. gray line, 10 *RAS* + *TBR* with a single tree/replication) eventually increases the times needed to produce equivalent results. The average times needed for a single instance of each search routine are *1 RAS* + *TBR* × *50* = 0.0291, *10 RAS* + *TBR* × *5* = 0.569, and *30 RAS* + *TBR* × *1* = 0.389. Note that *10 RAS* + *TBR* × *5* is the routine that takes longest, but the trees found are (on average) so much shorter than those found by the other routines that it decreases the overall time needed for finding a given length. **(h)** The effect of collapsing (black line) zero-length branches during searches, when doing multiple random addition sequences saving few trees per replication, is that equivalent results are produced in shorter times than when zero-length branches are not eliminated (white line). **(i)** Example showing that the accuracy of the estimated strict consensus of the *MPT*s is increased much more rapidly by doing additional independent hits to the minimum length (black line) than by saving numerous trees from a single starting point via TBR (white line).

```
B       1 0 1 0
C       0 1 1 0
D       0 1 0 1
```

characters 0–1 determine the tree (X((AB)(CD))), and characters 2–3 (X((BC)(AD))), each tree requiring 6 steps. The reader can verify that there is no way to cut a single branch and reinsert it at a different position or rooting such that one tree can be converted into the other. Any *TBR* search starting from a single tree will find only one of the trees and miss the other. In a modification of the example (from Goloboff 2014a),

```
        01 23 45 678
W       00 00 00 000
E       11 00 00 111
F       11 00 11 000
G       00 11 11 000
H       00 11 00 111
```

one of the trees, (W((EH)(FG))), with 13 steps, is shorter than the other, (W((EF)(GH))), with 14. Any other tree has 15 or more steps. As seen for the preceding example, the two trees (W((EH)(FG))) and (W((EF)(GH))) are not connected via direct *TBR* rearrangements. Thus, if *TBR* branch-swapping finds the tree of 14 steps, it will get stuck at that tree and never find the optimal tree of 13 steps—the tree of 14 steps is a local optimum for *TBR*. If a perfectly randomized Wagner tree is used as starting point for perfectly randomized *TBR* (randomly choosing sequences of addition and insertion in the Wagner trees and of clipping, rerooting, and reinserting in the swapping; Goloboff and Simmons 2014), the probability of finding the optimal tree is 0.75. If a random tree is instead the starting point of a perfectly randomized *TBR*, the probability of finding the optimal tree is 0.5—half the searches will get stuck at the suboptimal tree (W((EF)(GH))). Branch-swapping in actual computer programs always falls short of perfect randomization (e.g. if the implementation reinserts clipped branches in a specific sequence instead of randomly, which it may do to speed up calculations; in addition, pseudorandom number generators are not perfectly random). TNT approximates the theoretical results with Wagner trees but finds the optimal tree either more or less frequently, depending on the sequence of taxa in the file, when using random trees as the starting point.

5.3.5 Escaping Local Optima

The character conflict in the datasets just discussed is carefully chosen to produce perfect ties in tree scores. In real datasets, tree islands occur rarely in fewer than 15 or 20 taxa. Beyond those numbers of taxa, it is necessary to try strategies designed to escape local optima; just initiating a search from a single starting point and branch-swapping is not enough to produce optimality.

The basic methods to escape local optima can be grouped in three main types. The first type is based on using *different starting points to begin different searches*. Rather than expecting a single starting point to be located on the way to the optimum (as in Figure 5.1a), multiple points (e.g. each of the balls, in Figure 5.1b) used to begin independent searches with branch-swapping are likely to be placed on one or more of the slopes (e.g. the ball with a black arrow in Figure 5.1b) leading to the globally highest peak. Therefore, at least some of the separate, independent searches will probably end up reaching the global optimum (white arrow in Figure 5.1b). Obviously, the probability of reaching a global optimum is increased as more starting points are tried. For this, it is important that the initial points be spread along the space of possible trees. The initial points could well be random trees, but they are too far from optimal and then require a significant amount of time for branch-swapping to approach optimality. As showed prior, different addition sequences for Wagner trees produce different results if there is conflict in the data. Thus, Wagner trees with different, randomized addition sequences (*RAS*) work very well in practice.

The second type of method is based on *saving and swapping suboptimal trees*. A search by means of rearrangements can be allowed to descend into valleys in the tree landscape if suboptimal rearrangements are accepted, saved to the tree buffer, and eventually swapped as well. This can be seen graphically as starting to move one of the balls (black arrow, in Figure 5.1c) placed on the slopes of a suboptimal peak but allowing the ball to descend into lower heights along the way. This may eventually allow the ball to traverse valleys and find the slope leading to the highest peak (white arrow in Figure 5.1c). Goloboff (2014a) showed that this method may work well in specific datasets, but it requires that the sequence in which the trees are clipped for swapping is randomized (i.e. the branches to reinsert must be cut in a random sequence) and different for every tree being swapped (otherwise, the same shorter tree will be "rediscovered" again from the suboptimal tree it led to; see Goloboff 2014a: 125). Saving and swapping enough trees, sufficiently suboptimal, will guarantee eventually finding the optimal tree(s), but in practice this requires longer times than trying numerous starting points. The option of saving suboptimal trees is thus not considered further.

The third type of approach to avoid the problem of local optima is *using better search algorithms*. The concepts of local optima and islands of most parsimonious trees only apply for a specific search algorithm. It may well be that *SPR* is unable to find better trees, but application of *TBR* may succeed—after all, *TBR* tries all the *SPR* rearrangements and many more. The tree landscape, that is, actually looks different when viewed through the lens of different rearrangement algorithms. Using *TBR* instead of *SPR* is greatly advantageous in the case of parsimony, not only because it looks at many more rearrangements but because it can look at them in a more efficient way (i.e. using less time to evaluate each rearrangement, as already discussed). The method of multiple-cut swapping implemented in Nona was based on this idea, although in practice it produced results that tended to be inferior (in the tradeoff results/time) to those of standard *TBR*.

5.3.6 Comparing Efficiency of Search Algorithms

As should be evident from the preceding discussion, the comparison of different search algorithms is far from simple. Some algorithms are less likely to find optimal or near-optimal trees and run very quickly (e.g. a Wagner tree); others get close to optimality but are more time consuming (e.g. branch-swapping). The utility of an algorithm should also be defined by reference to specific goals—if we can be satisfied with results that are within (say) 10% of optimality, then a Wagner tree may be preferable to branch-swapping. Such results may be useful under several circumstances—for example, when trying to speed up estimations of group support by means of resampled matrices (see Chapter 8). And, as always, the method that works best for a type of dataset may not be the one that works best for another. Consider a dataset with perfectly congruent binary characters, with no missing entries, which determines a completely resolved tree. A Wagner tree will be the fastest way to obtain results for such dataset: it will always find the optimal tree; any subsequent branch-swapping is a waste of time. Of course, empirical morphological datasets never have so much congruence, and then Wagner trees by themselves are insufficient to get sufficiently close to optimality to be useful even as a means to obtain approximate results. The same reasoning applies to *SPR* and *TBR*. A full cycle of *TBR* may take longer than one of *SPR* because it examines many more rearrangements.[2] Thus, *TBR* guarantees finding shorter trees than *SPR* if run sufficiently longer (or for a sufficient number of starting points), but the time needed to reach specific target scores might be shorter for either method.

A question that remains open, in datasets where *SPR* takes shorter than *TBR*, is that of how it is preferable to invest the time available for computing: doing more shorter cycles of *SPR* or fewer longer cycles of *TBR*? The answer to that question depends on (a) the average probability of finding a tree within x steps of optimality and (b) the average time ratios between the two search algorithms. Consider a hypothetical dataset where the probability $P_{(SPR,x)}$ of finding a tree within x steps of optimality is a third of the probability $P_{(TBR,x)}$; for example $P_{(SPR,x)} = 0.0333$ and $P_{(TBR,x)} = 0.1$. Further assume that the average time t to complete *SPR* is $t_{(SPR)} = 1$ sec and $t_{(TBR)} = 2$ sec. If we need to independently find trees of x or fewer steps 10 times, we will need to run *SPR* an average of 10/0.03333 = 300 times, which will take 300 sec, or *TBR* an average of 10/0.10 = 100 times, which will take 200 sec. For this example, it is therefore preferable to use *TBR* instead of *SPR*; it will be 50% faster. The expected time ratio between the two algorithms to find the same number of trees within x steps of optimality is exactly $P_{(TBR,x)}/(P_{(SPR,x)} \times t_{(TBR)}/t_{(SPR)})$. Obtaining reliable estimations of the efficiency of different algorithms requires running the same routine many times, changing the random seed (ideally, developers of programs should use this type of comparison when working on new algorithms and implementations). Figure 5.1d shows (for the dataset of Goloboff 1993b, with 71 characters and 42 mygalomorph spiders) the cumulative probabilities for different numbers of steps beyond optimality. *RAS* + *TBR* is over 9 times more likely than *RAS* + *SPR* to find optimal trees (0.42/0.047), and it takes 45% longer to run (speed ratio *TBR/SPR* is 1.45). Thus, to find optimal trees a given number of times (Figure 5.1e), *RAS* + *TBR* will be 0.42 / (0.047 × 1.45), or over 6 times, faster than *RAS* + *SPR*. Of course, if we were to be satisfied with trees 10 steps longer than optimal (which a single *RAS* + *TBR* finds with probability 1.0 and *RAS* + *SPR* with 0.90), we are better off using *RAS* + *SPR*—it would take about 0.75 of the time than if using *RAS* + *TBR* (see Figure 5.1e).

For harder datasets (more taxa, more homoplasy), the difference between *SPR* and *TBR* may be much larger. Figure 5.1f shows that the time ratio between *RAS* + *SPR* and *RAS* + *TBR* for the dataset of Liebherr and Zimmerman (1998: 206 characters and 170 carabid beetles) is over 40 times in favor of *RAS* + *TBR* for trees of 6 or fewer steps beyond the best.

Although calculating the expected time ratios for a certain number of steps beyond optimality is not the only way to compare search algorithms, it is the most general and useful one. Subsequent comparisons of search algorithms therefore refer to these ratios.

5.4 DATASETS OF MEDIUM DIFFICULTY: MULTIPLE STARTING POINTS IN DEPTH

Running multiple random addition sequences is, for most datasets, a very effective method. A number of additional considerations help in making better use of this option.

5.4.1 Wagner Trees vs. Random Trees

The goal of multiple starting points is to sample tree landscape. That could be done just as well using random trees (instead of random addition sequence Wagner trees) as starting points. However, those random trees will be very far from optimality, and the initial stages of branch-swapping will need to go through many rearrangements to approach optimality. Therefore, the results will generally be similar but take a longer time with random trees as starting points (but see Goloboff 2014a: figure 3, and p. 121).

5.4.2 Saving Reduced Numbers of Trees per Replication

The larger the number of trees that are saved and swapped, the higher the probability of finding the optimal tree, *if* swapping can be run to completion. That is a big "*if*"—it may require swapping through tens of thousands of trees, thus taking a very long time. Saving instead lower numbers of trees per replication will allow running a larger number of starting points, even if the vicinity around each starting point is explored less thoroughly. As the starting points provided by Wagner trees with a *RAS* are likely to be more or less evenly spread across the landscape (e.g. as in Figs. 5.1b–c), this may provide a better sample of the tree space than using few starting points and using a lot of effort in moving around the close vicinity of those starting points. As the rearrangements obtained by branch-swapping (even *TBR*) are instead much more similar to each other, they represent relatively minor variations of the trees. Using *R* replications saving up to *N* trees for each will take a time roughly comparable to running *N* replications saving up to *R* trees for each, but when the effort is placed into doing more starting points (using less time in the minor variations produced by branch-swapping), the results tend to be better.

Exact numbers that always provide the best efficiency are hard to determine, but in empirical datasets, good results are rarely obtained if every replication saves more than 20 to 50 trees. An example is provided in Figure 5.1g for Liebherr and Zimmerman's (1998) dataset. Finding trees close to optimal saving and swapping 50 trees in a single replications takes, on average, about 2.5 times longer than saving and swapping 5 trees for each of 10 replications. Note that 10 *RAS* + *TBR* with 5 trees/replication takes almost twice as long (the initial stages of the search, to go from the strongly suboptimal Wagner tree to near optimality, use a significant amount of time), with 0.569 sec for every set of 10 replications saving up to 5 trees, instead of 0.291 sec per starting point saving 50 trees. However, as every set of 10 replications can explore many more starting points, it more than makes up for that extra time, and the net effect is finding (near) optimal trees much more efficiently than a single *RAS* + *TBR* saving up to 50 trees (Figure 5.1g).

There is, of course, a limit to how few trees we can save per replication and still have an effective search routine. If too few trees are saved, the chances for every individual replication to find an optimal tree are decreased too much. For example, Figure 5.1g shows that finding optimal trees by repeatedly saving a single tree for each of 30 *RAS* takes, on average, about 3 times longer than if repeating 10 *RAS* + *TBR* saving 5 trees. That is, saving too many trees per replication decreases efficiency, but so does saving too few. The default number of trees to save per replication in TNT is set to 10, which is a number that works well for most empirical datasets, but if you really need to fine-tune your search, you may need to change this number.

5.4.3 Effect of Full Tree Buffer

Saving reduced numbers of trees, in addition to allowing a wider exploration of tree space, has another beneficial side effect: it allows exploring the *TBR* neighborhoods with increased efficiency. This happens because of two different reasons. The first is that, when a tree buffer is full (i.e. no more equally parsimonious trees are to be stored in the buffer, only shorter trees, if found), tree-length calculations for a rearrangement can be abandoned sooner. Tree-length calculations can be abandoned as soon as the tree is confirmed to be of the *same length* as the best tree found so far, instead of abandoning it only when it is discovered that the tree is *longer*. Recall that, when using indirect calculation of tree lengths, every rearrangement is scored by making a comparison between two nodes for each character. The initial length of the rearrangement is the length of the two subtrees being joined, and if the length of the best tree is exceeded when having looked at (say) half the characters, the evaluation of that rearrangement requires half the time (relative to having to look at all the characters). But if rearrangements can be discarded when confirmed to be of equal length (instead of longer), then that length will be exceeded having looked at even fewer characters (say, 25% of the characters). Thus, every rearrangement can be discarded a little faster when the tree buffer is full.

The second factor speeding up the evaluation is one that can be important when saving very large numbers of trees per replication. Branch-swapping is not an orderly process; it is like stabbing in the dark, and it can repeatedly find the same trees. Thus, when producing a rearrangement of the same length as the best trees, it is necessary to compare the resulting tree with all the preexisting trees to ensure that the tree just found is not a duplicate of a previously found tree. The more preexisting trees there are, the more time invested in this comparison.[3] To make things worse, branches of the trees with no synapomorphies have to be removed, which requires calculating final state sets for the rearrangement (and this, of course, takes time). When the tree buffer is full, finding a tree of the same length requires nothing—it can be discarded right away.

The effect of saving fewer trees per replications is then that the majority of the search time is under the favorable conditions of a full tree buffer, thus giving the search a little additional push.

5.4.4 Convergence and Choice of Search Settings

No heuristic method can guarantee finding the optimal tree in an empirical dataset. Does this mean we can never have any idea of how likely we are to have found the optimal tree(s)? Or that any heuristic method we choose to use is fine, since, after all, none provides a guarantee of optimality? The answer to both questions is a resounding "no".

Comparing the results of multiple searches with different starting points (or different random seeds, in randomized processes like *RAS* Wagner trees) gives useful information on the tree landscape when viewed from the perspective of the search algorithm used. If repeating 100 different *RAS* + *TBR* always produces the same, single *MPT*, we can feel confident that the tree space is a simple one, with a single peak of optimality reachable from every point in tree space. If repeating 100 different searches produces the best known score in only 5 cases (producing different scores in the rest of cases), it seems obvious that the tree landscape is very rugged and the rearrangement algorithm in use has a difficult time traveling it. The degree to which different starting points produce a similar answer is known as *convergence*, and it can refer to the extent to which different searches converge to trees of the same topology (*topological convergence*) or the same scores (*convergence in optimality*). When convergence is hard to obtain (e.g. in the second case, only 5 out of 100 replicates producing best known score), one will probably want to perform some additional *RAS* + *TBR* until the minimum known length is independently found a larger number of times.

Some computer programs (e.g. TNT, PAUP*) produce reports on the number of times the best score was hit during a heuristic search. Publications describing results of searches should include this information, both in terms of total numbers of replications completed and numbers of hits to minimum known length, because this gives readers an idea of how likely you are to have indeed found all the optimal trees for your dataset.

We can always set the number of replications to tens of thousands of starting points and thousands of trees to be saved per replication and let the computer run for days or weeks. Once the search finishes, we can be pretty certain we have found optimal trees. But developing a morphological matrix requires many changes in coding as new specimens and characters are studied and added to the dataset. Even a change in a single matrix cell can change the results, so the dataset needs to be analyzed again. If we use the brute-force approach of huge numbers of replications and trees to save per replication, the process of developing a morphological dataset will take forever, and we will probably have lost our interest (and funding) to continue on the project before finishing. Some way to speed up this process, or at least be able to obtain approximate results with some degree of reliability, is necessary.

A much more rational approach is trying to make a good use of the computing time. The degree to which searches converge to optimality can be used to quickly decide on a good search strategy for the dataset at hand. Running 10 or 20 replications and observing whether the best score is found repeatedly, or consistently across sets of 10–20 replications repeated with different random seeds, gives an idea of whether the optimal score is likely to be found with current settings. If 10 out of 10 of those exploratory replications produce the same single tree, we are, in fact, done; there is no need to continue increasing the strength of the search. If that is not the case, increasing the number of trees saved per replication will make every replication more likely to find the best scores. Although there are no magic numbers in this (but see Section 5.7), the search should be set so that the optimal score is found at least 10–20 times independently. If obtaining such results with traditional search strategies proves unfeasible (i.e. it will require performing so many *RAS* + *TBR* that the analysis would take hours or days), then it is advisable to switch to more elaborate search schemes, explained under "Difficult Datasets" (Section 5.6).

5.4.5 Collapsing of Zero-Length Branches and Search Efficiency

Search algorithms work on binary trees mostly because they are easier to handle and algorithms for scoring trees can be simplified. Of course, a dataset may well not contain enough information to support the full resolution of the tree. The unsupported groups are eliminated a posteriori in the search and rearrangement process. If an optimal (binary) tree contains a branch along which there are no unambiguous state changes (i.e. synapomorphies) in any character, that branch can be eliminated, forming a trichotomy of the same length as the original binary resolution. Each of the three alternative binary resolutions of that trichotomy (including the original one) will then be of the same length. Although any of the four trees (the collapsed or the binary ones) is an equally good conclusion about relationships, with no grounds of preference for any one tree over the others in terms of fit to data, displaying only the polytomous tree is much more economic and compact. It includes (at least in theory; see Chapter 6) the possibility of the other three trees.

The different ways in which zero-length branches are treated during searches, even if setting the same criterion to decide which branches should be eliminated, is an occasional source of discrepancy in the behavior of TNT and PAUP*. PAUP*, to store in the tree buffer an *MPT* found during a search, will eliminate the unsupported branches, leaving no trace as to which original resolution had been polytomized. As branch-swapping only works on binary trees, when the tree is subsequently retrieved from the tree buffer to be swapped, it needs to be dichotomized again. PAUP* does this using generators of pseudorandom numbers to increase the chances that polytomized trees are not always resolved in the same way. It has the *advantage* that if some resolutions of a polytomy but not others can lead to better or additional trees, it is likely those will eventually be found. It has the *problem* that the tree may become longer when collapsing several unsupported groups at once (see Chapter 6), and since the program does not keep track of the original resolution, resolving the polytomies at random may produce a binary tree longer than the tree that had been polytomized to be saved in the tree buffer. This can potentially decrease the effectiveness of the search. TNT instead, when figuring out which branches are of zero length, uses a dual representation that keeps track of both the original binary and the polytomized tree. TNT stores this dual representation; as a new tree is found during the search, before storing it in the tree buffer, TNT makes sure that the trees are different as collapsed. But since the representation used is dual, when the tree is brought back from the tree buffer to be swapped, the original resolution which had produced the collapsed tree is reinstated. Advantages and problems are the opposite of those of PAUP*.

Because it is possible that, for a polytomous tree, only some of the binary resolutions lead to better or additional trees, one of the recommendations of Swofford and Bell (2017: 170) in the PAUP* manual is "not to use [elimination of zero-length branches] in conjunction with random replicates. This may greatly increase computation time, but will avoid the problem of arbitrarily missing parts of islands". That advice is correct in itself but only if very large numbers of trees are going to be saved and if we are willing to invest much longer times in the tree search. Except for small and simple datasets producing few trees, searches will be most effective if doing multiple *RAS* followed by *TBR* saving a limited number of trees (as shown in Figure 5.1g). In that case, saving the trees distinct as binary will quickly fill the buffer with trees that are (in terms of the groups that the dataset actually supports) very similar to each other. In being very similar, they will all be more or less equally likely to lead to better trees. When zero-length branches are removed instead, the few trees that are saved to the buffer will be more different from each other—they remain as distinct after a more stringent criterion is used to compare them. For that reason, it is more likely that one of the trees saved to the buffer will lead to finding a better tree (Goloboff 1996: 213–214; Davis et al. 2005 reached similar conclusions). An individual replicate will take somewhat longer to complete when removing zero-length branches (memory buffer will take longer to be filled, and removing zero-length branches requires finding final states for candidate trees, many of which get rejected anyway by virtue of being—when collapsed—a duplicate of some previously found tree). For the example shown in Figure 5.1h, every replicate with zero-length branches removed takes about 25% longer than without. But every replicate is so much more likely to find a shorter tree when zero-length branches are removed that the net effect is a significant speed up. Removing zero-length branches thus makes *RAS* + *TBR* with reduced number of trees saved per replication much more effective, particularly as more stringent criteria are used for the removal of unsupported branches (see Chapter 6 for options). Figure 5.1h shows how dramatic the effect of tree collapsing may be when saving only 10 trees per replication: the search takes on average 30+

times faster to find optimal trees due to the increased effectiveness of each replicate. While Swofford and Bell's (2017) advice may be useful under some circumstances, in most cases it is better avoided.

5.4.6 Many Hits or Many Trees?

If saving reduced numbers of trees per replication, it is obvious that the search is likely to miss some of the optimal trees. When possible, saving all the trees that are most parsimonious is desirable. As discussed at the introduction of this chapter, for determining whether a group is unsupported, it is necessary to consider whether the group is present in all of the optimal trees. In TNT, after a search with multiple *RAS* + *TBR* concludes, the program reports whether some of the individual replications had filled the memory buffer ("overflowed"). If so, then subjecting the resulting trees to a round of global *TBR* (saving as many trees as possible) will probably find many more additional trees. In small or very clean datasets, this two-step strategy allows effectively finding all equally parsimonious trees. When the multiple *RAS* + *TBR* finish without overflows noted at the end, that means none of the replications which found the best trees filled the allotted tree buffer (the tree buffer may have been filled in a replication that *failed* to find *MPT*s, which does not need to worry us). Therefore, when no overflow is reported at the end, subjecting the trees produced by *RAS* + *TBR* to a global round of *TBR* will produce no further trees.[4]

In large, ambiguous datasets, saving all the possible optimal trees may be unfeasible due to both time and memory constraints. Keep in mind that the number of trees can increase combinatorially. If each of 15 distant parts of a tree can be resolved in 3 possible ways due to character conflict (which still makes for a very well-resolved consensus tree), then the number of equally parsimonious trees will be 3^{15}, or over 14 million trees. Assuming that enough memory to store all those trees is available, confirming the actual number of trees via branch-swapping would require prohibitive amounts of time. As more trees accumulate in the tree buffer, swapping would become slower and slower because every new tree to be saved in the tree buffer would have to be confirmed as different from millions of preexisting trees—which eventually takes more time than the length calculations themselves! But the main goal of phylogenetic analysis is not finding and counting all *MPT*s; in most cases, the goal is instead only to find out which groups are supported by the data—which groups are present in each and every one of the *MPT*s. Those groups are the ones present in the strict consensus of all *MPT*s, and if there are means to calculate the same consensus, the actual number of trees input to the consensus procedure is irrelevant. In the example prior, even when 14 million *MPT*s exist, it is possible to have the correct strict consensus tree with just two trees, if every one of the 15 trichotomies is resolved differently in the two trees. If the two trees have been found by searches from independent starting points (e.g. *RAS* + *TBR*), having each of the trichotomies resolved differently is unlikely—a third of them will be matched in the two trees just by chance. But increasing the number of independent hits to minimum length, the probability that any one of the trichotomies is resolved equally in all the trees quickly approaches zero. With just 10 hits in TNT, there is a chance of about

1% of having a consensus with a spurious node; increasing the number of hits to 15, that probability decreases to less than 1/10,000). Note that saving 10 or 15 trees via *TBR* will not accomplish the same effect because *TBR* finds new trees in an *orderly* fashion—it will first save three trees for the first polytomy, then three for the second, and so on. By the time *TBR* gets to clip one of the nodes that resolve the fifth polytomy, it will have accumulated 15 trees (3 × 5) and filled the memory buffer with trees which must have the remaining 10 trichotomies resolved in the same way. Even saving 40 trees will leave the last of the 15 trichotomies resolved identically in all 42 trees (because 14 × 3 = 42). The lesson is that, to obtain accurate consensuses, we are better off spending our search time obtaining additional hits to maximum parsimony than trying to find as many *MPT*s as possible via branch-swapping. This is because—as discussed before in different contexts—the trees produced by independent hits in *RAS* + *TBR* tend to be much more different from each other than the trees found by *TBR* swapping from a single tree. This is illustrated in Figure 5.1i, which shows results for Pei et al.'s (2020) paleontological dataset (164 taxa, for *Velociraptor* and relatives), with much ambiguous resolution. If using trees from independent *RAS* + *TBR*, the correct consensus (with 110 nodes) is calculated with the first 10 hits to minimum length (grey line in Figure 5.1i); further hits to minimum length continue producing exactly the same consensus (because no *MPT* lacks any of those groups). Although only 10 hits to minimum length suffice to produce the correct consensus in this case, saving trees via *TBR* starting from a single tree (white line in Figure 5.1i) continues producing a consensus with 115 nodes with more than 50 trees saved. The consensus continues having 114 nodes (i.e. 4 wrong nodes) even after having found and saved 500 trees. That is, the actual number of equally parsimonious trees is a useless number; what matters is the strict consensus tree, which can be obtained most effectively with enough independent hits to optimality, not with many equally *MPT*s that are only minor variations of the same theme.

The situation described in the preceding paragraph refers to a case where there is actual character conflict producing different groupings. When the alternative binary resolutions can result from lack of information (lack of characters or ambiguity in character optimization), eliminating zero-length branches is an even more effective way to get rid of groups that are not truly supported by the data. Whether the trees are produced by many hits with *RAS* + *TBR* or by *TBR* from a few *MPT*s, collapsing them more stringently (e.g. by stripping off branches with a minimum possible length of zero) produces more accurate consensuses. Alternative ways to collapse zero-length branches are discussed in much more detail in Chapter 6.

5.5 SEARCHES UNDER CONSTRAINTS; BACKBONE TOPOLOGIES

In some contexts, finding the best trees which display or do not display a certain group can be very useful. The analysis of a dataset may reject a prior hypothesis of monophyly of a group. Calculating the number of steps required to force monophyly of the group gives an idea of how strongly the data contradict the monophyly of the group. Likewise with groups that are present in every one of the *MPT*s. Just finding a tree which displays the group to be forced for monophyly or lacks the group forced for non-monophyly is not enough—what is needed to properly test that alternative hypothesis is the shortest possible trees having the group or lacking the group. That cannot be done manually but only by applying searches where trees are properly filtered.

In TNT the reference groups can be defined as groups constrained to be present or absent. It is possible to define some group(s) to be present and other group(s) to be absent, simultaneously. Obviously, defining the same group to be simultaneously present and absent triggers an error message. Searches may also fail to find a tree satisfying the constraints. For example, for four taxa (A, B, C, and D, where A is used to root the tree), constraining three groups to be absent from the result (B+C, B+D, and C+D) fails to find any trees; only two of those groups can be constrained for non-monophyly at the same time, for otherwise every possible tree is forbidden.

Constraints for groups to be present can take the form of trees (called *tree constraints*) or lists of groups (called *positive constraints*); no group in the positive or tree constraints can contradict any other group. Constraints for groups to be absent (*negative constraints*) are given only as lists. Only in the case of positive and negative constraints is it possible to define floating taxa (i.e. a terminal may be inside or outside of an otherwise monophyletic group), to define a skeleton for the tree.

Tree constraints are checked (during a search) by reference to a tree and special lists of rearrangements to generate. When TNT performs branch-swapping under tree constraints, the rearrangements that produce trees violating the constraints are never attempted, thanks to the use of a special data structure. The number of rearrangements reported by the program includes the trees violating the constraints (as if they had been effected and rejected), but they are never truly done. Thus, tree constraints for several groups at the same time tend to speed up branch-swapping, often by a factor of two or more.

Positive and negative constraints, instead, are checked only after the length of a candidate tree has been found to be acceptable. The check internally uses a matrix of group variables that represent group membership; if optimizing the variable in the candidate tree requires more than a single step in the case of one of the positive constraints or fewer than two steps in the case of negative constraints, the tree is rejected. Given that this test is done subsequently to having generated the rearrangement and calculated its length, positive and negative constraints do not speed up searches (and if they imply strongly suboptimal trees, they slow searches down).

An important aspect of using negative constraints together with positive or tree constraints is that the addition sequence in Wagner trees cannot be fully randomized. For negative constraints (e.g. XYZ must be non-monophyletic), the first opportunity to make the group non-monophyletic is used; violation of monophyly of group XYZ may become impossible if delayed. Consider the case where X and Y have been placed together in the tree, Z remains to be added to the tree, and YZ is also forced to be a monophyletic group. If Z is the last taxon to add, then there is no branch where to place it such that both the positive and negative constraints are obeyed at the same time. Hence the need to obey negative constraints as soon as possible when forming the Wagner tree, making sure in this example that X and Y are not placed as sister groups. But this also requires that some addition sequences are banned—for example, for the first two taxa added in the sequence Y, Z, X it would be impossible to obey monophyly of YZ and at the same time violate monophyly of XYZ. The addition sequences in TNT are then modified, when negative constraints are used together with positive or tree constraints, to make sure such problematic cases are avoided.

5.6 DIFFICULT DATASETS: BASIC IDEAS AND METHODS

For datasets of medium size (i.e. 30 to 50 taxa), the standard methods described in previous sections work very well. However, as datasets become larger and have strong conflict between characters, those methods struggle to obtain optimal trees, or at least to obtain them in reasonably short runs. The basis of those methods is (a) to use independent starting points and (b) to use each of those starting points to separately begin branch-swapping. Eventually, one of the starting points will lead (via branch-swapping) to the global optimum. The mechanics are designed so that, if one of the starting points does not yield optimal trees after branch-swapping, the partial results obtained in that replicate are discarded, thus effectively making all the search effort in that replicate unproductive.

5.6.1 Composite Optima

The situation can be visualized by thinking of the phylogenetic problem as one of *composite optima* (Goloboff 1999). In real datasets, multiple islands or local optima begin to appear often at 30–40 taxa or so. When we have a large tree, there can be many parts of that size in the tree, each with its own multiple islands. Figure 5.2a shows a hypothetical large tree, where there are 10 separate portions of the tree with local optima. Assume for the sake of the argument that (a) for every portion, the probability of a *RAS* + *TBR* search to find the optimal resolution for the portion is 0.5, and (b) whether a given portion has been resolved optimally is more or less independent of whether other portions have been resolved optimally. Under those conditions, searching from a given starting point will land on optimal configurations in some parts of the tree (e.g. those marked in black in Figure 5.2a) but on suboptimal configurations in the others (marked in gray). The standard approach to searches amounts to throwing away all those results and starting all over again from scratch. Every attempt we make will be resolved optimally for about half the parts. The hope is that, eventually, we will be lucky enough to land on a tree with all 10 sectors optimally resolved—under the conditions described, that will happen with probability about 0.5^{10}, or once every 1,024 replications. This illustrates the complexity of finding *MPT*s as the number of taxa increases. Assumption (b), that the optimality in the local resolution of different parts of the tree is independent, is unlikely to be met in practice, as interactions between parts are likely to occur. This makes the problem even more difficult.

The idea of trees divided in parts helps us approach the problem from a useful perspective, different from the one embodied in *RAS*s and global *TBR*. Rather than starting all over every time a replicate failed to produce optimal trees, one should

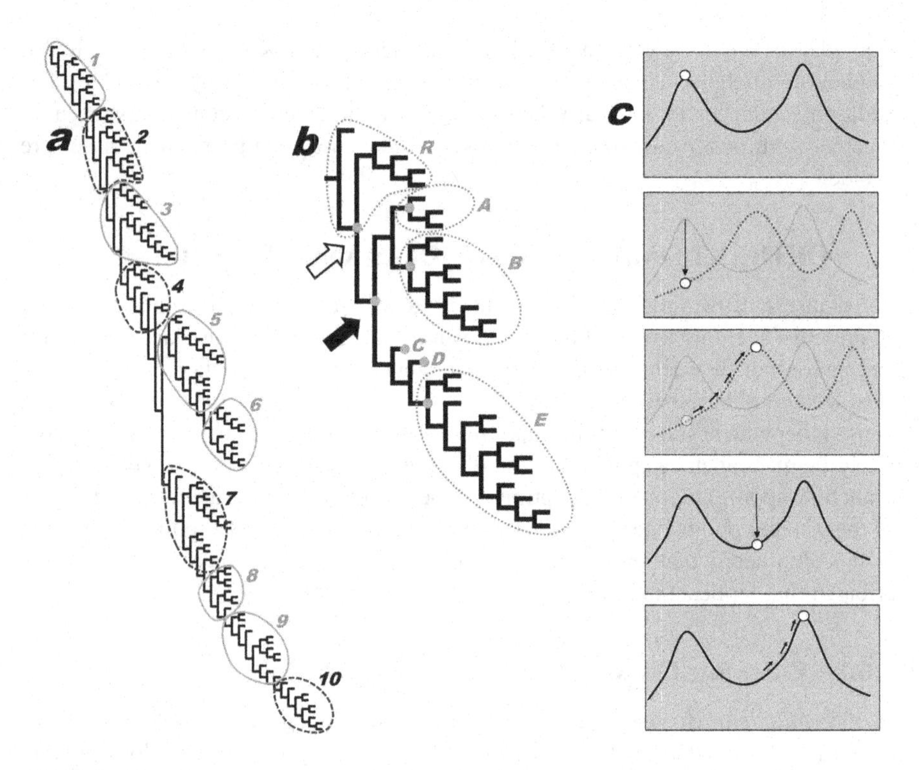

FIGURE 5.2 **(a)** A large tree can be thought of as composed of multiple subtrees, some of which are in optimal configurations (black) and some of which are not (gray). A globally optimal tree needs to have all the subtrees in an optimal configuration simultaneously. **(b)** A reduced dataset can be used to represent a subtree, using down-pass ancestral state reconstructions for internal nodes (i.e. the clades marked as A, B, or E). The root node must consist of the down-pass states for the remaining tree (R), if rerooted at the branch marked with arrows. With the reduced dataset created in this manner (and symmetric transformation costs), any length difference between trees for the reduced dataset will be matched by the same length difference in the corresponding configuration of the whole tree. **(c)** The parsimony ratchet, via random character reweighting, alters the tree landscape so that a point in a local optimum may now be at the base of a different optimum for the modified landscape. A quick branch-swapping search may now leave the point, when returning to the original landscape, at the foothill of a higher optimum.

strive to improve the tree one part at a time, without making the resolution of the other parts suboptimal in the process. There are four main methods implemented in TNT: sectorial searches, ratchet, tree drifting, and tree fusing (Goloboff 1999; Nixon 1999).

5.6.2 Sectorial Searches (*SS*)

Sectorial searches (Goloboff 1999) are the most obvious solution to the problem of composite optima. If *RAS* + *TBR* is generally capable of solving problems of 40–50 taxa, then one can select portions of the tree and perform a simpler analysis on each reduced subproblem (or *sector*). This creates a new reduced matrix to represent the sector (black arrow in Figure 5.2b), with "terminal" taxa A–E. Some of the terminals in the reduced matrix (A, B, E) are hypothetical ancestors representing groups (reconstructed via a down-pass character optimization within the pending clade), and some are truly terminals (C, D). The tree could be rerooted at the node marked with a black arrow, and then the group R would form a clade (instead of being paraphyletic); under such rooting, the down-pass states of R (i.e. the node indicated with a white arrow) are used as a terminal in the reduced matrix. In that case, the difference in length for the reduced matrix between two trees for R, A–E will be the same as the length difference for the original matrix produced by rearranging the groups represented by those terminals in the equivalent positions. Using the hypothetical ancestors, that is, allows creating a reduced matrix with no loss of information and no changes in the search landscape. Any reduced tree found that is better than the original resolution can just be plugged into the original tree to save the same number of steps. In the case of step-matrix characters, some special considerations are required. The most obvious one is that rerooting the tree so group R becomes a clade instead of a paraphyletic group may change tree lengths; when asymmetric step characters are present, therefore, the sector always extends down the root of the tree (in that case, it is advisable to make sure sectors are bigger; the program does not automatically extend sector size).

When creating the reduced matrix, the hypothetical ancestors (down pass) for all nodes in the original tree are saved and used so that successive selections of sectors do not require reoptimizing the entire tree; if the large tree is modified, hypothetical ancestors are changed only for nodes below the terminals and groups included in the sector. All the characters that become uninformative for the relationships between R, A–E are excluded from the reduced matrix, saving further time.

A vaguely similar method, under the name of Rec-I-DCM3, was proposed by Roshan et al. (2004). That method selects disjoint sets of terminals, and although Roshan et al. (2004) claimed that using TNT together with Rec-I-DCM3 improved the speed of searches, they only compared their method with basic settings of TNT. Goloboff and Pol (2007) reexamined Rec-I-DCM3 and showed that it produces results inferior to those of TNT alone with proper settings. In addition, Goloboff and Pol (2007) showed that (by virtue of selecting terminals instead of *HTU*s and thus changing the search landscape), Rec-I-DCM3 effectively behaves as a "perturbate-and-search" method more than as a true "divide-and-conquer" one (as publicized by Roshan et al. 2004).

5.6.2.1 Types of Sector Selection

Of course, in any empirical case, we have no idea what the conflictive parts of the tree are. The best course of action is then to select several sectors so as to cover the entire tree. That leads to several ways in which the sectors can be selected:

a) *Random sectorial search (RSS).* In the absence of any information about tree parts, an obvious choice is making repeated selections, every time making a random selection of connected nodes. For sectors of size S, out of T taxa, a number of selections $2T/S$ to $3T/S$ will on average cover the whole tree two to three times. Even finding the optimal resolution for each one of the reduced problems does not guarantee finding an optimal tree. First, finding a better global configuration may require moving terminals or groups (e.g. simultaneously) more than S nodes away in the tree, which can never be captured by selections of S nodes. Second, sectors are not fully independent; the configuration of each of two separate sectors can be suboptimal yet may turn out to be a better resolution when combined. Both of these problems are solved (or ameliorated) by selecting larger sectors. TNT can perform successive rounds of *RSS*, increasing the size for each one.

b) *Exclusive sectorial search (XSS)*. The tree is partitioned in exclusive sectors of an even size. This guarantees covering all the tree. In this case, instead of setting a size of sector to be analyzed by itself, the user determines a number of partitions. You should just make sure that the partitions are not too small (unlikely to contain significant conflict). Depending on the difficulty of the dataset (i.e. the amount of conflict and homoplasy, the ruggedness of the landscape), the number P of partitions should be chosen so that the resulting sectors have at least 40 terminals (i.e. $T/P > 40$). A series of partitions can also be used, e.g. for 160 taxa, one could do a round with 4 partitions (sectors of size ~40), followed by one with 3 (size 50–55), ending in a round with 2 (size ~80).

c) *Constrained sectorial search (CSS)*. It is possible to use sectors of the tree created from tree constraints using the constraints as reference. When the search is made under constraints and the reference tree contains polytomies, *CSS* creates reduced datasets corresponding to the sectors around each of those polytomies. TNT allows creating constraints on the fly, during multiple replications, by calculating the strict consensus of the *RAS* Wagner tree (or, optionally, the final tree produced by *TBR*) in one replication and the best tree found at the end of the preceding replication. In this way, the constraints to use (1) are determined from the data at hand, not prior hypotheses; (2) are likely to identify regions of dubious resolution (well-supported parts of the tree will be found by both independent replicates so that polytomies indicate less supported parts of the tree); and (3) change as the search proceeds, using different constraints for every new replication (so that they are unlikely to create a persistent bias in the search). A huge advantage of this *autoconstraint* option is that it speeds up the phase of the search that goes from a *RAS* Wagner tree to a *TBR* optimal (constrained) tree—recall that tree constraints make *TBR* searches proceed faster. Obviously, *within* the sector selected on the basis of polytomies in the constraint tree, the search can proceed without the use of constraints (which are effectively external to the reduced dataset). In TNT, *CSS* creates sectors only from polytomies of a degree above a certain limit (e.g. a trichotomy is unlikely to contain conflict that cannot be resolved by *TBR*; default limit is 10). Once the *CSS* is concluded, the constraints are abandoned and the replication continues—if the temporary autoconstraints happened to prevent some optimal resolution, hopefully that will be corrected later when the replication continues without constraints.

d) *User-defined sector (DSS).* Rarely used, this allows specifying a node around which the sector is to be created. It can be used to search in certain parts of the tree in more detail when using scripts or special routines.

5.6.2.2 Analysis of Reduced Datasets

Once the reduced dataset is created, TNT automatically submits it to a tree-search routine. If the reduced dataset is small enough, it can be analyzed with a simple routine such as multiple *RAS* + *TBR*. The default setting in TNT uses *RAS* + *TBR* for sectors within 75 taxa, running three initial replications and checking the lengths of the resulting trees. If the resulting trees are all of the same length as the original resolution (i.e. suggesting that the reduced dataset contains no significant conflict), the analysis of the sector is finished. If they have different lengths, an additional three *RAS* + *TBR* is performed, returning the best solution found at the end (obviously, if the original resolution is best, that one is kept).

As the reduced dataset is larger, it is unlikely that the optimal tree can be found by just *RAS* + *TBR*. TNT can then search using tree-drifting (see Section 5.6.4), using six cycles by default (starting from original resolution), when the number of taxa in the reduced dataset is more than 75 (or above the limit set by the user). For even more complex datasets or larger sectors (i.e. above a user-specified limit), the analysis of the sector can be done using a combination of multiple *RAS* + *TBR*, each replication followed by tree-drifting, at the end submitting the pool of the original tree and the best trees found to tree-fusing (see Section 5.6.5). All these settings can be changed with simple options of the command for sectorial searches. Last, for fine-tuning the search for the reduced dataset, *recursive sectorial searches* allow giving TNT the list of search commands to use on each sector (it can also be the name of a file with instructions for searching). This allows using *any* search routine on the reduced datasets, possibly including further subdivisions in subsectors (by calling the command for sectorial searches as part of the search instructions).

Every certain number of substitutions (or cycles, in the case of *XSS*) to the tree, a global round of *TBR* is performed. Once a better resolution for a sector is reinserted back into the big tree, the resolution of other sectors that seemed optimally resolved before the reinsertion could be changed to obtain further improvements. At the end, if any local changes to the tree were made, a round of global *TBR* is done to guarantee that the resulting tree is at least *TBR* optimal.

5.6.2.3 Performance

By virtue of improving the tree one part at a time, and without worsening the remaining parts, sectorial searches produce vast improvements over the use of *RAS* and *TBR*. The differences in times needed to find near-optimal trees depend, of course, on the dataset, but in empirical datasets with 100 or more taxa, the improvement is evident even without detailed measurements. For Liebherr and Zimmerman's dataset, running a cycle of *RSS* (default parameters) in addition to a simple *RAS* + *TBR* takes 3.28 times longer than the *RAS* + *TBR* alone (see Figure 5.3b). However, the

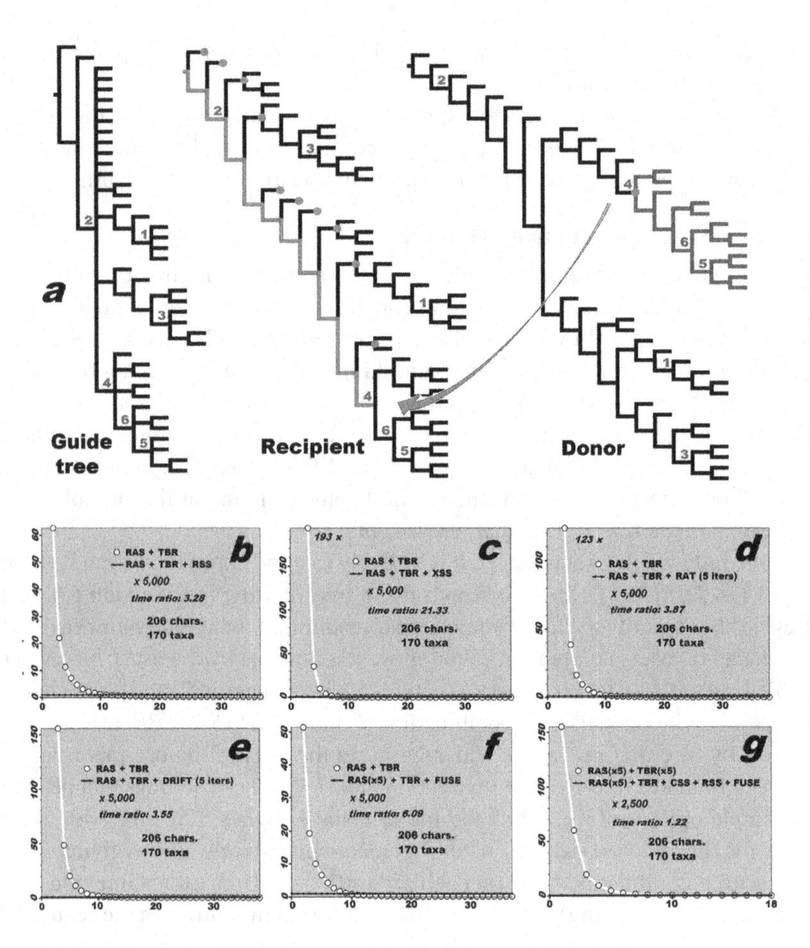

FIGURE 5.3 **(a)** In tree fusing, the insertion of resolutions of subtrees of a donor tree into a recipient tree are guided by the groups in a strict consensus of the two trees (i.e. any group present in the consensus is matched by groups of identical composition in donor and recipient). By using the down-pass states of the donor clade (marked as 4 in the example), and effecting a down pass from the equivalent point 4 in the recipient tree, the length resulting from the exchange can be calculated quickly (using the pre-calculated down-pass states of the successive sister groups of node 4 in recipient tree). **(b)** Expected time ratios (y-axis) to find trees of a given number of steps beyond minimum (x-axis), with just a *RAS* + *TBR*, relative to the times of a *RAS* + *TBR* and a random sectorial search. On average, a single *RAS* + *TBR* + *RSS* takes 3.28 times longer than a single *RAS* + *TBR*, but the trees it finds are so much shorter that trees close to optimality are found at least 60 times faster when adding *RSS*. **(c)** A comparison similar to that in (b) but using exclusive sectorial search (*XSS*, with `mult = rep 1 hold; sectsch = xss 4-2+4-1 gocom 5 combst 3`). The *XSS* routine takes 21.33 times longer than *RAS* + *TBR* but finds trees close to optimality almost 200 times faster. **(d)** Expected time ratios (y-axis) to find trees of a given number of steps beyond minimum (x-axis), with just a *RAS* + *TBR*, relative to the times of a *RAS* + *TBR* followed by 5 iterations of ratchet. On average, a single *RAS* + *TBR* plus five rounds of ratchet takes 3.87 times longer than a single *RAS* + *TBR*, but the trees it finds are so much shorter that trees close to optimality are found over 120 times faster when adding ratchet. **(e)** Tree drifting works similarly to

the ratchet (see d), finding trees close to optimality at least 150 times faster than a plain *RAS* + *TBR* (even when the process takes 3.55 times longer with the addition of drifting). **(f)** Tree fusing synergistically combines the trees from searches that failed to find optimal trees and produces better trees in very short times. Using the results of five *RAS* + *TBR* as input for tree fusing takes little additional time (only 6.09 times longer than for a single *RAS* + *TBR*), but the trees it finds are so short that trees close to optimality are found at least 50 times faster than if using only *RAS* + *TBR*. **(g)** Multiple *RAS* + *TBR* saving low numbers of trees per replication (i.e. the recommended traditional strategy) takes at least 150 times longer to find trees close to optimality than a combined search using multiple starting points, sectorial searches, and tree fusing.

tree scores obtained are so much lower that they can very rarely be attained with three rounds of *RAS* + *TBR*; the net result is that near-optimal trees are found, on average, over 60 times faster when using *RSS* on top of a *RAS* + *TBR*.

In the case of *XSS*, the improvement can be even more dramatic. Using four cycles of *XSS* on top of *RAS* + *TBR* (as in Figure 5.3c), starting with division of the tree in four pieces and ending in two and using for the analysis of each sector five *RAS* + *TBR* followed by tree drift and fusing at the end), takes 21 times longer than *RAS* + *TBR* alone. The tree scores found by *XSS* are so good, however, that the extra time more than pays off, and the longer routine finds near-optimal trees over 190 times faster than a plain *RAS* + *TBR*. This is an inversion of the old maxim: "more is less".

5.6.3 Ratchet

This method (Nixon 1999) uses alternate cycles of perturbation and search. A search cycle performs *TBR* on a single tree. A perturbation cycle performs *TBR* to obtain a tree with some deviations from current tree, which can be subsequently used as input to *TBR* in the next search cycle. One of Nixon's (1999) ideas was to be able to perform *TBR* from many starting trees but without having to spend all the time needed in the initial stages of a *RAS* + *TBR* search (i.e. to go from a strongly suboptimal Wagner tree to a *TBR*-optimal tree). The other idea was to use small perturbations of the data to be able to mildly deviate from the best tree found so far, but in a way that depends on character conflict. The perturbation is then done by changing the weights of some characters (chosen at random), searching with *TBR*, and using the resulting tree as a starting point for *TBR* on the original weights. The procedure is easy to implement with almost any program which can do branch-swapping, and that is one of the reasons for the popularity of the ratchet (the other being, of course, how well it works). With such a design, if the perturbation phase goes away from the pre-perturbation tree it used as starting point, it can only be because *some alternative characters support groups conflicting with groups present in the pre-perturbation tree*. Local optima are determined by character conflict, and thus the ratchet addresses the problem at its heart by the simple process of reweighting a randomly chosen subset of characters. An alternative but complementary view is that the ratchet, during the perturbation phase, alters the tree landscape (Figure 5.2c) so that the current solution is now placed at the foothills of a different peak, and when resetting weights and returning to the original landscape, we may now be placed at the slope of a higher peak than before (Figure 5.2c, bottom).

In TNT, the implementation of the ratchet uses some modifications of the procedure originally proposed by Nixon (1999) which produce additional improvements. The most important one is that the perturbation phase accepts rearrangements of equal score, instead of just those that improve the score under modified weights. As this could cycle back and forth between two or more trees of the same score, there is no completion in *TBR*; *TBR* is then stopped when a certain number of changes to the pre-perturbation tree have been accepted. The perturbation phase is not intended to find the best possible tree under the modified weights; the only purpose of the reweighting is to have a tree which differs in several rearrangements from the pre-perturbation tree, with groupings supported by alternative characters, and as soon as that is accomplished, the perturbation phase can stop. The number of changes to accept depends on the number of taxa; accepting 5 rearrangements is a radical change to the tree if it contains 10 taxa but insignificant if it contains 500. The default is using *T/8* (but no less than 20 or more than 200). In addition, if 99% of the clippings (with their reinsertions) have been effected (which could well happen without having reached the minimum number of changes to the tree if the data are very strongly structured), the perturbation phase is abandoned (this percentage can be changed by the user). This makes the ratchet even faster because it uses very little time in the perturbated search. Given that equally optimal rearrangements are accepted during the perturbation phase, the probability of reweighting characters (4% by default in TNT) is much smaller than the 50% originally proposed by Nixon (1999), or otherwise the perturbation phase tends to go too far away from the pre-perturbation tree (because of the acceptance of trees of equal optimality for the changed weights, which Nixon 1999 did not do).

5.6.3.1 Performance

The increase in performance by the ratchet is dramatic. For Liebherr and Zimmerman's dataset, adding five cycles of ratchet to the results of a *RAS+TBR* take almost four times longer. The resulting scores are so much better, however, that near-optimal lengths can be found with the ratchet over 120 times faster (on average) than if using just *RAS* + *TBR* (see Figure 5.3d).

5.6.4 Tree Drifting

Like the ratchet, tree drifting consists of alternating cycles of perturbation and search. The perturbation phase performs *TBR* accepting suboptimal rearrangements with a certain probability; the more suboptimal the rearrangement, the less likely to be accepted. Rearrangements producing equal score are always accepted, which allows the method to "drift" around tree space. The suboptimality of each rearrangement is determined with both the absolute fit difference in score, *AFD*, relative to current solution, and the *relative fit difference*, *RFD* (see Chapter 8). *RFD* compares character scores on the two trees, defining *C* as the sum of (weighted) step differences in all the characters fitting the rearranged tree better and *F* as the sum of (weighted) step differences in all the characters fitting the rearranged tree worse. While absolute score difference is simply $F - C$, *RFD* is $(F - C)/F$ and better takes into account character conflict (i.e. relative numbers of characters for and against a group; see Chapter 8). The *RFD* can be calculated exactly during branch-swapping by keeping track of the score differences for each character (i.e. without the need to optimize de novo). To further speed up calculations, tree drifting uses predefined acceptance limits, AFD_{lim} and RFD_{lim}. When clipping the tree, the decrease in total score is X; when evaluating a reinsertion, the increase in score will be Y. Note that $F - C = Y - X$ and $F \leq Y$ (F can be less than Y when some characters have their steps decreased when clipping and then increased again when reinserting). Therefore, only those rearrangements fulfilling $(Y - X)/Y < RFD_{lim}$, which is to say $Y < X/(1 - RFD_{lim})$, can be accepted. So, many rearrangements can be rejected only on the basis of the length increase when reinserting clipped clade, with no need to calculate *RFD*. If *RFD* for the rearrangement is less than RFD_{lim}, the acceptance or rejection is decided by generating a value $V = (random[0 - 100]/(100 \times (Y - X) + (2 \times f))$, where f is a factor measuring how far from the original score the drift has gone (f is initialized at 0 and then increased in $Y - X$ every time a rearrangement is accepted; this ensures that the perturbation does not drift to longer and longer trees). Note that $Y - X$ measures the length increase produced by the rearrangement (so that rearrangements producing longer trees have a lower average probability of acceptance). Then, the rearrangement is rejected if $RFD > V + (V \times Uf)$, where U is a user-defined acceptance factor (default is 3, larger values increase probability of accepting a suboptimal rearrangement). By using the *RFD*, tree drifting takes into account character conflict. The stopping rule for a cycle of perturbation is as in the modified ratchet procedure of TNT, based on either an absolute number of rearrangements accepted or a proportion of clippings effected.

In addition to accepting suboptimal rearrangements, tree drifting alternates phases where only equally optimal rearrangements are accepted (again, stopping on the same rules). This is also another aspect which allows tree drifting to examine tree space much more effectively than saving and swapping multiple trees via *TBR*. Consider the mechanics of saving and swapping multiple trees as implemented in most computer programs. Starting from an initial tree (e.g. the first tree of the best score found so far), swapping begins, and a number N of trees with equal score can be saved. Every one of these trees is only *one TBR* move away from the original, and to begin swapping on those, we need to complete swapping on the first tree. Assume that swapping each of these N trees also produces, in every case, another N trees. This leaves the memory buffer with N^2 trees that are *two* moves away from the original, but to begin swapping on those, we need to complete swapping on each of the preceding N trees that are one move away. As we continue saving N trees for each of these N^2 trees, we will produce N^3 trees *three TBR* moves away from the original, and to begin swapping on those, we need to complete swapping on the preceding N^2 trees. The progression is clear: to begin swapping on trees that are m moves away from the original tree, when every tree swapped produces approximately N new trees, we need to swap to completion about N^{m-1} trees. This explains why saving many multiple trees with *TBR* is not too efficient. In the case of tree drifting, in the phase that accepts only equally optimal rearrangements, with just a few replacements to the tree, we can end up with a tree differing in 20 or more moves from the original almost instantly—without even completing swapping on a single tree. This results in a vastly improved exploration of the tree space: equally optimal trees that are more different are more likely to lead to better trees. Given that this drifting through suboptimal trees increases the effectiveness of the search, TNT also uses it in the ratchet (Nixon's 1999 original description of the ratchet only accepted rearrangements of better score in a perturbation phase).

In computer science, search algorithms with some similarities to tree drifting have long been known as simulated annealing (from a metallurgic process of heating and cooling). Some experimental programs (e.g. Barker 2004; Richer et al. 2012) have made attempts to adapt simulated annealing for phylogenetic searches. Those programs rely (Goloboff 2015) on worse search implementations than TNT and use a very generic adaptation of the simulated annealing algorithms (e.g. only using absolute score differences to determine the probability of acceptance). By virtue of strategies more specific to phylogeny (e.g. the use of *RFD* in addition to absolute differences in score, alternation between perturbation and search phases, and specific stopping rules for the perturbation phase), the implementation of tree drifting in TNT is generally much more efficient than those programs (Goloboff 2015).

One of the drawbacks of tree drifting is that the user needs to determine a limit for the *AFD* (the limit for *RFD* is less problematic). In the case of datasets with very large numbers of characters (e.g. phylogenomic datasets), small differences in optimality produce no acceptable trees (all trees in a *TBR* neighborhood close to any local optimum may be many steps away). In that case, it is necessary to set AFD_{lim} to a large value (remembering to keep RFD_{lim} to a relatively low value; only conflicts between roughly equal amounts of evidence, meaning values of RFD_{lim} from 0.1 to 0.2 at the most, are relevant for determining local optima). Tree drifting, on the other hand, has the advantage over the ratchet that internal memory handling is somewhat easier (because character weights are not altered).

5.6.4.1 Performance

The performance of tree drifting is similar to that of the ratchet. For Liebherr and Zimmerman's carabid dataset, it finds near-optimal trees about 150 times faster than just *RAS* + *TBR* (Figure 5.2e).

5.6.5 Tree Fusing (*TF*)

Thinking of large trees as composed of several sections, each of which can have its own islands, allows approaching the problem from a different perspective. Consider the case of a tree with 10 sections, a probability of 0.5 of finding the optimal resolution for each individual section, and the final trees for five different *RAS* + *TBR*. There is probably no tree which optimally resolves all the sections simultaneously, but every one of the sections is very likely to be optimally resolved in at least one of the trees. In that case, if we put together pieces from different trees, taking the good parts from each, it is quite likely that we will be able to assemble a globally optimal tree (or at least one which is much closer than each of the individual trees). The method of *tree fusing* (*TF*, Goloboff 1999) performs exactly that kind of combination.

In computer science, *genetic* algorithms for optimization attempt to mimic evolution by natural selection, where "fitness" is given by the scores of solutions and "crossover" is given by mixing parts of candidate solutions (e.g. making solutions more likely to "breed" if they have better scores). Several programs in addition to TNT use different kinds of genetic algorithms (e.g. Moilanen 1999; Goëffon et al. 2006; Lin 2008; Ribeiro and Vianna 2009), but none of them is widely used; they generally have a strong random component for the exchange between tree parts. *TF* has some similarities with genetic algorithms, but it differs (Goloboff 2015) in using a much more deterministic process because it exchanges pieces in such a way that the resulting scores can be evaluated quickly so that only exchanges producing improvements are effected.

To perform a round of *TF* (Figure 5.3a), a recipient tree (*rt*) is chosen first (by default, the best tree in the set of trees to be fused; optionally, a tree randomly chosen from the set). Each of the remaining trees will then be evaluated as donor trees (*dt*) for parts. To fuse *rt* and *dt*, the groups shared by both trees are identified first; this is easily done with a strict consensus of both trees as a guide. In Figure 5.3a, the equivalent groups (i.e. those shared by *rt* and *dt*) are numbered. Using the guide tree and node numbering, a group in *rt* can be removed and replaced with the equivalent but differently resolved group from *dt* without affecting the taxon composition of the tree. Further, if the states for the down-pass character optimization are stored for each of the nodes of *rt* and *dt*, together with the lengths of each pending subtree, the length of the combined tree can be calculated quickly by just moving down the tree from the node corresponding to the exchange (as indicated in Figure 5.3a, with a gray path from the point of exchange to the root of the tree). In the implementation of TNT, the nodes to be exchanged are tried in a down pass. Once the exchange of a node is accepted, all the nodes ancestral to that one are eliminated from the list of exchanges to attempt (it is better to try the exchange of the smaller pieces first, only some of which may improve resolution, than trying to exchange the big subtree first, which may contain some poorly resolved parts). Further, equivalent nodes that form part of polytomies of degree p or less (default is $p = 3$) in the guide tree are skipped. For a node split in the same two subclades, it is preferable to exchange each subclade separately, and polytomies of degree < 3 are easily resolved by *TBR* (and thus unlikely to imply local optima).

With those restrictions, only few exchanges need to be tried, and each can be evaluated quickly. At the end, the recipient tree is submitted to a round of global *TBR*. This is needed because the exchanges of subtrees might make some further improvement possible; to increase the chances of this happening, it is also possible to optionally accept all the exchanges between subtrees that produce the same score in the recipient tree (instead of just better scores). Then, a new tree is chosen as recipient, and all the remaining trees are tried as donors. Eventually, after some rounds of exchanges (TNT uses three rounds as default), all the parts that can be combined have been attempted for exchange, and the process produces no further improvements.

The obvious advantage of *TF* is making it possible to reuse the good parts of trees found by separate searches, trees that would otherwise be discarded completely, thus wasting all the search effort invested in finding those trees that came close to optimality but did not quite reach it. With *TF*, just a little additional time invested in mixing the trees produces a much better result almost instantly. Another advantage of *TF* is that, to some extent, it can provide another criterion (in addition to mere convergence) for deciding whether optimality has been reached: when mixing numerous independent hits to minimum known length fails to produce better trees, this makes it more likely that such length is indeed the minimum possible for the dataset at hand.

5.6.5.1 Performance

Submitting the trees resulting from five *RAS* + *TBR* to *TF* requires only an additional 20% of time and, when compared to a single *RAS* + *TBR* without saving multiple trees, near-optimal scores are found (for Liebherr and Zimmerman's dataset) over 50 times faster (Figure 5.3f).

5.7 DIFFICULT DATASETS: COMBINED METHODS AND DRIVEN SEARCHES

Every one of the methods discussed in the previous section, by itself, produces much better results than just application of *RAS* + *TBR*. However, most methods have significant differences, and where a method is stuck at a local optimum, another may be able to go through. Thus, the best results are obtained when different replicates of *RAS* + *TBR* are each followed by sectorial searches and ratchet or drifting and the final trees from each replicate are submitted to *TF*. This is exemplified with Liebherr and Zimmerman's dataset (Figure 5.3g), where the combined methods produce near-optimal trees over 150 times faster than five *RAS* + *TBR* saving up to five trees per replication (recall that, as shown in Figure 5.1g, saving a few trees per replication produces near-optimal trees about three times faster than saving a single one).

When the best tree length is known and fusing the first set of trees fails to produce trees of such length, rather than discarding all the trees and starting a whole new set of replications with *RAS* + *TBR* + *SS* + *drift*, the trees can be kept, adding replications of *RAS* + *TBR* + *SS* + *drift* only a few at a time. Periodically submitting the enlarged set of trees to *TF* will eventually produce an optimal tree; this could happen with just a few more replications, taking advantage of the results from numerous previous replications, instead of just discarding them. This is called a *driven search*, where the search continues until a tree of the best known length is found. With a driven search, instead of setting TNT to use a specific number of replications, the program is set to do a specific number of *hits* to minimum length. A given hit may produce several trees, e.g. via fusing, which are not completely independent from each other. But different hits are fully independent because the entire process starts from scratch for every hit. The length is, by default, updated as the program runs. If the user has requested 10 hits to best length and the first five hits produce trees of (say) 100 steps and the sixth hit finds a tree of 99 steps, then the counter of hits is set again to 0 and TNT continues searching until it finds trees of 99 steps independently 10 times.

5.7.1 Searching for a Stable Consensus

As discussed in previous sections, numerous independent hits to optimal scores are much more useful, to produce accurate consensus trees, than the numerous trees found by subjecting trees from a single hit to global *TBR*. The problem remains of how many hits to minimum length should be used. But, just as the criterion of convergence offered a guide (in the context of standard searches using only *RAS* + *TBR*) on when to consider that minimum length had probably been found, the same criterion can be used to decide whether the consensus obtained is correct. Thus, a search can be driven, not toward finding optimal trees a specific number of times but instead toward finding the point at which additional hits do not further decrease the resolution of the strict consensus. That point can be determined by comparing the strict consensus of the trees resulting from increasing numbers of hits. In TNT, this checking is done periodically every time a significant number of hits is produced. Obviously, the consensus of the first 20 hits will probably be identical to the consensus produced by just one more hit. A substantial number of additional hits needs to be produced for the stability test to be meaningful. In TNT, the default is to check stability at the first 5 hits and then when the number of additional hits is at least 75% of the number of previous hits (but no more than 12 or less than 5).

Two settings help further speed up the process of calculating a stable consensus. First, the minimum length may be known in advance (e.g. by having performed aggressive searches); then, the driven search can give up a hit as soon as that length is found (e.g. without the need to do all initial replications or *TF*). Second, collapsing the trees more strictly generally makes the consensus stable with fewer hits (see Chapter 6 for discussion). The best results are obtained when the trees are collapsed by means of *TBR* (Goloboff 1999; Goloboff and Farris 2001). For this, *TBR* is performed on each tree, and every time a rearrangement produces the same score, the nodes intermediate between the two positions (taking into account rerooting) are recorded for collapsing. This amounts to producing the strict consensus of all the equally optimal trees that can be found by doing *TBR* on the tree, but without actually saving those trees to memory. For increased certainty about the correctness of the strict consensus calculated, the entire process can be repeated, calculating the grand consensus of the stabilized consensuses (and, of course, finding that every stabilization produced the same consensus increases the confidence that the settings are appropriate for the dataset at hand).

5.7.2 Strengths and Weaknesses of the Different Algorithms

Depending on the characteristics of the dataset being analyzed, some of the basic algorithms may have advantages over the others. Obviously, in a dataset with almost perfectly congruent characters, where every group is well supported, nothing will beat the plain *TBR* algorithm—or perhaps even Wagner trees—in terms of efficiency to find optimal trees. Empirical datasets are rarely that clean, and *SS*, drifting, ratchet, and *TF* are very useful there.

Both *SS* and *TF* assume well-defined sections or portions in the tree. This may not be true when a dataset has too much character conflict (e.g. random data, with equiprobability of 0, 1 in every matrix cell). In the case of *TF*, trees produced by independent starting points for such data are unlikely to have any groups in common, and thus *TF* cannot effect any exchanges between groups. In the case of *SS*, improving the tree may require simultaneously changing distant parts of the tree so that selecting smaller sectors never produces improvements. In larger, better structured datasets where distinct sectors can exist, *SS* is very advantageous. By doing a number of searches on small datasets, the time needed to complete *SS* on large trees increases more or less lineally with number of taxa, as opposed to global *TBR*, for which the time increases with T^2 to T^3 (in datasets with high or low congruence, respectively). As ratchet and drifting mostly rely on *TBR*, *SS* tends to outperform them in datasets with several hundreds of taxa.

Both ratchet and drifting, during the perturbation phase, can go far away from the initial tree if character conflict determines so. Relatively small datasets (i.e. 100 or fewer taxa) will be difficult to analyze mostly when they have poorly supported topologies, and in that case, the search should emphasize the ratchet or drift search (e.g. doing many more cycles of perturbation/search), using lighter options for *SS* and *TF*. In the case of drifting, it is important to remember that a pre-specified maximum score difference determines which trees can potentially replace the current tree during the perturbation phase. By default, this difference is one step, but in datasets with very large numbers of characters (e.g. molecular datasets), the perturbation phase may find no alternative trees within such small difference, making it necessary to set the value of acceptable suboptimality to a larger number. An alternative is to determine this automatically from the branch lengths (with the command `subopt`, see the following).

5.7.2.1 Alternatives to the Algorithms Described

In very poorly structured datasets, ratchet or drifting alone may be insufficient to reliably find near-optimal trees. Other search algorithms of TNT can be useful in those cases. Goloboff (2015) described methods for tree hybridization and tree rebuilding.

Tree *hybridization* takes two trees and identifies a partition in each tree which splits similar subsets of about half the taxa (with a maximum number of unshared taxa allowed). Then, pruning first the excess taxa in each group, it inserts one part of a tree into the other, then reinserting the pruned taxa and performing *TBR* swapping on the completed tree. It does this in both directions. The exchanges between trees are not deterministic, as in the case of *TF*, so that most of the changes do not improve the trees. However, some do (particularly when coupled with the subsequent reinsertion of pruned taxa at their optimal locations and global *TBR*).

Tree *rebuilding* prunes from the tree some randomly chosen pre-specified proportion of taxa, reinserts them at the best available locations, and subjects the new tree to *TBR*. By simultaneously pruning several taxa, this allows jumping between islands, using in every case a new starting point that is closer to optimality than a Wagner tree. This is a type of perturbated search where the perturbation can be more severe than in the case of the ratchet. Another type of perturbation algorithm (described by Goloboff 2015, inspired on Roshan et al.'s 2004 rec-I-DCM3; see discussion in Goloboff and Pol 2007) is the `pfijo` command (*piñón fijo*, Spanish for "fixed gear", a play on words for the ratchet). The *piñón fijo* analyzes reduced datasets, but instead of creating them as in *SS*, creates them by selecting terminals or unconnected hypothetical ancestors. Unlike the *HTU*s created for *SS*, this selection of terminals for the reduced dataset alters the tree landscape so that a quick *TBR* search for the reduced dataset will move away from the current tree. Subsequent *TBR* from this tree (now on the complete dataset) may then explore tree space moving across local optima, much like the ratchet. Goloboff (2015) and Goloboff et al. (2021a) found these three algorithms (hybridization, rebuilding, and *piñón fijo*) to be effective on very poorly structured datasets, such as random or near-random datasets, or datasets mixing characters modeled from different trees (i.e. different "signals").

5.7.3 Challenges Posed by Morphological Datasets

Morphological datasets typically include a couple hundred taxa at the most. Unlike molecular or phylogenomic datasets, morphological datasets also tend to include fewer characters, rarely above 1,000. With lower numbers of characters, much of the difficulty in finding most parsimonious trees is often not just one of conflict or islands but also one of a flat landscape: it is common for datasets to often allow (tens of) thousands of *MPT*s. This makes it even more important to calculate consensus trees by multiple independent hits to minimum length, instead of by saving all possible *MPT*s. In addition, sometimes a flat portion of the landscape (i.e. many trees of the same score) may contain a single isolated peak with one or a few trees. This is not an island in Maddison's (1991) sense because by swapping all possible suboptimal trees of equal score, the peak is eventually found. While possible in principle, the numbers of trees that may need to be saved to locate the single peak may be too many; theoretical examples of such flat landscapes are given in Goloboff (2014a). In such datasets, the homoplasy (i.e. character conflict) may be relatively low, so ratchet or drifting need to emphasize the phase where trees of equal score (i.e. without perturbation) are accepted. To do this, the number of changes to effect to the tree before returning to the search phase can be set to a larger number. The limit of absolute and relative score differences (in the case of drifting) or the proportion of characters whose weight is altered (in the case of the ratchet) must be set to lower values so that the perturbation phase (which will then accept more substitutions) does not end up in trees deviating too much from optimal. In the case of *SS*, accepting alternative resolutions of equal score for reduced subproblems also allows moving around the space of equally optimal trees. These modifications allow exploring flat landscapes with more efficiency than just saving huge numbers of equally parsimonious trees with *TBR*. See Goloboff (2014a) for additional discussion of these and related problems.

5.8 APPROXIMATE SEARCHES AND QUICK CONSENSUS ESTIMATION

Of course, there are datasets where even the most sophisticated algorithms known fail to consistently produce optimal scores. In those cases, there are methods of analysis that produce approximate answers. Farris et al. (1996), in a seminal paper, laid out the original idea for such an approximation. Their insight was noting that, to find well-supported groups, quick superficial searches should suffice—a group supported by many characters and contradicted by few or none is unlikely to be missed by any type of search algorithm. The problem with a tree found in a search not reaching minimum length is not the problem of lacking well-supported groups but instead that it will also display some groups that are not supported at all. What is needed, then, is a way to eliminate those unsupported groups. Farris et al. (1996) also noted that it is unlikely to have the same unsupported groups found repeatedly by independent searches, each proceeding from a randomized starting point. Farris et al. (1996) used that idea in connection with measures for group supports, but the idea can be extended (Goloboff and Farris 2001) into methods for quickly estimating the strict consensus trees (that is, for finding groups with any degree of support). This method is implemented in the TNT command `qnelsen`.

Thus, when a group is present in all (or a high proportion) of the trees produced by independent searches, it is likely that the group is actually supported by the data. Goloboff and Farris (2001) proposed using a few independent searches, finally displaying a *double consensus*, with only those groups present in both a high proportion of replicates (they used 85%, thus allowing a supported group to be missed in one or a few searches) and every one of the best trees (they used 25% of the best trees). To the extent that searches are more superficial, e.g. only Wagner trees, the trees produced will be less similar to each other and thus share fewer groups. The consensus estimations produced by using a Wagner tree are much less resolved than the actual consensus, but the few groups that they display are correct. If the searches are more exhaustive, e.g. *TBR*, then every search is less likely to miss a group and the estimations produced are better resolved, displaying more groups. In essence, this approach is based on obtaining some information on how the strict consensus of optimal trees could look like, but without ever actually finding any optimal trees. Goloboff and Farris (2001) observed that, in practice, more exhaustive searches increase the resolution of the double consensus—that is, the number of correct groups discovered. The quick estimation often displayed a low number of groups that were actually unsupported, and somewhat surprisingly, these were not random groups. Goloboff and Farris (2001) noted that (just like in measures of group supports; see Chapter 8), when a group is present in the majority of optimal or near-optimal trees but absent from one or more optimal trees, the group is unsupported. If a search from a given (randomized) starting point is equally likely to land in any of the optimal or near-optimal trees (that is, the search is *unbiased*; see Chapter 8 for further discussion and examples), then it is more likely to land on a tree having the frequent group. Those frequent but unsupported groups thus tend to be found repeatedly in the independent searches. Goloboff and Farris (2001) demonstrated that making the individual searches more exhaustive—i.e. to come closer to optimality—improved the resolution of the estimate but did *not* decrease the average number of spurious groups found. This is shown in Figure 5.4a, and while seemingly counterintuitive, it is expected: most spurious groups result from groups present in some but not all the optimal trees. The only way to decrease the number of spurious groups is by better taking into account ambiguity—that is, the existence of trees tied in optimality or

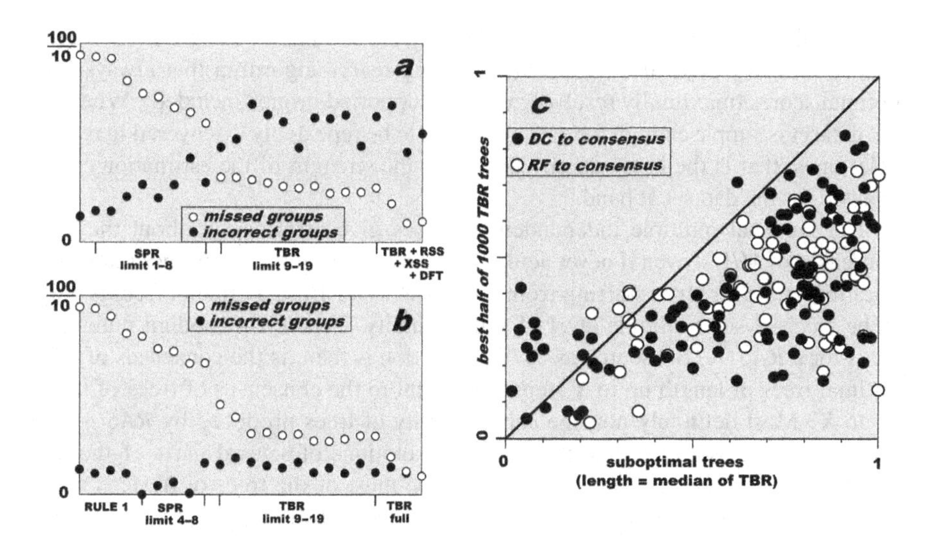

FIGURE 5.4 (**a**) The results of using a double-consensus without collapsing the trees (using 15 independent starting points, with the groups in common between the strict consensus of the best 25% and the 85% majority rule consensus of the final 15 trees) on the dataset of Pei et al. (2020). For this dataset, the correct strict consensus has 110 nodes. The search strength (always saving a single tree) increases along the x-axis. The number of missed groups (i.e. estimated consensus is poorly resolved) is shown on a 0–100 scale and the number of incorrect groups (extra groups in the estimated consensus) on a scale of 0–10. Increasing the search strength misses fewer correct groups but does not decrease the number of incorrect groups present in the estimated consensus. (**b**) A double consensus, with summary and search strategy similar to (a) but collapsing the trees with different methods (increasing along x-axis). As the trees are collapsed more strictly and search strength increases, the number of incorrect groups remains low [instead of increasing with search strength, as in (a)]. Tree collapsing reduces the number of incorrect groups without significantly reducing the number of correct groups. (**c**) Simulations (60 taxon datasets, generated with `xread+ 0 85 100 2`) showing that consensuses estimated with trees suboptimal by the median degree of suboptimality in a *RAS + TBR* are farther from the correct strict consensus (x-axis, as measured by the Robinson-Foulds distance and the symmetric distortion coefficient, see Chapter 6) than estimates based on the strict consensus of the best half of 10,00 replications with *RAS + TBR*, even if both estimates consider trees of similar degrees of suboptimality.

near optimality—which a single tree cannot do. An easy way to take into account ambiguity is using *SPR* or *TBR* collapsing, and Figure 5.4b shows that this keeps the number of spurious groups at a much lower level for all levels of search strength. The problem of a group erroneously concluded as supported can also affect any type of search algorithm, even if unbiased, if the consensus is constructed from too few independent hits or too few trees.

To the extent that the individual searches come closer to actual optimality, the more resolved the consensus is—it will be missing fewer groups that are actually supported (as obvious in Figs. 5.4a–b). Thus, by varying the strength of the searches, it is possible to have a continuum between a quick estimation using only a Wagner tree (with few groups discovered) and a strong search algorithm that always finds optimal scores (maximally resolved, with all supported groups included). Whenever the dataset is simple enough for optimal scores to be repeatedly discovered in reasonable times, that is the best option. Otherwise, the strength of the estimation can be adapted for the dataset at hand.

The fact that multiple independent searches give information about the strict consensus of *MPT*s, even if never actually finding an *MPT*, raises another interesting question. A *TBR* search starting from a given *RAS* (or from a random tree) is going to be stuck at some middle level of suboptimality. Denote the median number of steps for a *RAS* + *TBR* search as ***X***. The question is then: is the consensus of *TBR*-optimal trees of length up to ***X*** steps equivalent to the consensus of trees of length up to ***X***? Most definitely not; the suboptimality of trees produced by *RAS* + *TBR* may come from simultaneously incorrect resolutions of several parts of the tree, summing up to a larger value. Put differently, most of the trees of ***X*** steps would not be *TBR* optimal; a tree of ***X*** steps can be a reliable estimator of the consensus if *TBR* optimal but far from it otherwise. A tree ***A*** can be one step off from optimality because the characters are tangled so that relatively simple rearrangements cannot move to a better tree. Another tree, ***B***, can be one step off because a group supported by one synapomorphy and contradicted by none is broken into two groups in the tree. Both trees are one step away from optimality, yet the second (lacking a group present in the consensus) is a case where the *TBR* algorithm would never get stuck. An implication of this is that interrupting *TBR* searches (on the basis of either time or number of rearrangements, as optional in some programs, TNT included) risks leaving the solution in a situation like that of tree ***B***, missing a group that would otherwise be present. Although *TBR* optimality (i.e. having completed *TBR* without finding a better tree) cannot guarantee global optimality, it does have a special meaning, and the trees produced by *TBR*, even if suboptimal at ***X*** steps, provide information about the *MPT*s that other trees of ***X*** steps cannot. This is illustrated in Figure 5.4c.

5.9 IMPLEMENTATION IN TNT

A general summary of the main commands and options is in Figure 5.5 and in Table 5.1. In TNT, searches can be done either by creating trees de novo or by improving preexisting trees. The trees to improve may be trees found by a previous search, created at random, read from a tree file, or created by the user. For saving multiple trees with branch-swapping, the maximum size of the tree buffer must be set prior to starting the search. All the search commands can take additional arguments for specific settings. Running the command without arguments (i.e. with a semicolon) effects the search with current settings. The command followed by = changes the settings (with options given after the = symbol) and runs; the command followed by a colon (:) plus options changes settings but does not run. A colon without further arguments (i.e. followed by a semicolon) reports current settings.

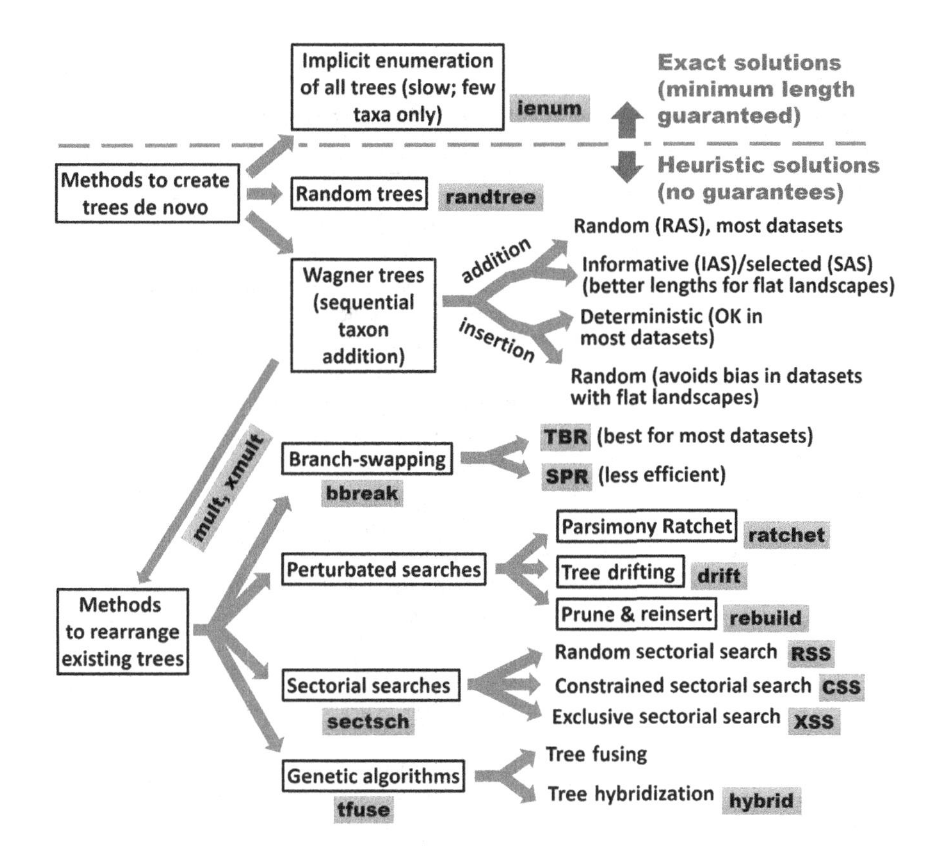

FIGURE 5.5 Summary of main search methods described in this chapter. The corresponding TNT commands (or options) are shown in gray boxes.

5.9.1 General Settings

In addition to search commands themselves, several other commands change options relating to searches. The command that sets the maximum number of trees to retain in memory during a search is `hold`; the maximum number of trees can also be set (in Windows *GUI* versions) from *Settings > Memory > Max.trees*. When searching, the trees in memory are modified; if you need to keep a copy of the trees prior to a search, you can create a "vault" where the trees can be placed and retrieved subsequent to the search. The tree vault is set by specifying the maximum number `N` of trees that will be placed in the vault in the `hold` command after the size `S` of the visible tree buffer: `hold S/N`. Once the vault is set, trees are placed in the vault with `tvault > L` (where `L` is the list of trees to place in the vault) and retrieved with `tvault < L` (where `L` is the list of trees to get using the vault's own numbering). Group membership of the trees (see Chapter 1, Section 1.11.4, "Lists and Ranges") is also saved to (and retrieved from) the vault.

TABLE 5.1
List of Search Commands and Settings for Searches in TNT

Command	Minimum Truncation	Action(s)
alltrees	al	Generate all possible trees (or tree shapes) for currently active taxa
bbreak	bb	Perform branch-swapping on trees now in memory
best	bes	Calculate scores for all trees in memory, retain best ones, and eliminate duplicate trees
collapse	col	Set collapsing level; set whether trees are retained as binary or collapsed after searches
constrain	cons	If constraints defined, enforce them during searches
drift	dr	Perform tree drifting starting from (first) best tree in memory
force	fo	Define constraints
hold	ho	Set size of tree buffer and tree vault
hybrid	hy	Hybridize trees
ienum	i	Perform an exact calculation of *MPT*s (implicit enumeration)
mult	mu	Perform multiple *RAS* + *TBR*
pfijo	pf	Perform *piñón fijo* (an algorithm for perturbated searches)
qcollapse	q	Set a shortcut for quick pre-collapsing of trees on (=) and off (–)
qnelsen	qn	Estimate quick consensus
randtrees	ra	Generate random trees (all trees equiprobable)
ratchet	rat	Perform a ratchet search from trees in memory
rebuild	re	Perform rebuilding, for trees in memory
report	rep	Change options for reporting status
riddup	ri	Deactivate taxa with duplicate character states to speed up searches; they can be subsequently reactivated and inserted next to their similar taxa
rseed	rs	Change seed for randomizations (0 = time)
sectsch	se	Perform a sectorial search on each of the existing trees in memory
subopt	su	Change value of absolute and relative suboptimal
tchoose	tc	Select some of the tree(s) in memory
tfuse	tf	Perform tree fusing on the trees in memory
timeout	ti	Change timeout
tvault	tv	Place (or retrieve) trees in (or from) the vault
xmult	xm	Do a search, combining multiple *RAS*, branch-swapping, sectorial searches, ratchet, drifting, and tree fusing

The default in TNT is to produce periodic status reports to screen, with only a final report saved to the text buffer (or log file) at the end. The real-time report is changed with `report` (`report-` disconnects the reporting, `report=` connects it back). To save periodic reports, use the `report+` option (accessed with *Settings > Report Levels > Report Progress*); `report+s/r/t` will save a report every `s` seconds, `r` replications completed (or their equivalent), or `t` trees swapped or processed. This can be used to produce a more detailed monitoring of the progress of a search and the scores and numbers of trees found by each replication. Setting to 0 (or skipping) any of `s`, `r`, or `t` turns off the recording for that option. The search commands can take a long time to run, and it is possible to set a `timeout` (in the `hh:mm:ss` format) so that each search command runs for no more than the specified time.

Any command that does some randomizations will use the current random seed, changed with `rseed`. Setting `rseed 0` will use the computer clock time every time a randomizing process is called so that repeating searches may produce different outcomes (as long as the repetition is not done within a second). `Rseed *` will change the random seed (generating a new random seed at random); you can use this to repeat short routines, making sure that a different randomization is used every time. Most search commands only retain trees that are different if zero-length branches (i.e. branches without synapomorphies; see Chapter 6) are ignored. The criterion to use for elimination of zero-length branches (including *TBR* collapsing) is set with the `collapse` command or with *Settings > Collapsing Rules*.

5.9.2 Exact Searches

Results guaranteed as optimal are obtained with the `ienum` command (invoked with *Analyze > Implicit Enumeration*). `Ienum` saves no more than the number of trees set with `hold`; the search proceeds faster when the memory buffer is full (because search paths are abandoned when best score is matched, not when it is worsened; `hold 1 ienum` proceeds the fastest). If constraints or `subopt` (see Section 5.9.3) are in effect, `ienum` will honor those. When allowing very large numbers of trees to be saved, if zero-length branches are being collapsed (see Chapter 6, Section 6.4.1), TNT will need to compare every tree found with previously found trees—two different binary trees may collapse to the same polytomized tree. When zero-length branches are not collapsed, `ienum` does not need to compare new with preexisting trees because the process generates every binary tree exactly once. This can save time in exact searches for some datasets where there are many unsupported branches.

5.9.3 Heuristic Searches—Basic Algorithms

The command `mult`, or *Analyze > Traditional Search* (then selecting *Wagner trees* as starting trees), does multiple *RAS*, with or without branch-swapping. The default swapper is *TBR*; the type of swapping is set with `mult=tbr` or `mult=spr`; use `mult=wagner` to skip the swapping (in Windows, select *none* as swapping algorithm). The maximum number `N` of trees to save per replication (if doing branch-swapping) is set with `mult=hold N` (in Windows, *trees to save per replication*). By default, TNT will retain only the trees from replications that succeeded in finding the best trees; to retain the best trees from each individual replication, use `mult=keepall` (or *keep all trees found*, in Windows); `mult=nokeepall` sets back the default. The number `R` of replicates (starting points) is controlled with `mult=replic R`. At the end of the search, TNT reports the number of hits to minimum length (to assess convergence) and whether any of the replications that succeeded in finding the best scores filled the partial memory buffer (to determine whether additional branch-swapping is likely to produce additional *MPTs*). When invoked, the `mult` command will check the trees in memory, retaining only the optimal trees that are distinct under the current criterion for elimination of zero-length branches. Note that (for reasons given in Chapter 6) the maximum severity of zero-length branch elimination during a search is collapsing on minimum length of zero ("rule 1"); if using a stricter criterion (e.g. *TBR* collapsing), rule 1 is enforced for filtering previous trees and during the search. Saving suboptimal trees with multiple *RAS* + *TBR* is generally a bad idea; unless very large numbers of trees per replication are being saved, this will make it more difficult to find optimal trees. If one needs suboptimal trees (e.g. for calculating Bremer supports; see Chapter 8), the best way to get them is generally by doing multiple *RAS* + *TBR* retaining all the best trees found in each replication and only then applying global *TBR* to save suboptimal trees.

If the `mult` command fails to produce trees as parsimonious as the preexisting ones, it will retain the best trees found, together with the preexisting trees (so that the optimal trees are not lost). If you want to make sure that the new trees found are wholly independent of preexisting trees, you may issue a `keep 0` command before the `mult` command (or tick the *replace existing trees* option, in *GUI* versions).

To perform branch-swapping on preexisting trees (e.g. those created by `mult`), use the `bbreak` command. For example:

```
tnt*>mult; hold 1000; bbreak;
```

will save up to 1,000 *MPTs* by doing *TBR* on the best trees resulting from `mult`. In Windows, the equivalent of `bbreak` is under *Analyze > Traditional Search*, selecting *trees from RAM*. By default, `bbreak` will save as many equally parsimonious trees (or trees within the difference specified with the `subopt` command) as set with `hold`; use `bbreak=nomulpars` to swap the tree without saving multiple trees. The only types of swapping implemented in TNT are *SPR* (`bbreak=spr`) and *TBR* (`bbreak=tbr`, the default). By default, every clipped clade will be reinserted at every possible remaining location (with every possible rerooting, in *TBR*); use `bbreak=limit N` to reinsert (or reroot) no more than `N` nodes away from the original. Doing this examines fewer rearrangements but also decreases the efficiency, so it is not generally recommended. As in any command that searches starting from trees in memory, if constraints are defined (with `force`) and enforced (with `constrain=`), the starting trees must obey constraints (if none does, swapping has no trees to start from).

5.9.4 Heuristic Searches—Special Algorithms

The command `sectsch` does sectorial searches on preexisting trees. Options are `rss`, `css` (this one requires definition of constraints and trees obeying them), and `xss`. Small sectors are analyzed by means of three *RAS* + *TBR*, followed by an additional three *RAS* + *TBR* if some of the initial replications produce trees of different lengths (number of initial *RAS* + *TBR* is changed with `sectsch:starts`). Sectors above a given size `D` (by default, 75, changed with sectsch:godrift D) are analyzed with tree drifting (by default, six cycles per reduced sector, changed with sectsch:drift). The option sectsch:gocomb S sets sectors of size S or more to be analyzed with combined strategies: N *RAS* + *TBR* (by default, four), tree drifting, and R rounds of tree fusing at the end (changed with sectsch: combstarts N fuse R); default S is 0 (=never use combined searches). If S set to a number different from 0, then any sector of S or more is analyzed by combined routines (if S is smaller than D, the size to switch to drifting, then drifting is never done; all sectors of size above S are analyzed with combined methods).

Recursive sectorial searches are also possible. If the options given to sectsch after the = symbol include some search commands within square brackets, then those commands are used to analyze each of the reduced sectors. The commands themselves (if complex, e.g. including some scripting commands to decide how to analyze the sector on the basis of its size or other characteristics) can be included in a file, called for each sector (e.g. [proc filename;]). Being able to run recursive sectorial searches requires that a sectsch:recurse L command (where L is the maximum level of recursion permitted during a sectorial search) is issued before reading the dataset (recursive sectorial searches require allocation of special memory buffers when reading the dataset). Other options (see online help of TNT) control the size of sectors, the number of substitutions to perform global *TBR*, etc. In Windows versions, those options are also controlled from *Analyze > New Technology Search*, selecting *settings* for *Sect. Search*.

The commands drift and ratchet perform their homonymous methods from preexisting trees. Since they operate similarly, some of their options are similar. The option iter N sets the number of iterations (default are 30 iterations for drifting and 50 for ratchet); numsubs N changes the number of substitutions to give up the perturbation phase, and giveup P sets the proportion of clippings (out of the $2T - 3$ possible clippings) to abandon the perturbation phase. For drifting, the acceptable absolute (A) and relative (R) score differences are set with drift: fitdiff A rfitdiff R. For the ratchet, ratchet:upfactor and downfactor set the probability of up—or down—weighting characters; tradrat enables the original ratchet described by Nixon (i.e. without accepting equally optimal trees during the perturbation phase; the default is set back with notradrat). In Windows versions, these options are controlled with the *settings* buttons for *Ratchet* and *Drift* in the dialog for *Analyze > New Technology Search*.

Tree fusing is effected with tfuse (only on the trees in the subsequent list or on all trees in memory if no tree list is provided). Some of the options are accepting rearrangements of equal score (tfuse=equal), the number of rounds to effect (tfuse=rounds N), and the minimum polytomy in the guide tree to attempt exchanges (tfuse=minfork N). The command tfuse also effects tree hybridization, useful only for very poorly structured datasets, with tfuse=hybrid (use nohybrid to set back to the default). For example, doing 10 random *RAS* + *TBR* and keeping the best tree from each replication, the resulting trees can be submitted to tree fusing:

```
tnt*>mult=repl 10 ho1 keep; tfuse;
```

For tree hybridization, it is preferable to use larger numbers of trees as input, as the process is less deterministic:

```
tnt*>hold 1000; mult=repl 100 hol keep; tfuse=hyb;
```

Finally, all the search algorithms described so far can be combined with the `xmult` command. With no arguments, it will just do a number of replications (five *RAS* by default), followed by *TBR*, *CSS* (with autoconstraints, i.e. a constraint created from the final tree of previous replication and the *RAS* of current; first replication is not autoconstrained and cannot use *CSS*), and *TF* at the end. All the search algorithms can be invoked with

```
tnt*>xmult=ratchet N drift N xss autoconst css rss fuse;
```

The settings (other than cycles of ratchet and drift, specified within the `xmult` command itself) for each algorithm are handled by the respective command (recall that the command followed by a colon changes settings but does not run). The option shown prior will only run, fuse the trees at the end, and stop, regardless of the scores found. To make `xmult` continue searching until a specific score `S` is found, use `xmult=target S`; to find score `S` with `N` independent hits, use `xmult=hits N`. To make a replication abandon (moving onto the next hit) as soon as score `X` is found, use `xmult=giveupscore X`. By default, the `target` and `giveupscore` will be updated if a shorter tree is found. To `giveup` at the score originally specified, without updating it, use `xmult=noupdate`. These options, in Windows versions, are changed from the main dialog for *Analyze > New Technology Search* (with *find min. length, use score bound of,* and *give up at score*).

Instead of searching until best length is found a specified number of times, the `xmult` command can search until the consensus becomes stable a number `N` of times, with `xmult=consense N`. The minimum number of hits `M` to check for consensus stabilization is set with `xmult=conbase M`, and the number `N` of additional hits over the hits done is `xmult=confactor N` (where `N` is a percentage). Remember that the consensus stabilizes with fewer hits when trees are being collapsed more strictly so that it is advisable to issue a `collapse tbr` command before running the `xmult` command in this manner. In Windows version, the consensus stabilization is invoked with *Analyze > New Technology Search*, selecting the *Stabilize consensus* option.

Although the default options of the `mult`, `bbreak`, and `xmult` commands suffice for the analysis of many datasets, the routine use of defaults for these commands is strongly discouraged. To make a thorough analysis, a much better approach is to make preliminary explorations of the behavior of the dataset at hand with different search algorithms, choosing the best settings for your final analysis.

NOTES

1 Rerooting a subtree does not change ancestral state assignments if the character has symmetric transformation costs.

2 This is not always the case, because *TBR* in large datasets can use less time per rearrangement due to the use of special shortcuts discussed prior. For datasets with many taxa and relatively good character congruence, with the implementation in TNT, completing *TBR* takes less time than *SPR*.

3 TNT uses a method known as *hash tables* to speed up comparison with preexisting trees by separating trees in groups based on *keys*. The keys used in TNT are *imperfect* in that they partition tree lists into 128 groups (i.e. two different trees can have the same key). Cutting down the number of trees to compare with greatly reduces computations, but in the case of many thousands of trees, the time needed for tree comparisons is still significant. Another time-saving shortcut used by TNT during searches is based on rejecting branch moves that produce trees of equal score but only cross branches unsupported under the collapsing criterion in effect (`qcollapse` shortcut; see Chapter 6, and Goloboff 1996); this saves time in avoiding calculation of zero-length branches for the rearranged tree.

4 This statement needs a qualification: a global round of *TBR* will produce no further trees *if the same binary trees found during the search are retained* (using, when zero-length branches are eliminated for tree-comparisons, the dual binary-collapsed representation of trees discussed prior). TNT can be set (with the collapse auto option) to actually remove the zero-length branches, thus losing track of the original resolution. In that case, TNT needs to re-dichotomize the trees for swapping and it is *possible* (if uncommon) that swapping again with *TBR* from some trees earlier found by *TBR* produces a different result (particularly if using different random seeds). The only behavior possible in PAUP* is the latter one (as PAUP* does not keep track of original binary resolutions), and thus it always has this behavior (in addition, PAUP* by default changes the random seed for every search, increasing the chances of this happening).

References

Agnarsson, I., and Miller, J. 2008. Is ACCTRAN better than DELTRAN? *Cladistics* 24, 1032–1038.

Akaike, H. 1973. Information theory as an extension of the maximum likelihood principle. In Petrov, B.N., and Csaki, F. (eds.), *Second International Symposium on Information Theory*. Akademiai Kiado. New York: Academic Press, pp. 267–281.

Anderson, C., Liang, L., Pearl, D., and Edwards, S. 2012. Tangled trees: The challenge of inferring species trees from coalescent and non-coalescent genes. In Anisimova, M. (ed.), *Evolutionary Genomics: Statistical and Computational Methods*. Vol. 2. New York: Humana Press, pp. 3–28.

Archie, J. 1985. Methods for coding variable morphological features for numerical taxonomic analysis. *Syst. Zool.* 34, 326–345.

Arnold, M. 2016. *Divergence with Genetic Exchange*. Oxford: Oxford University Press, 251 pp.

Bandelt, H. 1990. Recognition of tree metrics. *SIAM J. Discrete Math.* 3, 1–6.

Barker, D. 2004. LVB: Parsimony and simulated annealing in the search for phylogenetic trees. *Bioinformatics* 20, 274–275.

Baum, D., and Smith, S. 2013. *Tree Thinking: An Introduction to Phylogenetic Biology*. Greenwood Village, CO: Roberts and Co. Publishers.

Beatty, J. 1982. Classes and cladists. *Syst. Zool.* 31, 25–34.

Berlocher, S., and Swofford, D. 1997. Searching for phylogenetic trees under the frequency parsimony criterion: An approximation using generalized parsimony. *Syst. Biol.* 46, 211–215.

Bertelli, S. 2017. Advances on Tinamou phylogeny: An assembled cladistic study of the volant palaeognathous birds. *Cladistics* 33, 351–374.

Boden, R., Cleland, D., Green, P., Katayama, Y., Uchino, Y., Murrell, J., and Kelly, D. 2012. Phylogenetic assessment of culture collection strains of *Thiobacillus thioparus*, and definitive 16S rRNA gene sequences for *T. thioparus*, *T. denitrificans*, and *Halothiobacillus neapolitanus*. *Arch Microbiol.* 194, 187–195.

Bond, J., and Hedin, M. 2006. A total evidence assessment of the phylogeny of North American euctenizine trapdoor spiders (Araneae, Mygalomorphae, Cyrtaucheniidae) using Bayesian inference. *Mol Phylogenet Evol.* 41, 70–85.

Bond, J., Hendrixson, B., Hamilton, C., and Hedin, M. 2012. A reconsideration of the classification of the spider infraorder Mygalomorphae (Arachnida: Araneae) based on three nuclear genes and morphology. *Plos One* 7(6), e38753. https://doi.org/10.1371/journal.pone.0038753.

Bookstein, F. 1994. Can biometrical shape be a homologous character? In Hall, B.K. (ed.), *Homology: The Hierarchical Basis of Comparative Biology*. San Diego: Academic Press, pp. 197–227.

Borkent, A. 1978. Upper Oligocene fossil pupae and larvae of *Chaoborus tertiarius* (von Heyden) (Chaoboridae, Diptera) from West Germany. *Quaest. Entomol.* 14, 491–496.

Borkent, A. 2018. The state of phylogenetic analysis: Narrow visions and simple answers—examples from the Diptera (flies). *Zootaxa* 4374, 107–143.

Boussau, B., and Gouy, M. 2006. Efficient likelihood computations with nonreversible models of evolution. *Syst. Biol.* 55, 756–768.

Boyden, A. 1935. Genetics and homology. *Quart. Rev. Biol.* 10, 448–451.

Boyden, A. 1947. Homology and analogy. A critical review of the meanings and implications of these concepts in biology. *Amer. Midl. Nat.* 37, 648–669.

Brady, R. 1982. Theoretical issues and "pattern cladistics". *Syst. Zool.* 31, 286–291.
Brady, R. 1983. Parsimony, hierarchy, and biological implications. In Platnick, N., and Funk, V. (eds.), *Advances in Cladistics Vol. 2. Proceedings of the Second Meeting of the Willi Hennig Society.* New York: Columbia University Press, pp. 49–60.
Brady, R. 1985. On the independence of systematics. *Cladistics* 1, 113–126.
Brazeau, M., Guillerme, T., and Smith, M. 2019. An algorithm for morphological phylogenetic analysis with inapplicable data. *Syst. Biol.* 68, 619–631. Vignettes at https://tguillerme.github.io/inapp.html.
Bremer, K., and Wanntorp, H. 1979. Hierarchy and reticulation in systematics. *Syst. Zool.* 28, 624–627.
Broda, S., and Raymond, K. 2014. On distributions of ratios. *Tinbergen Institute Discussion Paper* 13–211/III, available at http://dx.doi.org/10.2139/ssrn.2375638.
Brower, A. 2000. Evolution is not a necessary assumption of cladistics. *Cladistics* 16, 143–154.
Brower, A. 2015. Transformational and taxic homology revisited. *Cladistics* 31, 197–201.
Brower, A. 2018. Fifty shades of cladism. *Biol Philos* 33, 1–11.
Brower, A. 2019. Background knowledge: The assumptions of pattern cladistics. *Cladistics* 35, 717–731.
Brower, A. 2020. Oh, what a tangled web we weave. *Cladistics* 36, 341–344.
Brower, A., and de Pinna, M. 2012. Homology and errors. *Cladistics* 28, 529–538.
Brower, A., DeSalle, R., and Vogler, A. 1996. Gene trees, species trees, and systematics: A Cladistic Perspective. *Annu. Rev. Ecol. Syst.* 27, 423–450.
Brundin, L. 1966. Transantarctic relationships and their significance, as evidenced by Chironomid midges. *Kungl. Svenska Vetenskap* 11, 1–472.
Bryant, D., Galtier, N., and Poursat, A.-M. 2005. Likelihood calculation in molecular phylogenetics. In Gascuel, O. (ed.), *Mathematics of Evolution and Phylogeny.* Oxford: Oxford University Press, pp. 33–62.
Bull, J., Badgett, M., Wichman, H., Huelsenbeck, J., Hillis, D., Gulati, A., Ho, C., and Molineux, I. 1997. Exceptional convergent evolution in a virus. *Genetics* 147, 1497–1507.
Burnham, K., and Anderson, D. 2002. *Model Selection and Multimodel Inference—a Practical Information-Theoretic Approach.* New York: Springer Verlag, 488 pp.
Camin, J., and Sokal, R. 1965. A method for deducing branching sequences in phylogeny. *Evolution* 19, 311–326.
Camp, C. 1923. Classification of the lizards. *Bull. Am. Mus. Nat. Hist.* 48, 289–481.
Carpenter, J., Goloboff, P., and Farris, J. 1998. PTP is meaningless, T-PTP is contradictory: A Reply to Trueman. *Cladistics* 14, 105–116.
Catalano, S.A., Goloboff, P., and Giannini, N. 2010. Phylogenetic morphometrics (I): The use of landmark data in a phylogenetic framework. *Cladistics* 26, 539–549.
Catalano, S.A., Segura, V., and Candioti, F. 2019. PASOS: A method for the phylogenetic analysis of shape ontogenies. *Cladistics* 35, 671–687.
Chang, J. 1996. Full reconstruction of Markov models on evolutionary trees: Identifiability and consistency. *Math. Biosci.* 137, 51–73.
Chen, M.-H., Kuo, L., and Lewis, P. 2014. *Bayesian Phylogenetics: Methods, Algorithms, and Applications.* Oxfordshire: CRC Press, Taylor & Francis Group, 358 pp.
Clemente, J, Ikeo, K., Valiente, G., and Gojobori, T. 2009. Optimized ancestral state reconstruction using Sankoff parsimony. *BMC Bioinformatics* 10, 51. 10.1186/1471-2105-10-51.
Cunningham, C., Zhu, H., and Hillis, D. 1998. Best-fit maximum likelihood models for phylogenetic inference: Empirical tests with known phylogenies. *Evolution* 52, 978–987.
Darwin, C. 1859. *On the Origin of Species by Means of Natural Selection, or the Preservation of Favoured Races in the Struggle for Life.* London: John Murray.

Davis, C. 1938. Studies in Australian Embioptera. Part III: Revision of the genus Metoligotoma, with descriptions of new species, and other notes on the family Oligotomidae. *Proc. Linn. Soc. N.S.W.* 63, 226–272.
Davis, C. 1940. Taxonomic notes of the Order Embioptera. Part XX: The distribution and comparative morphology of the Order Embioptera. *Proc. Linn. Soc. N.S.W.* 65, 533–542.
Davis, J., Nixon, K., and Little, D. 2005. The limits of conventional cladistic analysis. In Albert, V. (ed.), *Parsimony, Phylogeny, and Genomics*. Oxford: Oxford University Press, pp. 119–147.
de Beer, G. 1971. *Homology, an Unsolved Problem*. Oxford: Oxford University Press.
De Laet, J. 2005. Parsimony and the problem of inapplicables in sequence data. In Albert, V. (ed.), *Parsimony, Phylogeny, and Genomics*. Oxford: Oxford University Press, pp. 81–116.
De Laet, J. 2015. Parsimony analysis of unaligned sequence data: Maximization of homology and minimization of homoplasy, not minimization of operationally defined total cost or minimization of equally weighted transformations. *Cladistics* 31, 550–567.
De Laet, J. 2018. *Anagallis: A Program for Parsimony Analysis of Character Hierarchies*. Computer Program and Documentation, available at www.anagallis.be/anagallis/index.html.
De Laet, J. 2019. A note on Brazeau et al.'s (2017) algorithm for characters with inapplicable data, illustrated with an analysis of their Fig. 3d using anagallis, a program for parsimony analysis of character hierarchies. https://doi.org/10.31219/osf.io/hp8vs.
De Laet, J., and Smets, E. 1998. On the three-taxon approach to parsimony analysis. *Cladistics* 14, 363–381.
de Pinna, M. 1991. Concepts and tests of homology in the cladistic paradigm. *Cladistics* 7, 367–394.
de Queiroz, K. 1985. The ontogenetic method for determining character polarity and its relevance to phylogenetic systematics. *Syst. Zool.* 34, 280–299.
de Queiroz, K., and Donoghue, M. 1988. Phylogenetic systematics and the species problem. *Cladistics* 4, 317–338.
de Queiroz, K., and Poe, S. 2001. Philosophy and phylogenetic inference: A comparison of likelihood and parsimony methods in the context of Karl Popper's writings on corroboration. *Syst. Biol.* 50, 305–321.
Decae, A. 1984. A theory on the origin of spiders and the primitive function of spider silk. *J. Arachnol.* 12, 21–28.
Doolittle, W., and Brunet, T. 2016. What is the tree of life? *PLoS Genet.* 12, e1005912.
Doyle, J. 1992. Gene trees and species trees: Molecular systematics as one-character taxonomy. *Syst. Bot.* 17, 144–163.
Drummond, A., and Rambaut, A. 2007. BEAST: Bayesian evolutionary analysis by sampling trees. *BMC Evol. Biol.* 7, 214. https://doi.org/10.1186/1471-2148-7-214.
Drummond, A., Suchard, M., Xie, D., and Rambaut, A. 2012. Bayesian phylogenetics with BEAUti and the BEAST 1.7. *Mol. Biol. Evol.* 29, 1969–1973.
Duchêne, S., Molak, M., and Ho, S. 2014. ClockstaR: Choosing the number of relaxed-clock models in molecular phylogenetic analysis. *Bioinformatics* 7, 1017–1019.
Ebach, M., Carvalho, M., and Williams, D. 2011. Opening Pandora's molecular box. In Carvalho, M.R. de and Craig, M.T. (eds.), Morphological and molecular approaches to the phylogeny of fishes: Integration or conflict? *Zootaxa* 2946, 1–142.
Ebach, M., Williams, D., and Vanderlaan, T. 2013. Implementation as theory, hierarchy as transformation, homology as synapomorphy. *Zootaxa* 3641, 587–594.
Edwards, A. 1992. *Likelihood*, 2nd ed. Baltimore: John Hopkins University Press.
Edwards, S.V. 2009. Is a new and general theory of molecular systematics emerging? *Evolution* 63, 1–19.

Eldredge, N., and Cracraft, J. 1980. *Phylogenetic Patterns and the Evolutionary Process.* New York: Columbia University Press.
Emanuel, K. 2020. The relevance of theory for contemporary research in atmospheres, oceans, and climate. *AGU Adv.* 1, e2019AV000129. https://doi.org/10.1029/2019AV000129.
Engelmann, G., and Wiley, E. 1977. The place of ancestor-descendant relationships in phylogeny reconstruction. *Syst. Zool.* 26, 1–11.
Fan, Y., Wu, R., Chen, M.-H., Kuo, L., and Lewis, P. 2011. Choosing among partition models in Bayesian phylogenetics. *Mol. Biol. Evol.* 28, 523–532.
Farris, J. 1966. Estimation of conservatism of characters by constancy within biological populations. *Evolution* 20, 587–591.
Farris, J. 1967. The meaning of relationship and taxonomic procedure. *Syst. Zool.* 16, 44–51.
Farris, J. 1970. Methods for computing Wagner trees. *Syst. Zool.* 19, 83–92.
Farris, J. 1972. Estimating phylogenetic trees from distance matrices. *Am. Natur.* 106, 645–668.
Farris, J. 1977. On the phenetic approach to vertebrate classification. In Hecht, M.K., Goody, P.C., and Hecht, B.M. (eds.), *Major Patterns in Vertebrate Evolution.* NATO Advanced Study Institute Series, no. 14. New York: Plenum Press.
Farris, J. 1979a. On the naturalness of phylogenetic classification. *Syst. Zoo*l. 28, 200–214.
Farris, J. 1979b. The information content of the phylogenetic system. *Syst. Zool.* 28, 483–519.
Farris, J. 1980a. Naturalness, information, invariance, and the consequences of phenetic criteria. *Syst. Zool.* 29, 360–381.
Farris, J. 1980b. The efficient diagnoses of the phylogenetic system. *Syst. Zool.* 29, 386–401.
Farris, J. 1981. Distance data in phylogenetic analysis. In Funk, V.A., and Brooks, D.R. (eds.), *Advances in Cladistics: Proceedings of the First Meeting of the Willi Hennig Society.* Bronx: New York Botanical Garden, pp. 3–23.
Farris, J. 1982a. Outgroups and parsimony. *Syst. Zool.* 31, 328–334.
Farris, J. 1982b. Simplicity and informativeness in systematics and phylogeny. *Syst. Zool.* 31, 413–444.
Farris, J. 1983. The logical basis of phylogenetic analysis. In Platnick, N.I., and Funk, V.A. (eds.), *Advances in Cladistics.* New York: Columbia University Press, pp. 7–36.
Farris, J. 1985. Distance data revisited. *Cladistics* 1, 67–85.
Farris, J. 1986a. On the boundaries of phylogenetic systematics. *Cladistics* 2, 14–27.
Farris, J. 1986b. Distances and statistics. *Cladistics* 2, 144–157.
Farris, J. 1988. *Hennig86.* Program and documentation, distributed by the author. New York.
Farris, J. 1990. Phenetics in camouflage. *Cladistics* 6, 91–100.
Farris, J. 2008. Parsimony and explanatory power. *Cladistics* 24, 825–847.
Farris, J. 2012. 3ta sleeps with the fishes. *Cladistics* 28, 422–436.
Farris, J. 2014a. "Pattern cladistics" really means paraphyly. *Cladistics* 30, 236–239.
Farris, J. 2014b. Homology and misdirection. *Cladistics* 30, 555–561.
Farris, J. 2014c. "Taxic homology" is neither. *Cladistics* 30, 113–115.
Farris, J., Albert, V., Källersjö, M., Lipscomb, D., and Kluge, A. 1996. Parsimony jackknifing outperforms neighbor-joining. *Cladistics* 12, 99–124.
Farris, J., Kallersjo, M., Albert, V., Allard, M., Anderberg, A., Bowditch, B., Bult, C., Carpenter, J., Crowe, T., de Laet, J., Fitzhugh, K., Frost, D., Goloboff, P., Humphries, C., Jondelius, U., Judd, D., Karis, P., Lipscomb, D., Luckow, M., Mindell, D., Muona, J., Nixon, K., Presch, W., Seberg, O., Siddall, M., Struwe, L., Tehler, A., Wenzel, J., Wheeler, Q., and Wheeler, W. 1995. Explanation. *Cladistics* 11, 211–218.
Farris, J., and Kluge, A. 1997. Parsimony and history. *Syst. Biol.* 46, 215–218.
Felsenstein, J. 1973. Maximum likelihood and minimum-steps methods for estimating evolutionary trees from data on discrete characters. *Syst. Zool.* 22, 240–249.
Felsenstein, J. 1978a. The number of evolutionary trees. *Syst. Zool.* 27, 27–33.

Felsenstein, J. 1978b. Cases in which parsimony and compatibility methods will be positively misleading. *Syst. Zool.* 27, 401–410.

Felsenstein, J. 1979. Alternative methods of phylogenetic inference and their interrelationship. *Syst. Zool.* 28, 49–62.

Felsenstein, J. 1981. A likelihood approach to character weighting and what it tells us about parsimony and compatibility. *Biol. J. Linn. Soc.* 16, 183–196.

Felsenstein, J. 1982. Numerical methods for inferring evolutionary trees. *Q. Rev. Biol.* 57, 379–404.

Felsenstein, J. 1992. Phylogenies from restriction sites: A maximum-likelihood approach. *Evolution* 46, 159–173.

Felsenstein, J. 2004. *Inferring Phylogenies.* Sunderland, MA: Sinauer Associates, 664 pp.

Fink, W. 1982. The conceptual relationship between ontogeny and phylogeny. *Paleobiol.* 8, 254–264.

Fisher, D. 2008. Stratocladistics: Integrating temporal data and character data in phylogenetic inference. *Ann. Rev. Ecol. Syst.* 39, 365–385.

Fitch, W. 1971. Toward defining the course of evolution: Minimum change for a specific tree topology. *Syst. Zool.* 20, 406–416.

Fitch, W., and Margoliash, E. 1967. Construction of phylogenetic trees. *Science* 155, 279–284.

Fitch, W., and Markowitz, E. 1970. An improved method for determining codon variability in a gene and its application to the rate of fixation of mutations in evolution. *Biochem. Genet.* 4, 579–593.

Forey, P., and Kitching, I. 2000. Experiments in coding multistate characters. In Scotland, R., and Pennigton, T. (eds.), *Homology and Systematics, Coding Characters for Phylogenetic Analysis.* Systematics Association Special 58, 54–80.

Foulds, L., and Graham, R. 1982. The steiner problem in phylogeny is NP-complete. *Adv. Appl. Math.* 3, 43–49.

Fox, D., Fisher, D., and Leighton, L. 1999. Reconstructing phylogeny with and without temporal data. *Science* 284, 1816–1819.

Gauthier, J., Kluge, A., and Rowe, T. 1988. Amniote phylogeny and the importance of fossils. *Cladistics* 4, 105–209.

Goëffon, A., Richer, J.M., and Hao, J.K., 2006. A distance-based information preservation tree crossover for the maximum parsimony problem. *Lect. Notes Comp. Sci.* 4193, 761–770.

Goldberg, E., and Igic, B. 2008. On phylogenetic tests of irreversible evolution. *Evolution* 62–11, 2727–2741.

Goldman, N., 1990. Maximum likelihood inference of phylogenetic trees, with special reference to a Poisson process model of DNA substitution and to parsimony analyses. *Syst. Zool.* 39, 345–361.

Goloboff, P. 1982. Nota sobre algunas Ctenizidae (Araneae) de la Argentina. *Physis* secc. C, 40, 75–79.

Goloboff, P. 1991a. Homoplasy and the choice among cladograms. *Cladistics* 7, 215–232.

Goloboff, P. 1991b. Random data, homoplasy, and information. *Cladistics* 7, 395–406.

Goloboff, P. 1993a. Estimating character weights during tree search. *Cladistics* 9, 83–91.

Goloboff, P. 1993b. Character optimization and calculation of tree lengths. *Cladistics* 9, 433–436.

Goloboff, P. 1993c. A reanalysis of Mygalomorph spider families. *Amer. Mus. Novitates No.* 3056, 1–32.

Goloboff, P. 1993d. *NONA: A Tree-Searching Program.* Program and Documentation, available at www.lillo.org.ar/phylogeny/Nona-PeeWee.

Goloboff, P. 1993e. *Pee-Wee: Parsimony and Implied Weights.* Program and Documentation, available at www.lillo.org.ar/phylogeny/Nona-PeeWee.

Goloboff, P. 1995. Parsimony and weighting: A reply to Turner and Zandee. *Cladistics* 10, 91–104.

Goloboff, P. 1996. Methods for faster parsimony analysis. *Cladistics* 12, 199–220.

Goloboff, P. 1997. Self-weighted optimization: Character state reconstructions and tree searches under implied transformation costs. *Cladistics* 13, 225–245.

Goloboff, P. 1998a. Principios básicos de cladística. *Soc. Argent. de Botánica*, Buenos Aires.

Goloboff, P. 1998b. Tree searches under Sankoff parsimony. *Cladistics* 14, 229–237.

Goloboff, P. 1999. Analyzing large data sets in reasonable times: Solutions for composite optima. *Cladistics* 15, 415–428.

Goloboff, P. 2002. Optimization of polytomies: State set and parallel operations. *Mol. Phyl. Evol.* 22, 269–275.

Goloboff, P. 2003. Parsimony, likelihood, and simplicity. *Cladistics* 19, 91–103.

Goloboff, P. 2014a. Hide and vanish: Data sets where the most parsimonious tree is known but hard to find, and their implications for tree search methods. *Mol. Phyl. Evol.* 79, 118–131.

Goloboff, P. 2014b. Oblong, a program to analyse phylogenomic data sets with millions of characters, requiring negligible amounts of RAM. *Cladistics* 30, 273–281.

Goloboff, P. 2015. Computer science and parsimony: A reappraisal, with discussion of methods for poorly structured data sets. *Cladistics* 31, 210–225.

Goloboff, P., and Arias, J.S. 2019. Likelihood approximations of implied weights parsimony can be selected over the Mk model by the Akaike information criterion. *Cladistics* 35, 695–716.

Goloboff, P., and Catalano, S. 2012. GB-to-TNT: Facilitating creation of matrices from GenBank and diagnosis of results in TNT. *Cladistics* 28, 503–513.

Goloboff, P., and Catalano, S. 2016. TNT version 1.5, including a full implementation of geometric morphometrics. *Cladistics* 32, 221–238.

Goloboff, P., Catalano, S., Mirande, M., Szumik, C., Arias, J.S., Källersjö, M., and Farris, J.S. 2009. Phylogenetic analysis of 73,060 taxa supports major evolutionary groups. *Cladistics* 25, 211–230.

Goloboff, P., Catalano, S., and Torres, A. 2021a. Parsimony analysis of phylogenomic datasets (II): Evaluation of PAUP*, MEGA, and MPBoot. *Cladistics*. https://doi.org/10.1111/cla.12476

Goloboff, P., De Laet, J., Ríos-Tamayo, D., and Szumik, C. 2021b. A reconsideration of inapplicable characters, and an approximation with step-matrix recoding. *Cladistics* 37, 596–629.

Goloboff, P., and Farris, J. 2001. Methods for quick consensus estimation. *Cladistics* 17, S26–S34.

Goloboff, P., Farris, J., and Nixon, K. 2003. *TNT: Tree Analysis Using New Technology*. Program and Documentation, available at www.lillo.org.ar/phylogeny/tnt.

Goloboff, P., Farris, J., and Nixon, K. 2008. TNT, a free program for phylogenetic analysis. *Cladistics* 24, 774–786.

Goloboff, P., Mattoni, C., and Quinteros, S. 2006. Continuous characters analyzed as such. *Cladistics* 22, 589–601.

Goloboff, P., Pittman, M., Pol, D., and Xu, X. 2018. Morphological data sets fit a common mechanism much more poorly than DNA sequences and call into question the Mkv model. *Syst. Biol.* 68, 494–504.

Goloboff, P., and Pol, D. 2005. Parsimony and Bayesian phylogenetics. In Albert, V. (ed.), *Parsimony, Phylogeny, and Genomics*. Oxford: Oxford University Press, pp. 148–159.

Goloboff, P., and Pol, D. 2007. On divide-and-conquer strategies for parsimony analysis of large data sets: Rec-i-dcm3 vs. TNT. *Syst. Biol.* 56, 485–495.

Goloboff, P., and Sereno, P. 2021. Comparative cladistics: Identifying the sources for differing phylogenetic results between competing morphology-based datasets. *J. Syst. Palaeont.* 19, 761–786.

Goloboff, P., and Simmons, M. 2014. Bias in tree searches and its consequences for measuring groups supports. *Syst. Biol.* 63, 851–861.
Goloboff, P., Torres, A., and Arias, J. 2017. Weighted parsimony outperforms other methods of phylogenetic inference under models appropriate for morphology. *Cladistics* 34, 407–437.
Goloboff, P., Torres, A., and Arias, J. 2018. Parsimony and model-based phylogenetic methods for morphological data: Comments on O'Reilly et al. (2017). *Palaeontology* 61, 625–630.
Goloboff, P., and Wilkinson, M. 2018. On defining a unique phylogenetic tree with homoplastic characters. *Mol. Phylogenet. Evol.* 122, 95–101.
Goodman, M., Czelusniak, J., William Moore, G., Romero-Herrera, A., and Matsuda, G. 1979. Fitting the gene lineage into its species lineage, a parsimony strategy illustrated by cladograms constructed from globin sequences. *Syst. Biol.* 28, 132–163.
Gower, J. 1975. Generalized Procrustes analysis. *Psychometrika* 40, 33–51.
Granara de Willink, M., and Szumik, C. 2007. Phenacoccinae de Centro y Sudamérica (Hemiptera: Coccoidea: Pseudoccidae): Sistemática y filogenia. *Rev. Soc. Entomol. Argent.* 66, 29–129.
Grant, T., and Kluge, A. 2004. Transformation series as an ideographic character concept. *Cladistic* 20, 23–31.
Groh, S., Upchurch, P., Barrett, P., and Day, J. 2019. The phylogenetic relationships of neosuchian crocodiles and their implications for the convergent evolution of the longirostrine condition. *Zool. J. Linn. Soc.* 188, 473–506.
Gruenheit, N., Lockhart, P., Steel, M., and Martin, W. 2008. Difficulties in testing for covarion-like properties of sequences under the confounding influence of changing proportions of variable sites. *Mol. Biol. Evol.* 25, 1512–1520.
Guindon, S., and Gascuel, O. 2003. A simple, fast and accurate algorithm to estimate large phylogenies by maximum likelihood. *Syst. Biol.* 52, 696–704.
Haas, O., and Simpson, G.G. 1946. Analysis of some phylogenetic terms, with attempts at redefinition. *Proc. Amer. Phil. Soc.* 90, 319–349.
Hamilton, A. 2014. *The Evolution of Phylogenetic Systematics.* Species and Systematics, Vol. 5. Berkeley, CA: University of California Press, 308 pp.
Harrison, L., and Larsson, H. 2014. Among-character rate variation distributions in phylogenetic analysis of discrete morphological characters. *Syst. Biol.* 64, 307–324.
Hartigan, J. 1973. Minimum mutations fit to a given tree. *Biometrics* 29, 53–65.
Hauser, D., and Presch, W. 1993. The effect of ordered characters on phylogenetic reconstruction. *Cladistics* 7, 243–265.
Hennig, W. 1966. *Phylogenetic Systematics.* Urbana: University of Illinois Press.
Herbst, L., Li, H., and Steel, M. 2019. Quantifying the accuracy of ancestral state prediction in a phylogenetic tree under maximum parsimony. *J. Math. Biol.* 78, 1953–1979.
Hillis, D., Bull, J., White, M., Badgett, M., and Molineux, I. 1992. Experimental phylogenetics: Generation of a known phylogeny. *Science* 255, 589–592.
Höhna, S., Landis, M., Heath, T., Boussau, B., Lartillot, N., Moore, B., Huelsenbeck, J., and Ronquist, F. 2016. RevBayes: Bayesian phylogenetic inference using graphical models and an interactive model-specification language. *Syst. Biol.* 65, 726–736.
Holder, M., Lewis, P., and Swofford, D. 2010. The Akaike information criterion will not choose the no common mechanism model. *Syst. Biol.* 59, 477–485.
Hormiga, G. 1994. Cladistics and the comparative morphology of linyphiid spiders and their relatives (Araneae, Araneoidea, Linyphiidae). *Zool. J. Linn. Soc.* 111, 1–71.
Hovenkamp, P. 1999. Unambiguous data or unambiguous results? *Cladistics* 15, 99–102.
Hoyal Cuthill, J. 2015. The morphological state space revisited: What do phylogenetic patterns in homoplasy tell us about the number of possible character states? *Interface Focus* 5, 20150049. http://dx.doi.org/10.1098/rsfs.2015.0049.

Hubbs, C. 1944. Concepts of homology and analogy. *Amer. Nat.* 78, 289–307.
Huelsenbeck, J., Alfaro, M., and Suchard, M. 2011. Biologically inspired phylogenetic models strongly outperform the no common mechanism model. *Syst. Biol.* 60, 225–232.
Huelsenbeck, J., Ane, C., Larget, B., and Ronquist, F. 2008. A Bayesian perspective on a non-parsimonious parsimony model. *Syst. Biol.* 57, 406–419.
Humphries, P., and Wu, T. 2013. On the neighborhoods of trees. *IEEE/ACM Trans. Comp. Biol. Bioinf.* 10, 721–728.
Huson, D., Rupp, R., and Scornavacca, C. 2011. *Phylogenetic Networks: Concepts, Algorithms, and Applications*. Cambridge: Cambridge University Press, 362 pp.
Huson, D., and Steel, M. 2004. Distances that perfectly mislead. *Syst. Biol.* 53, 327–332.
Huxley, J. 1942. *Evolution: The Modern Synthesis*. London: Allen and Unwin.
Jukes, T., and Cantor, C. 1969. Evolution of protein molecules. In Munro, N. (ed.), *Mammalian Protein Metabolism*. Vol. 3. New York: Academic Press, pp. 21–132.
Källersjö, M., Albert, V., and Farris, J. 1999. Homoplasy *increases* phylogenetic structure. *Cladistics* 15, 91–93.
Kendall, D. 1977. The diffusion of shape. *Adv. Appl. Prob.* 9, 428–430.
Kernighan, B., and Ritchie, D. 1988. *The C Programming Language*. Englewood Cliffs, NJ: Prentice Hall.
Kim, J. 1996. General inconsistency conditions for maximum parsimony: Effects of branch lengths and increasing numbers of taxa. *Syst. Biol.* 45, 363–374.
Kimura, M. 1968. Evolutionary rate at the molecular level. *Nature* 217, 624–626.
Kimura, M. 1980. A simple method for estimating evolutionary rates of base substitutions through comparative studies of nucleotide sequences. *J. Mol. Evol.* 16, 111–120.
Kingman, J. 1982. On the genealogy of large populations. *J. App. Prob.* 19, 27–43.
Klingenberg, C.P. 2011. MorphoJ: An integrated software package for geometric morphometrics. *Mol. Ecol. Resources* 11, 353–357.
Klingenberg, C.P., and Gidaszewski, N. 2010. Testing and quantifying phylogenetic signals and homoplasy in morphometric data. *Syst. Biol.* 59, 245–261.
Klopfstein, S., Vilhelmsen, L., and Ronquist, F. 2015. A nonstationary Markov model detects directional evolution in Hymenopteran morphology. *Syst. Biol.* 64, 1089–1103.
Kluge, A. 1985. Ontogeny and phylogenetic systematics. *Cladistics* 1, 13–27.
Kluge, A. 1989. A concern for evidence and a phylogenetic hypothesis of relationships among Epicrates (Boidae, Serpentes). *Syst. Zool.* 38, 7–25.
Kluge, A. 1994. Moving targets and shell games. *Cladistics* 10, 403–413.
Kluge, A. 1997. Testability and the refutation and corroboration of cladistic hypotheses. *Cladistics* 13, 81–96.
Kluge, A. 1998. Total evidence or taxonomic congruence: Cladistics or consensus classification. *Cladistics* 14, 151–158.
Kluge, A. 1999. The science of phylogenetic systematics: Explanation, prediction, and test. *Cladistics* 15, 429–436.
Kluge, A. 2003. The repugnant and the mature in phylogenetic inference: Atemporal similarity and historical identity. *Cladistics* 19, 356–368.
Kluge, A. 2005. What is the rationale for 'Ockham's razor' (a.k.a. parsimony) in phylogenetic inference. In Albert, V. (ed.), *Parsimony, Phylogeny, and Genomics*. Oxford: Oxford University Press, pp. 15–42.
Lakatos, I. 1980. *The Methodology of Scientific Research Programmes: Volume 1: Philosophical Papers*. Cambridge: Cambridge University Press.
Lanfear, R., Frandsen, P., Wright, A., Senfeld, T., and Calcott, B. 2017. PartitionFinder 2: New methods for selecting partitioned models of evolution for molecular and morphological phylogenetic analyses. *Mol. Biol. Evol.* 34, 772–773.

Lankester, E. 1870. On the use of the term homology in modern zoology, and the distinction between homogenetic and homoplastic agreements. *Ann. Mag. Nat. Hist.* 6, 34–43.
Larget, B., and Simon, D. 1999. Markov chain Monte Carlo algorithms for the Bayesian analysis of phylogenetic trees. *Mol. Biol. Evol.* 16, 750–759.
Lartillot, N., and Philippe, H. 2006. Computing Bayes factors using thermodynamic integration. *Syst. Biol.* 55, 195–207.
Lee, D.-C., and Bryant, H. 1999. A reconsideration of the coding of inapplicable characters: Assumptions and problems. *Cladistics* 15, 373–378.
Lemoine, F., Domelevo Entfellner, J.-B., Wilkinson, E., Correia, D., Dávila Felipe, M., De Oliveira, T., and Gascuel, O. 2018. Renewing Felsenstein's phylogenetic bootstrap in the era of big data. *Nature* 556, 452–456.
Lewis, P. 2001a. A likelihood approach to estimating phylogeny from discrete morphological character data. *Syst. Biol.* 50, 913–925.
Lewis, P. 2001b. Phylogenetic systematics turns over a new leaf. *TREE* 16, 30–37.
Li, G., Steel, M., and Zhang, L. 2008. More taxa are not necessarily better for the reconstruction of ancestral character states. *Syst. Biol.* 57, 647–653.
Liebherr, J., and Zimmerman, E. 1998. Cladistic analysis, phylogeny, and biogeography of the Hawaiian Platynini (Coleoptera: Carabidae). *Syst. Entomol.* 23, 101–136.
Lin, Y.M. 2008. *Tabu Search and Genetic Algorithms for Phylogeny Inference.* PhD thesis, North Carolina University, Faculty of Operations Research, Raleigh, NC, 119 pp.
Lin, Y.M., Nag, S., and Thorne, J. 2007. A tabu search algorithm for maximum parsimony phylogeny inference. *Europ. J. Op. Res.* 176, 1908–1917.
Lipscomb, D. 1992. Parsimony, homology, and the analysis of multistate characters. *Cladistics* 8, 45–65.
Lorenz, K. 1958. The evolution of behavior. *Sci. Am.* 199, 67–78.
Lovtrup, S. 1987. On species and other taxa. *Cladistics* 3, 157–177.
Mabee, P. 1989. An empirical rejection of the ontogenetic polarity criterion in phylogenetic inference. *Syst. Zool.* 37, 106–141.
Mabee, P. 2000. The usefulness of ontogeny in interpreting morphological characters. In Wiens, J. (ed.), *Phylogenetic Analysis of Morphological Data.* Washington: Smithsonian Institution Press, pp. 84–114.
Maddison, D. 1991. The discovery and importance of multiple islands of most-parsimonious trees. *Syst. Zool.* 40, 315–328.
Maddison, D., and Maddison, W. 2000. *McClade 4.* Sunderland, MA: Sinauer Associates.
Maddison, D., Swofford, D., and Maddison, W. 1997. NEXUS: An extensible file format for systematic information. *Syst. Biol.* 46, 590–621.
Maddison, W. 1989. Reconstructing character evolution on polytomous cladograms. *Cladistics* 5, 365–377.
Maddison, W. 1993. Missing data versus missing characters in phylogenetic analysis. *Syst. Biol.* 42, 576–581.
Maddison, W. 1997. Gene trees in species trees. *Syst. Biol.* 46, 523–536.
Maddison, W., Donoghue, M., and Maddison, D. 1984. Outgroup analysis and parsimony. *Syst. Zool.* 33, 83–103.
Maddison, W., and Maddison, D. 2018. *Mesquite: A Modular System for Evolutionary Analysis.* Version 3.51, available at www.mesquiteproject.org.
Marsaglia, G. 1965. Ratios of normal variables and ratios of sums of uniform variables. *J. Am. Stat. Assoc.* 60, 193–204.
Marsaglia, G. 2006. Ratios of normal variables. *J. Stat. Soft.* 16, 1–10.
Maslin, T. 1952. Morphological criteria of phyletic relationships. *Syst. Zool.* 1, 49–70.

Matos-Maraví, P., Wahlberg, N., Freitas, A., DeVries, P., Antonelli, A., and Penz, C. 2021. Mesoamerica is a cradle and the Atlantic forest is a museum of neotropical butterfly diversity: Insights from the evolution and biogeography of Brassolini (Lepidoptera: Nymphalidae). *bioRxiv* 762393. https://doi.org/10.1101/762393.

Mau, B., and Newton, M. 1997. Phylogenetic inference for binary data on dendrograms using Markov chain Monte Carlo. *J. Comput. Graph. Stat.* 6, 122–131.

Mayr, E. 1982. *The Growth of Biological Thought: Diversity, Evolution, and Inheritance.* Cambridge: Harvard University Press.

Mickevich, M. 1982. Transformation series analysis. *Syst. Zool.* 31, 461–478.

Mickevich, M., and Weller, S. 1992. Evolutionary character analysis: Tracing character change on a cladogram. *Cladistics* 6, 137–170.

Minelli, A. 2009. *Perspectives in Animal Phylogeny and Evolution.* New York: Oxford University Press, 345 pp.

Mirande, J. 2017. Combined phylogeny of ray-finned fishes (Actinopterygii) and the use of morphological characters in large-scale analyses. *Cladistics* 33, 333–350.

Mirarab, S., and Warnow, T. 2015. ASTRAL-II: Coalescent-based species tree estimation with many hundreds of taxa and thousands of genes. *Bioinformatics* 31, i44–i52.

Mishler, B. 2005. The logic of the data matrix in phylogenetic analysis. In Albert, V. (ed.), *Parsimony, Phylogeny, and Genomics.* New York: Oxford University Press, pp. 57–70.

Moilanen, A. 1999. Searching for most parsimonious trees with simulated evolutionary optimization. *Cladistics* 15, 39–50.

Mongiardino, N., Soto, I., and Ramírez, M. 2015. Overcoming problems with the use of ratios as continuous characters for phylogenetic analyses. *Zool. Scripta* 44, 463–474.

Monteiro, L. 2000. Why morphometrics is special: The problem with using partial warps as characters for phylogenetic inference. *Syst. Biol.* 49, 796–800.

Moody, S. 1985. Charles L. Camp and his 1923 classification of Lizards: An early cladist? *Syst. Zool.* 34, 216–222.

Mooi, R., and Gill, A. 2010. Phylogenies without synapomorphies—A crisis in fish systematics: Time to show some character. *Zootaxa* 2450, 26–40.

Mooi, R., Williams, D., and Gill, A. 2011. Numerical cladistics, an unintentional refuge for phenetics—a reply to Wiley et al. In Carvalho, M. de, and Craig, M. (eds.), *Morphological and Molecular Approaches to the Phylogeny of Fishes: Integration or Conflict?* pp. 17–28. *Zootaxa* 2946, 1–142.

Mossel, E., and Vigoda, E. 2005. Phylogenetic MCMC algorithms are misleading on mixtures of trees. *Science* 309, 2207–2209.

Nakhleh, L., Warnow, T., Linder, C., and St. John, K. 2005. Reconstructing reticulate evolution in species— theory and practice. *J. Comp. Biol.* 12, 796–811.

Needleman, S., and Wunsch, C. 1970. A general method applicable to the search for similarities in the amino acid sequences of two proteins. *J. Mol. Biol.* 48, 443–453.

Nelson, G. 1978. Ontogeny, phylogeny, paleontology, and the biogenetic law. *Syst. Zool.* 27, 324–345.

Nelson, G. 1983. Reticulation in cladograms. In Platnick, N.I., and Funk, V.A. (eds.), *Advances in Cladistics.* New York: Columbia University Press, pp. 105–111.

Nelson, G. 1985. Outgroups and ontogeny. *Cladistics* 1, 29–46.

Nelson, G. 1989a. Cladistics and evolutionary models. *Cladistics* 5, 275–289.

Nelson, G. 1989b. Species and taxa: Systematics and evolution. In Otte, D., and Endler, J. (eds.), *Speciation and Its Consequences.* Sunderland, MA: Sinauer Associates, Inc., pp. 60–84.

Nelson, G. 1994. Homology and systematics. In Hall, B.K. (ed.), *Homology: The Hierarchical Basis of Comparative Biology.* San Diego: Academic Press, pp. 101–149.

Nelson, G. 2004. Cladistics: Its arrested development. In Williams, D., and Forey, P. (eds.), *Milestones in Systematics*. Systematics Association Special Volume Series 67. London: CRC Press, pp. 127–147.

Nelson, G., and Platnick, N. 1981. *Systematics and Biogeography: Cladistics and Vicariance*. New York: Columbia University Press.

Nelson, G., and Platnick, N. 1991. Three-taxon statements: A more precise use of parsimony? *Cladistics* 7, 351–366.

Newton, M., and Raftery, A. 1994. Approximate Bayesian inference by the weighted likelihood bootstrap. *J. Roy. Stat. Soc.*, Series B 56, 3–48.

Neyman, J. 1971. Molecular studies of evolution: A source of novel statistical problems. In Gupta, S., and Yackel, J. (eds.), *Statistical Decision Theory and Related Topics*. New York: Academic Press, pp. 1–17.

Nguyen, L., Schmidt, H., von Haeseler, A., and Minh, B. 2015. IQ-TREE: A fast and effective stochastic algorithm for estimating maximum likelihood phylogenies. *Mol. Biol. Evol.* 32, 268–274.

Nixon, K. 1999. The parsimony ratchet a new method for rapid parsimony analysis. *Cladistics* 15, 407–414.

Nixon, K. 2002. *WinClada ver. 1.00.08*. Ithaca, NY: The Author. Available at http://www.lillo.org.ar/phylogeny/winclada.

Nixon, K., and Carpenter, J. 2012a. On homology. *Cladistics* 28, 160–169.

Nixon, K., and Carpenter, J. 2012b. More on homology. *Cladistics* 28, 225–226.

Nixon, K., and Davis, J. 1991. Polymorphic taxa, missing values, and cladistic analysis. *Cladistics* 7, 233–241.

Nylander, J., Ronquist, F., Huelsenbeck, J., and Nieves-Aldrey, J. 2004. Bayesian phylogenetic analysis of combined data. *Syst. Biol.* 53, 47–67.

Opatova, V., Hamilton, C., Hedin, M., Montes de Oca, L., Král, J., and Bond, J. 2020. Phylogenetic systematics and evolution of the spider infraorder Mygalomorphae using genomic scale data. *Syst. Biol.* 69, 671–707.

O'Reilly, J., and Donoghue, P. 2017. The efficacy of consensus tree methods for summarizing phylogenetic relationships from a posterior sample of trees estimated from morphological data. *Syst. Biol.* 67, 354–362.

O'Reilly, J., Puttick, M., Parry, L., Tanner, A., Tarver, J., Fleming, J., Pisani, D., and Donoghue, P. 2016. Bayesian methods outperform parsimony but at the expense of precision in the estimation of phylogeny from discrete morphological data. *Biol. Lett.* 12, 1–5.

O'Reilly, J., Puttick, M., Pisani, D., and Donoghue, P. 2018. Empirical realism of simulated data is more important than the model used to generate it: A reply to Goloboff et al. *Palaeontology* 61, 631–635.

Owen, R. 1843. *Lectures of the Comparative Anatomy and Physiology of the Invertebrate Animals, Delivered at the Toyal College of Surgeons in 1843*. London: Longman, Brown, Green, Longmans, 392 pp.

Page, R. 1994. Maps between trees and cladistic analysis of historical associations among genes, organisms, and areas. *Syst. Biol.* 43, 58–77.

Page, R. 1996. Tree View: An application to display phylogenetic trees on personal computers. *Bioinformatics* 12, 357–358.

Panchen, A. 1992. *Classification, Evolution, and the Nature of Biology*. Cambridge: Cambridge University Press, 403 pp.

Parins-Fukuchi, C., and Brown, J. 2017. What drives results in Bayesian morphological clock analyses? *bioArxiv*. http://dx.doi.org/10.1101/219048.

Parins-Fukuchi, C., Greiner, E., MacLatchy, L., and Fisher, D. 2019. Phylogeny, ancestors and anagenesis in the hominin fossil record. *bioRxiv*. https://doi.org/10.1101/434894.

Patterson, C. 1982a. Classes and cladists or individuals and evolution. *Syst. Zool.* 31, 284–286.

Patterson, C. 1982b. Morphological characters and homology. In Joysey, K.A., and Friday, A.E. (eds.), *Morphological Characters and Homology*. London: Academic Press, pp. 21–74.

Pei, R., Pittman, M., Goloboff, P., Dececchi, T., Habib, M., Kaye, T., Larsson, H., Norell, M., Brusatte, S., and Xu, X. 2020. Potential for powered flight neared by most close avialan relatives but few crossed its thresholds. *Current Biology* 30, 1–14.

Pickett, K., and Randle, C. 2005. Strange Bayes indeed: Uniform topological priors. *Mol. Phyl. Evol.* 34, 203–211.

Platnick, N. 1977. Cladograms, phylogenetic trees, and hypothesis testing. *Syst. Zool.* 26, 438–442.

Platnick, N. 1979. Philosophy and the transformation of cladistics. *Syst. Zool.* 28, 537–546.

Platnick, N. 1982. Defining characters and evolutionary groups. *Syst. Zool.* 31, 282–284.

Platnick, N. 1993. Character optimization and weighting: Differences between the standard and three-taxon approaches to phylogenetic inference. *Cladistics* 9, 267–272.

Platnick, N. 2012. Less on homology. *Cladistics* 29, 10–12.

Platnick, N., Humphries, C., Nelson, G., and Williams, D. 1996. Is Farris optimization perfect? Three-Taxon statements and multiple branching. *Cladistics* 12, 243–252.

Poe, S., and Wiens, J. 2000. Character selection and the methodology of morphological phylogenetics. In Wiens, J. (ed.), *Phylogenetic Analysis of Morphological Data*. Washington: Smithsonian Institution Press, pp. 20–36.

Prendini, L. 2000. Phylogeny and classification of the superfamily Scorpionoidea Latreille 1802 (Chelicerata, Scorpiones): An exemplar approach. *Cladistics* 16, 1–78.

Puttick, M., O'Reilly, J., Pisani, D., and Donoghue, P. 2018. Probabilistic methods outperform parsimony in the phylogenetic analysis of data simulated without a probabilistic model. *Palaeontology* 62, 1–17.

Puttick, M., O'Reilly, J., Tanner, A., Fleming, J., Clark, J., Holloway, L., Lozano-Fernández, J., Parry, L., Tarver, J., Pisani, D., and Donoghue, P. 2017. Uncertain-tree: Discriminating among competing approaches to the phylogenetic analysis of phenotype data. *Proceedings of the Royal Society B* 284, 20162290.

Puttick, M., Thomas, G., and Benton, M. 2016. Dating Placentalia: Morphological clocks fail to close the molecular fossil gap. *Evolution* 70, 873–886.

Quinn, A. 2017. When is a cladist not a cladist? *Biol. Phil.* 32, 581–598.

Rae, T. 1998. The logical basis for the use of continuous characters in phylogenetic systematics. *Cladistics* 14, 221–228.

Rambaut, A. 2009. FigTree, version 1.4.3. Computer program, distributed by the author, available at http://tree.bio.ed.ac.uk/software/figtree/.

Ramírez, M. 2007. Homology as a parsimony problem: A dynamic homology approach for morphological data. *Cladistics* 23, 1–25.

Randle, E., and Sansom, R. 2017. Exploring phylogenetic relationships of Pteraspidiformes heterostracans (stem-gnathostomes) using continuous and discrete characters. *J. Syst. Paleont.* 5, 583–599.

Raven, R. 1985. The spider infraorder Mygalomorphae: Cladistics and systematics. *Bull. Am. Mus. Nat. Hist.* 182, 1–180.

Remane, A. 1952. *Die Grundlagen des Naturlichen Systems der Vergleichenden Anatomie und der Phylogenetik*. Leipzig, Germany: Geest and Portig.

Ribeiro, C.C., and Vianna, D.S. 2009. A hybrid genetic algorithm for the phylogeny problem using path-relinking as a progressive crossover strategy. *Int. Trans. Oper. Res.* 16, 641–657.

Ridley, M. 1986. *Evolution and Classification: The Reformation of Cladism*. Harlow: Longman.

Richer, J.M., Rodriguez-Tello, E., and Vazquez-Ortiz, K.E. 2012. Maximum parsimony phylogenetic inference using simulated annealing. Proceedings of the EVOLVE 2012, Mexico City, Mexico. *Adv. Intell. Soft Comp.* 175, 189–203.

Rieppel, O. 1985. Ontogeny and the hierarchy of the types. *Cladistics* 1, 234–246.

Rieppel, O. 2003. Popper and systematics. *Syst. Biol.* 52, 259–271.

Rieppel, O. 2013. The early cladogenesis of cladistics. In Hamilton, A. (ed.), *The Evolution of Phylogenetic Systematics*. Berkeley: University of California Press, pp. 117–137.

Rieppel, O., and Kearney, M. 2002. Similarity. *Biol. J. Linn. Soc.* 75, 59–82.

Rinsma, I., Hendy, M., and Penny, D. 1990. Minimally colored trees. *Math. Biosci.* 98, 201–210.

Robinson, D., and Foulds, L. 1981. Comparison of phylogenetic trees. *Math. Biosci.* 53, 131–147.

Roch, S., and Warnow, T. 2015. On the robustness to gene tree estimation error (or lack thereof) of coalescent-based species tree methods. *Syst. Biol.* 64, 663–676.

Rogers, J. 1972. Measures of genetic similarity and genetic distance. Studies in genetics VII. *Univ. Texas Publ.* 7213, 145–153.

Rogers, J. 1997. On the consistency of maximum likelihood estimation of phylogenetic trees from nucleotide sequences. *Syst. Biol.* 46, 354–357.

Rogers, J. 2001. Maximum likelihood estimation of phylogenetic trees is consistent when substitution rates vary according to the invariable sites plus Gamma distribution. *Syst. Biol.* 50, 713–722.

Rohlf, J. 2000. On the use of shape spaces to compare morphometric methods. *Hystrix It. J. Mamm.* 11 (1). https://doi.org/10.4404/hystrix-11.1-4134.

Rohlf, J. 2002. Geometric morphometrics and phylogeny. In MacLeod, N., and Forey, P. (eds.), *Morphology, Shape and Phylogeny*, Systematic Association Special Volume Series 64. London: Taylor & Francis, pp. 175–193.

Ronquist, F. 1998. Fast Fitch-parsimony algorithms for large data sets. *Cladistics* 14, 387–400.

Ronquist, F., Huelsenbeck, J., and Teslenko, M. 2011. MrBayes version 3.2 manual: Tutorials and model summaries, available at http://mrbayes.sourceforge.net/mb3.2_manual.pdf.

Ronquist, F., Larget, B., Huelsenbeck, J., Kadane, J., Simon, D., and van der Mark, P. 2006. Comment on 'Phylogenetic MCMC algorithms are misleading on mixtures of trees'. *Science* 312, 367.

Ronquist, F., Teslenko, M., van der Mark, P., Ayres, D., Darling, A., Höhna, S., Larget, B., Liu, L., Suchard, M., and Huelsenbeck, J. 2012. MrBayes 3.2: Efficient Bayesian phylogenetic inference and model choice across a large model space. *Syst. Biol.* 61, 539–542.

Rosa, B., Melo, G., and Barbeitos, M. 2019. Homoplasy-based partitioning outperforms alternatives in Bayesian analysis of discrete morphological data. *Syst. Biol.* 68, 657–671.

Roshan, U., Moret, B., Williams, T., and Warnow, T. 2004. Rec-I-DCM3: A fast algorithmic technique for reconstructing large phylogenetic trees. *Proceedings 3rd IEEE Computational Systems Bioinformatics Conference* (CSB 2004), 98–109.

Roth, V. 1984. On homology. *Biol. J. Linn. Soc.* 22, 13–29.

Rydin, C., and Källersjö, M. 2002. Taxon sampling and seed plant phylogeny. *Cladistics* 18, 485–513.

Saitou, N., and Nei, M. 1987. The neighbor-joining method: A new method for reconstructing phylogenetic trees. *Mol. Biol. Evol.* 4, 406–425.

Sanderson, M., and Kim, J. 2000. Parametric phylogenetics? *Syst. Biol.* 49, 817–829.

Sankoff, D., and Cedergren, R. 1983. Simultaneous comparison of three or more sequences related by a tree. In Sankoff, D., and Kruskal, B. (eds.), *Time Warps, String Edits, and Macromolecules: The Theory and Practice of Sequence Comparison*. Reading, MA: Addison-Wesley, pp. 253–263.

Sankoff, D., and Rousseau, P. 1975. Locating the vertices of a Steiner tree in an arbitrary space. *Math. Program.* 9, 240–246.

Sarich, V.M., and Wilson, A.C. 1967. Immunological time scale for hominid evolution. *Science* 158, 1200–1203.

Sattler, R. 1984. Homology—a continuing challenge. *Syst. Bot.* 9, 382–394.

Sattler, R. 1994. Homology, homeosis, and process morphology in plants. In Hall, B. (ed.), *Homology, the Hierarchical Basis of Comparative Biology*. San Diego: Academic Press, pp. 423–475.

Schöniger, M., and von Haeseler, A. 1994. A stochastic model for the evolution of autocorrelated DNA sequences. *Mol. Phyl. Evol.* 3, 240–247.

Schröder, E. 1870. Vier kombinatoriche probleme. *Zeitschrift für Matemmatik und Physik* 15, 489–503.

Schuh, R. 1991. Phylogenetic, host and biogeographic analyses of the Pilophorini (Heteroptera: Miridae: Phylinae). *Cladistics* 7, 157–189.

Scott, B. 1996. Phylogenetic relationships of the Camaenidae (Pulmonata: Stylommatophora: Helicoidea). *J. Moll. Stud.* 62, 65–73.

Scott, E. 2005. A phylogeny of ranid frogs (Anura: Ranoidea: Ranidae), based on a simultaneous analysis of morphological and molecular data. *Cladistics* 21, 507–574.

Scott-Ram, N. 1990. *Transformed Cladistics, Taxonomy, and Evolution*. Cambridge: Cambridge University Press.

Semple, C., and Steel, M. 2003. *Phylogenetics*. Oxford: Oxford University Press.

Sereno, P. 2007. Logical basis for morphological characters in phylogenetics. *Cladistics* 23, 565–587.

Sereno, P. 2009. Comparative cladistics. *Cladistics* 25, 624–659.

Shultz, J. 1987. The origin of the spinning apparatus in spiders. *Biol. Rev.* 62, 89–113.

Sibley, C., and Ahlquist, J. 1984. The phylogeny of the hominoid primates, as indicated by DNA-DNA hybridization. *J. Mol. Evol.* 20, 2–15.

Simmons, M. 2012. Misleading results of likelihood-based phylogenetic analyses in the presence of missing data. *Cladistics* 28, 208–222.

Simmons, N. 1993. The importance of methods: Archontan phylogeny and cladistic analysis of morphological data. In McPhee, R. (ed.), *Primates and Relatives in Phylogenetic Perspective*. New York: Springer, pp. 1–61.

Simon, C. 1983. A new coding procedure for morphometric data with an example from periodical cicada wing veins. In Felsenstein, J. (ed.), *Numerical Taxonomy*. Proceedings of the NATO Advanced Study Institute. NATO ASI Series G (Ecological Sciences), No. 1. Berlin, Heidelberg: Springer-Verlag, pp. 378–382.

Simon, E. 1892. *Histoire Naturelle des Araignees*. Vol. 2. Paris, part 4, pp. 69–1080.

Siu-Ting, K., Pisani, D., Creevey, C., and Wilkinson, M. 2015. Concatabominations: Identifying unstable taxa in morphological phylogenetics using a heuristic extension to Safe Taxonomic Reduction. *Syst. Biol.* 64, 137–143.

Slowinski, J. 1993. "Unordered" versus "ordered" characters. *Syst. Biol.* 42, 155–165.

Slowinski, J., and Page, R. 1999. How Should Species Phylogenies Be Inferred from Sequence Data? *Syst. Biol.* 48, 814–825.

Sokal, R., and Rohlf, J. 1962. The comparison of dendrograms by objective methods. *Taxon* 11, 33–40.

Sokal, R., and Sneath, P. 1963. *Principles of Numerical Taxonomy*. San Francisco: Freeman & Co.

Smith, M. 2019. Bayesian and parsimony approaches reconstruct informative trees from simulated morphological datasets. *Biol. Lett.* 15, 20180632. http://dx.doi.org/10.1098/rsbl.2018.0632.

Springer, M., and Gatesy, J. 2018. On the importance of homology in the age of phylogenomics. *Syst. Biodiv.* 16, 210–228.
Stamatakis, A. 2014. RAxML version 8: A tool for phylogenetic analysis and post-analysis of large phylogenies. *Bioinformatics* 30, 1312–1313.
Stamatakis, A., Ludwig, T., and Meier, M. 2005. RAxML-III: A fast program for maximum likelihood-based inference of large phylogenetic trees. *Bioinformatics* 21, 456–463.
Steel, M. 1989. *Distributions on Bicoloured Evolutionary Trees.* PhD thesis, Massey University, New Zealand.
Steel, M. 1994. The maximum likelihood point for a phylogenetic tree is not unique. *Syst. Biol.* 43, 560–564.
Steel, M. 2001. Sufficient conditions for two tree reconstruction methods to succeed on sufficiently long sequences. *SIAM J. Disc. Math.* 14, 36–48.
Steel, M. 2005. Should phylogenetic models be trying to "fit an elephant"? *Trends Genet.* 21, 307–309.
Steel, M. 2011. Can we avoid "sin" in the house of "no common mechanism"? *Syst. Biol.* 60, 96–109.
Steel, M. 2013. Consistency of Bayesian inference of resolved phylogenetic trees. *J. Theo. Biol.* 336, 246–249.
Steel, M., and Penny, D. 2000. Parsimony, likelihood, and the role of models in molecular phylogenetics. *Mol. Biol. Evol.* 17, 839–850.
Steel, M., and Penny, D. 2004. Two further links between MP and ML under the Poisson model. *App. Math. Lett.* 17, 785–790.
Steel, M., and Pickett, K. 2006. On the impossibility of uniform priors on clades. *Mol. Phyl. Evol.* 39, 585–586.
Stevens, P. 1991. Character states, morphological variation, and phylogenetic analysis: A review. *Syst. Bot.* 16, 553–583.
Strait, D., Moniz, M., and Strait, P. 1996. Finite mixture coding: A new approach to coding continuous characters. *Syst. Biol.* 45, 67–78.
Strong, E., and Lipscomb, D. 1999. Character coding and inapplicable data. *Cladistics* 15, 363–371.
Swofford, D. 1985. *PAUP: Phylogenetic Analysis Using Parsimony.* User's Manual. Champaign, IL: Illinois Natural History Survey.
Swofford, D. 1993. *PAUP: Phylogenetic Analysis Using Parsimony, Version 3.1.* Washington: Program and documentation, Laboratory of Molecular Systematics, Smithsonian Institution.
Swofford, D. 2001. *PAUP*: Phylogenetic Analysis Using Parsimony (*and Other Methods).* Sunderland: Sinauer Associates.
Swofford, D., and Bell, C. 2017. PAUP* manual, available at https://paup.phylosolutions.com/.
Swofford, D., and Berlocher, S. 1987. Inferring evolutionary trees from gene frequency data under the principle of maximum parsimony. *Syst. Zool.* 36, 293–325.
Swofford, D., and Maddison, W. 1987. Reconstructing ancestral character states under Wagner parsimony. *Math. Biosc.* 87, 199–229.
Swofford, D., and Maddison, W. 1992. Parsimony, character-state reconstructions, and evolutionary inferences. In Mayden, R. (ed.), *Systematics, Historical Ecology, and North American Freshwater Fishes.* Stanford, CA: Stanford University Press, pp. 186–223.
Swofford, D., Olsen, G., Waddell, P., and Hillis, D. 1996. Phylogenetic inference. In Hillis, D., Moritz, C., and Mable, B. (eds.), *Molecular Systematics*, 2nd ed. Sunderland, MA: Sinauer, pp. 407–514.
Szumik, C. 1996. The higher classification of the Order Embioptera: A cladistic analysis. *Cladistics* 12, 41–64.

Telford, M., and Budd, G. 2003. The place of phylogeny and cladistics in Evo-Devo research. *Int. J. Dev. Biol.* 47, 479–490.

Thiele, K. 1993. The Holy Grail of the perfect character: The cladistic treatment of morphometric data. *Cladistics* 9, 275–304.

Tillier, E., and Collins, R. 1995. Neighbor joining and maximum-likelihood with RNA sequences—addressing the interdependence of sites. *Mol. Biol. Evol.* 12, 7–15.

Truszkowski, J., and Goldman, N. 2016. Maximum likelihood phylogenetic inference is consistent on multiple sequence alignments, with or without gaps. *Syst. Biol.* 65, 328–333.

Tuffley, C., and Steel, M. 1997. Links between maximum likelihood and maximum parsimony under a simple model of site substitution. *Bull. Math. Biol.* 59, 581–607.

Van Valen, L. 1982. Homology and causes. *J. Morphol.* 173, 305–312.

Varón, A., Vinh, L., and Wheeler, W. 2010. POY version 4: Phylogenetic analysis using dynamic homologies. *Cladistics* 26, 72–85.

Velasco, J. 2008. The prior probabilities of phylogenetic trees. *Biol. Philos.* 23, 455–473.

Viana, G., Gomes, F., Ferreira, C., and Meneses, C. 2007. Uma implementacão eficiente de uma heurıstica de busca local em multi-vizinhanças para um problema de filogenia. *Proceedings of the XXXIX Simposio Brasileiro de Pesquisa Operacional*, Fortaleza, Brasil, pp. 2045–2056.

Votsis, I. 2016. Ad Hoc hypotheses and the monsters within. In Müller, V.C. (ed.), *Fundamental Issues of Artificial Intelligence* (Synthese Library), Berlin: Springer, pp. 299–313.

Wägele, J. 2004. Hennig's phylogenetic systematics brought up to date. In Williams, D.M., and Forey, P.L. (eds.), *Milestones in Systematics.* The Systematics Association Special Volume Series 67. London: CRC Press, pp. 101–125.

Wägele, J. 2005. *Foundations of Phylogenetic Systematics*, 2nd ed. Munchen: Verlag.

Wagner, G. 1989. The biological homology concept. *Annu. Rev. Ecol. Syst.* 20, 51–69.

Wagner, P., Ruta, M., and Coates, M. 2006. Evolutionary patterns in early tetrapods. II. Differing constraints on available character space among clades. *Proc. R. Soc. B* 273, 2113–2118.

Wagner, W. 1961. Problems in the classification of ferns. In *Recent Advances in Botany.* Toronto: University Toronto Press, pp. 841–844.

Wang, Y., Chen, M., Kuo, L., and Lewis, P. 2018. A new monte Carlo method for estimating marginal likelihoods. *Bayesian Analysis* 13, 311–333.

Wang, Y., and Yang, Z. 2014. Priors in Bayesian phylogenetics. In Chen, M.-H., Kuo, L., and Lewis, P. (eds.), *Bayesian Phylogenetics: Methods, Algorithms, and Applications.* Oxfordshire: CRC Press, Taylor & Francis Group, pp. 5–23.

Watrous, L., and Wheeler, Q. 1981. The out-group comparison method of character analysis. *Syst. Zool.* 30, 1–11.

Wenzel, J. 1992. Behavioral homology and phylogeny. *Annu. Rev. Ecol. Syst.* 23, 361–381.

Wenzel, J. 1993. Application of the biogenetic law to behavioral ontogeny: A test using nest architecture in paper wasps. *J. Evol. Biol.* 6, 229–247.

Weston, P. 1988. Indirect and direct methods in systematics. In Humphries, C. (ed.), *Ontogeny and Systematics.* New York: Columbia University Press, pp. 27–56.

Wheeler, Q. 1990. Ontogeny and character phylogeny. *Cladistics* 6, 225–268.

Wheeler, Q. 2008. Toward the new taxonomy. In: Wheeler, Q. (ed.), *The New Taxonomy.* The Systematics Association Special Volumes Series 76. Boca Raton: CRC Press, pp. 1–17.

Wheeler, W. 1996. Optimization alignment: The end of multiple sequence alignment in phylogenetics? *Cladistics* 12, 1–9.

Wheeler, W. 2001a. Homology and DNA sequence data. In Wagner, G. (ed.), *The Character Concept in Evolutionary Biology.* New York: Academic Press, pp. 303–318.

Wheeler, W. 2001b. Homology and the optimization of DNA sequence data. *Cladistics* 17, S3–S11.

Wheeler, W. 2010. Trees, tree-shaped objects, super-trees, and support. *Cladistics* 26, 225–226.
Wheeler, W. 2015. Phylogenetic network analysis as a parsimony optimization problem. *BMC Bioinformatics* 16, 296.
Wheeler, W. 2016. *Systematics: A Course of Lectures*. Hoboken: John Wiley and Sons.
Wheeler, W., Coddington, J., Crowley, L., Dimitrov, D., Goloboff, P., Griswold, C., Hormiga, G., Prendini, L., Ramirez, M., Sierwald, P., Almeida-Silva, L., Alvarez-Padilla, F., Arnedo, M., Benavides Silva, L., Benjamin, S., Bond, J., Grismado, C., Hasan, E., Hedin, M., Izquierdo, M., Labarque, F., Ledford, J., Lopardo, L., Maddison, W., Miller, J., Piacentini, L., Platnick, N., Polotow, D., Silva-Davila, D., Scharff, N., Szüts, T., Ubick, D., Vink, C., Wood, H., and Zhang, J. 2016. The spider tree of life: Phylogeny of Araneae based on target-gene analyses from an extensive taxon sampling. *Cladistics* 33, 574–616.
Wheeler, W., and Pickett, K. 2008. Topology-Bayes versus clade-Bayes in phylogenetic analysis. *Mol. Biol. Evol.* 25, 447–453.
Whidden, C., and Matsen, F. 2015. Quantifying MCMC exploration of phylogenetic tree space. *Syst. Biol.* 64, 472–491.
Whiting, M., Bradler, S., and Maxwell, T. 2003. Loss and recovery of wings in stick insects. *Nature* 421(6920), 264–267.
White, W.T., and Holland, B.R. 2011. Faster exact maximum parsimony search with XMP. *Bioinformatics* 27, 1359–1367.
Wiens, J. 1995. Polymorphic characters in phylogenetic systematics. *Syst. Biol.* 44, 482–500.
Wiley, E. 1975. Karl R. Popper, systematics, and classification: A reply to Walter bock and other evolutionary taxonomists. *Syst. Zool.* 24, 233–243.
Wiley, E., Chakrabarty, P., Craig, M., Davis, M., Holcroft, N., Mayden, R., and Leo Smith, W. 2011. Will the real phylogeneticists please stand up? In Carvalho, M. de, and Craig, M. (eds.), *Morphological and Molecular Approaches to the Phylogeny of Fishes: Integration or Conflict?* pp. 7–16. *Zootaxa* 2946, 1–142.
Wiley, E., and Lieberman, B. 2011. *Phylogenetics: Theory and Practice of Phylogenetic Systematics*, 2nd ed. Hoboken: John Wiley and Sons.
Wilkins, J. 2009. *Species: A History of the Idea. Species and Systematics*. Vol. 1. Berkeley, CA: University of California Press, 303 pp.
Wilkinson, M. 1995. Coping with abundant missing entries in phylogenetic inference using parsimony. *Syst. Biol.* 44, 501–514.
Williams, D. 2004. Homologues and homology, phenetics and cladistics: 150 years of progress. In Williams, D., and Forey, P. (eds.), *Milestones in Systematics*. The Systematics Association Special Volume Series 67. London: CRC Press, pp. 191–224.
Williams, D., and Ebach, M. 2008. *Foundations of Systematics and Biogeography*. Berlin: Springer, 328 pp.
Williams, D., and Ebach, M. 2012a. Confusing homologs as homologies: A reply to 'On homology'. *Cladistics* 28, 223–224.
Williams, D., and Ebach, M. 2012b. 'Phenetics' and its application. *Cladistics* 28, 229–230.
Williams, D., and Ebach, M. 2014. Patterson's curse, molecular homology, and the data matrix. In Hamilton, A. (ed.), *The Evolution of Phylogenetic Systematics*, Species and Systematics. Vol. 5. Berkeley: University of California Press, pp. 151–187.
Williams, D., and Ebach, M. 2017. What is intuitive taxonomic practice? *Syst. Biol.* 66, 637–643.
Williams, D., and Ebach, M. 2018. A Cladist is a systematist who seeks a natural classification: Some comments on Quinn (2017). *Biol. Phil.* 33, 1–4.
Williams, D., Ebach, M., and Wheeler, Q. 2010. Beyond belief—the steady resurrection of phenetics. In Williams, D., and Knapp, S. (eds.), *Beyond Cladistics—The Branching of a Paradigm*. Berkeley: University of California Press, 352 pp.

Williams, D., and Siebert, D. 2000. Characters, homology, and three-item analysis. In Scotland, R., and Pennington, R. (eds.), *Homology and Systematics: Coding Characters for Phylogenetic Analysis.* The Systematics Association Special Volume Series 58. London: Taylor and Francis, pp. 183–208.

Winther, J. 2009. Character analysis in cladistics: Abstraction, reification, and the search for objectivity. *Acta Biotheor.* 57, 129–162.

Wolpert, D., and Macready, W. 1997. No free lunch theorems for optimization. *IEEE Trans. Evol. Comp.* 1, 67–82.

Wright, A. 2019. A systematist's guide to estimating Bayesian phylogenies from morphological data. *Insect Syst. Diversity* 3, 1–14.

Wright, A., and Hillis, D. 2014. Bayesian analysis using a simple likelihood model outperforms parsimony for estimation of phylogeny from discrete morphological data. *PLoS One* 9, e109210. https://doi.org/10.1371/journal.pone.0109210.

Wright, A., Lloyd, G., and Hillis, D. 2016. Modeling character change heterogeneity in phylogenetic analyses of morphology through the use of priors. *Syst. Biol.* 65, 602–611.

Xie, W., Lewis, P., Fan, Y., Kuo, L., and Chen, M.-H. 2011. Improving marginal likelihood estimation for Bayesian phylogenetic model selection. *Syst. Biol.* 60, 150–160.

Yang, Z. 1993. Maximum-likelihood estimation of phylogeny from DNA sequences when substitution rates differ over sites. *Mol. Biol. Evol.* 10, 1396–1401.

Yang, Z. 1994. Maximum likelihood phylogenetic estimation from DNA sequences with variable rates over sites: Approximate methods. *J. Mol. Evol.* 39, 306–314.

Yang, Z. 1996. Phylogenetic analysis using parsimony and likelihood methods. *J. Mol. Evol.* 42, 294–307.

Yang, Z. 2006. *Computational Molecular Evolution.* Oxford: Oxford University Press, 357 pp.

Yang, Z., and Rannala, B. 1997. Bayesian phylogenetic inference using DNA sequences: A Markov Chain Monte Carlo method. *Mol. Biol. Evol.* 14, 717–724.

Yang, Z., and Zhu, T. 2018. Bayesian selection of misspecified models is overconfident and may cause spurious posterior probabilities for phylogenetic trees. *P.N.A.S.* 115, 1854–1859.

Yu, Y., Dong, J., Liu, K., and Nakhleh, L. 2014. Maximum likelihood inference of reticulate evolutionary histories. *Proc. Natl. Acad. Sc.* 111, 16448–16453.

Zangerl, R. 1948. The methods of comparative anatomy and its contribution to the study of evolution. *Evolution* 2, 351–374.

Zuckerkandl, E., and Pauling, L. 1965. Evolutionary divergence and convergence in proteins. In Bryson, V., and Vogel, H.J. (eds.), *Evolving Genes and Proteins.* New York: Academic Press, pp. 97–166.

Zwickl, D. 2006. *Genetic Algorithm Approaches for the Phylogenetic Analysis of Large Biological Sequence Datasets Under the Maximum Likelihood Criterion.* PhD dissertation, The University of Texas, Austin.

Index

Note: Page numbers in *italic* indicate a figure and page numbers in **bold** indicate a table on the corresponding page.

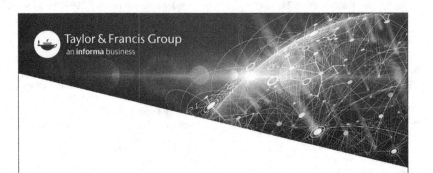

Printed in the United States
by Baker & Taylor Publisher Services